T0231538

PHYSICAL CHEMISTRY FOR ENGINEERING AND APPLIED SCIENCES

Theoretical and Methodological Implication

Innovations in Physical Chemistry: Monograph Series

PHYSICAL CHEMISTRY FOR ENGINEERING AND APPLIED SCIENCES

Theoretical and Methodological Implication

Edited by

A. K. Haghi, PhD
Cristóbal Noé Aguilar, PhD
Sabu Thomas, PhD
Praveen K. M.

Apple Academic Press Inc.
3333 Mistwell Crescent
Oakville, ON L6L 0A2
Canada

Apple Academic Press Inc.
9 Spinnaker Way
Waretown, NJ 08758
USA

© 2018 by Apple Academic Press, Inc.
First issued in paperback 2021
Exclusive worldwide distribution by CRC Press, a member of Taylor & Francis Group
No claim to original U.S. Government works
ISBN-13: 978-1-77463-066-2 (pbk)
ISBN-13: 978-1-77188-627-7 (hbk)

Library and Archives Canada Cataloguing in Publication

Physical chemistry for engineering and applied sciences : theoretical and methodological implication / edited by A.K. Haghi, PhD, Cristóbal Noé Aguilar, PhD Sabu Thomas, PhD, Praveen K.M.

(Innovations in physical chemistry : monograph series)
Includes bibliographical references and index.
Issued in print and electronic formats.
ISBN 978-1-77188-627-7 (hardcover).--ISBN 978-1-315-10972-5 (PDF)

1. Chemistry, Physical and theoretical. I. Haghi, A. K., editor.
II. Series: Innovations in physical chemistry. Monograph series

| QD453.3.P59 2018 | 541 | C2018-901671-X | C2018-901672-8 |

Library of Congress Cataloging-in-Publication Data

Names: Haghi, A. K., editor.
Title: Physical chemistry for engineering and applied sciences : theoretical
and methodological implication / editors, A.K. Haghi, PhD, Cristâobal Noâe Aguilar, PhD,
Sabu Thomas, PhD, Praveen K.M.
Description: Toronto ; Toronto ; New Jersey : Apple Academic Press, [2018] | Series:
Innovations in physical chemistry : monograph series | Includes
bibliographical references and index.
Identifiers: LCCN 2018012388 (print) | LCCN 2018013521 (ebook) |
ISBN 9781315109725 (ebook) | ISBN 9781771886277 (hardcover : alk. paper)
Subjects: LCSH: Chemistry, Physical and theoretical.
Classification: LCC QD453.3 (ebook) | LCC QD453.3 .P47 2018 (print) | DDC 541--dc23
LC record available at https://lccn.loc.gov/2018012388

Apple Academic Press also publishes its books in a variety of electronic formats. Some content that appears in print may not be available in electronic format. For information about Apple Academic Press products, visit our website at **www.appleacademicpress.com** and the CRC Press website at **www.crcpress.com**

ABOUT THE EDITORS

A. K. Haghi, PhD

A. K. Haghi, PhD, is the author and editor of 165 books, as well as 1000 published papers in various journals and conference proceedings. Dr. Haghi has received several grants, consulted for a number of major corporations, and is a frequent speaker to national and international audiences. Since 1983, he served as professor at several universities. He is currently Editor-in-Chief of the *International Journal of Chemoinformatics and Chemical Engineering* and *Polymers Research Journal* and on the editorial boards of many international journals. He is also a member of the Canadian Research and Development Center of Sciences and Cultures (CRDCSC), Montreal, Quebec, Canada. He holds a BSc in urban and environmental engineering from the University of North Carolina (USA), an MSc in mechanical engineering from North Carolina A&T State University (USA), a DEA in applied mechanics, acoustics and materials from the Université de Technologie de Compiègne (France), and a PhD in engineering sciences from the Université de Franche-Comté (France).

Cristóbal Noé Aguilar, PhD

Cristóbal Noé Aguilar, PhD, is Dean of the School of Chemistry at the Universidad Autónoma de Coahuila, México. Dr. Aguilar has published more than 160 papers in indexed journals, more than 40 articles in Mexican journals, and 250 contributions in scientific meetings. He has also published many book chapters, several Mexican books, four editions of international books, and more. Prof. Aguilar is a member level III of S.N.I. (National System of Researchers of Mexico). He has been awarded several prizes and awards; the most important are the National Prize of Research 2010 from the Mexican Academy of Sciences; the Prize "Carlos Casas Campillo 2008" from the Mexican Society of Biotechnology and Bioenegineering; National Prize AgroBio–2005 and the Mexican Prize in Food Science and Technology.

Dr. Aguilar is a member of the Mexican Academy of Science, the International Bioprocessing Association, Mexican Academy of Sciences, the Mexican Society for Biotechology and Bioengineering, and the Mexican

Association for Food Science and Biotechnology. He has developed more than 21 research projects, including six international exchange projects. He has been advisor for PhD, MSc, and BSc theses.

Dr. Aguilar earned his chemist designation from the School of Chemistry at the Universidad Autónoma de Coahuila, México. He attended the MSc Program in Food Science and Biotechnology that was held at the Universidad Autónoma de Chihuahua, México. His PhD in Fermentation Biotechnology was awarded by the Universidad Autónoma Metropolitana, Mexico.

Sabu Thomas, PhD

Sabu Thomas, PhD, is the Pro-Vice Chancellor of Mahatma Gandhi University and Founding Director of the International and Inter University Center for Nanoscience and Nanotechnology, Mahatma Gandhi University, Kottayam, Kerala, India. He is also a full professor of polymer science and engineering at the School of Chemical Sciences of the same university. He is a fellow of many professional bodies. Professor Thomas has (co-)authored many papers in international peer-reviewed journals in the area of polymer science and nanotechnology. He has organized several international conferences. Professor Thomas's research group has specialized in many areas of polymers, which includes polymer blends, fiber-filled polymer composites, particulate-filled polymer composites and their morphological characterization, ageing and degradation, pervaporation phenomena, sorption and diffusion, interpenetrating polymer systems, recyclability and reuse of waste plastics and rubbers, elastomeric crosslinking, dual porous nanocomposite scaffolds for tissue engineering, etc. Professor Thomas's research group has extensive exchange programs with different industries and research and academic institutions all over the world and is performing world-class collaborative research in various fields. Professor Thomas's Center is equipped with various sophisticated instruments and has established state-of-the-art experimental facilities, which cater to the needs of researchers within the country and abroad.

Professor Thomas has published over 750 peer-reviewed research papers, reviews, and book chapters and has a citation count of 31,574. The H index of Prof. Thomas is 81, and he has six patents to his credit. He has delivered over 300 plenary, inaugural, and invited lectures at national/international meetings over 30 countries. He is a reviewer for many international journals. He has received MRSI, CRSI, nanotech medals for his outstanding work in nanotechnology. Recently Prof. Thomas has been conferred an Honoris

Causa (DSc) by the University of South Brittany, France, and University Lorraine, Nancy, France.

Praveen K. M.

Praveen K. M. is an Assistant Professor of Mechanical Engineering at SAINTGITS College of Engineering, India. He is currently pursuing a PhD in Engineering Sciences at the University of South Brittany (Université de Bretagne Sud)—Laboratory IRDL PTR1, Research Center "Christiaan Huygens," in Lorient, France, in the area of coir-based polypropylene micro composites and nanocomposites. He has published an international article Applied Surface Science (Elsevier) and has also presented poster and conference papers at national and international conferences. He also has worked with the Jozef Stefan Institute, Ljubljana, Slovenia; Mahatma Gandhi University, India; and the Technical University in Liberec, Czech Republic. His current research interests include plasma modification of polymers, polymer composites for neutron shielding applications, and nanocellulose.

ABOUT THE INNOVATIONS IN PHYSICAL CHEMISTRY: MONOGRAPH SERIES

This book series aims to offer a comprehensive collection of books on physical principles and mathematical techniques for majors, non-majors, and chemical engineers. Because there are many exciting new areas of research involving computational chemistry, nanomaterials, smart materials, high-performance materials, and applications of the recently discovered graphene, there can be no doubt that physical chemistry is a vitally important field. Physical chemistry is considered a daunting branch of chemistry—it is grounded in physics and mathematics and draws on quantum mechanics, thermodynamics, and statistical thermodynamics.

Editors-in-Chief

A. K. Haghi, PhD
Editor-in-Chief, International Journal of Chemoinformatics and Chemical Engineering and Polymers Research Journal; Member, Canadian Research and Development Center of Sciences and Cultures (CRDCSC), Montreal, Quebec, Canada E-mail: AKHaghi@Yahoo.com

Lionello Pogliani, PhD
University of Valencia-Burjassot, Spain
E-mail: lionello.pogliani@uv.es

Ana Cristina Faria Ribeiro, PhD
Researcher, Department of Chemistry, University of Coimbra, Portugal
E-mail: anacfrib@ci.uc.pt

Books in the Series

- **Applied Physical Chemistry with Multidisciplinary Approaches**
  Editors: A. K. Haghi, PhD, Devrim Balköse, PhD, and Sabu Thomas, PhD

- **Chemical Technology and Informatics in Chemistry with Applications**
  Editors: Alexander V. Vakhrushev, DSc, Omari V. Mukbaniani, DSc, and Heru Susanto, PhD

- **Engineering Technologies for Renewable and Recyclable Materials: Physical-Chemical Properties and Functional Aspects**
  Editors: Jithin Joy, Maciej Jaroszewski, PhD, Praveen K. M., Sabu Thomas, PhD, and Reza Haghi, PhD

- **Engineering Technology and Industrial Chemistry with Applications**
  Editors: Reza Haghi, PhD, and Francisco Torrens, PhD

- **High-Performance Materials and Engineered Chemistry**
  Editors: Francisco Torrens, PhD, Devrim Balköse, PhD, and Sabu Thomas, PhD

- **Methodologies and Applications for Analytical and Physical Chemistry**
  Editors: A. K. Haghi, PhD, Sabu Thomas, PhD, Sukanchan Palit, and Priyanka Main

- **Modern Physical Chemistry: Engineering Models, Materials, and Methods with Applications**
  Editors: Reza Haghi, PhD, Emili Besalú, PhD, Maciej Jaroszewski, PhD, Sabu Thomas, PhD, and Praveen K. M.

- **Physical Chemistry for Chemists and Chemical Engineers: Multidisciplinary Research Perspectives**
  Editors: Alexander V. Vakhrushev, DSc, Reza Haghi, PhD, and J. V. de Julián-Ortiz, PhD

- **Physical Chemistry for Engineering and Applied Sciences: Theoretical and Methodological Implication**
  Editors: A. K. Haghi, PhD, Cristóbal Noé Aguilar, PhD, Sabu Thomas, PhD, and Praveen K. M.

- **Theoretical Models and Experimental Approaches in Physical Chemistry: Research Methodology and Practical Methods**
  Editors: A. K. Haghi, PhD, Sabu Thomas, PhD, Praveen K. M., and Avinash R. Pai

CONTENTS

LIST OF CONTRIBUTORS

L. E. Agibayeva
Al-Farabi Kazakh National University, Almaty, Republic of Kazakhstan

Yelda Akdeniz
MGM Material Technologies, Urla, İzmir, Turkey

R. Arunkumar
Department of Physics, School of Advanced Sciences, VIT University, Vellore, India

K. Anver Basha
P.G. & Research Department of Chemistry, C. Abdul Hakeem College (Autonomous), Melvisharam, Tamil Nadu 632509, India. E-mail: kanverbasha@gmail.com

Ravi Shanker Babu
Department of Physics, School of Advanced Sciences, VIT University, Vellore, India. E-mail: ravina2001@rediffmail.com

Devrim Balköse
Department of Chemical Engineering, İzmir Institute of Technology, Urla, İzmir, Turkey. E-mail: devrimbalkose@gmail.com

Tanmoy Chakraborty
Department of Chemistry, Manipal University, Jaipur 303007, Rajasthan, India. E-mail: tanmoy. chakraborty@jaipur.manipal.edu, tanmoychem@gmail.com

E. Chkhaidze
Faculty of Chemical Technology and Metallurgy, Department of Chemical and Biological Engineering, Georgian Technical University, 77 Kostava st., Tbilisi 0175, Georgia. E-mail: ekachkhaidze@yahoo. com

Mehmet Gönen
Department of Chemical Engineering, Engineering Faculty, Süleyman Demirel University, Batı Yerleşkesi, Isparta 32260, Turkey. E-mail: mehmetgonen@sdu.edu.tr

Sushma P. Ijardar
Present address: Department of Chemistry, Veer Narmad South Gujarat University, Surat 395007, India Salt and Marine Chemicals Division, CSIR—Central Salt and Marine Chemicals Research Institute (CSIR-CSMCRI), Council of Scientific & Industrial Research (CSIR), Bhavnagar, Gujarat 364002, India. E-mail: sushmaijardar@yahoo.co.in

T. K. Jumadilov
JSC "Institute of Chemical Sciences after A.B. Bekturov", Almaty, Republic of Kazakhstan. E-mail: jumadilov@mail.ru

S. Kalainathan
Department of Physics, School of Advanced Sciences, VIT University, Vellore, India

Porteen Kannan
Department of Veterinary Public Health and Epidemiology, Madras Veterinary College, Chennai 600007, India

D. Kharadze
Ivane Beritashvili Center of Experimental Biomedicine, 14 Gotua st. Tbilisi 0160, Georgia

V. I. Kodolov
Chemistry and Chemical Technology Department, M. T. Kalashnikov
Izhevsk State Technical University, Izhevsk, Russia

R. G. Kondaurov
JSC "Institute of Chemical Sciences after A.B. Bekturov", Almaty, Republic of Kazakhstan

Ajay Kumar
Department of Mechatronics, Manipal University, Jaipur 303007, Rajasthan, India

Shilna KV
Department of Physics, Central University of Kerala, Kasaragod, Kerala 671314, India.
E-mail: shilnakvnarayanan@gmail.com

Naved I. Malek
Applied Chemistry Department, Sardar Vallabhbhai National Institute of Technology, Ichchhanath,
Surat, Gujarat 395007, India
Institute of Chemistry, The University of São Paulo, Box 26077, São Paulo, SP 05513-970, Brazil.
E-mail: navedmalek@chem.svnit.ac.in, navedmalek@yahoo.co.in

Utkarsh U. More
Applied Chemistry Department, Sardar Vallabhbhai National Institute of Technology, Ichchhanath,
Surat, Gujarat 395007, India

Swapna S. Nair
Assistant Professor, Department of Physics, School of Mathematical and Physical Science, Central
University of Kerala, Kasaragod, Kerala 671314, India. E-mail: swapnasharp@gmail.com

N. Nepharidze
Faculty of Chemical Technology and Metallurgy, Department of Chemical and Biological Engineering,
Georgian Technical University, 77 Kostava st., Tbilisi 0175, Georgia

Sukanchan Palit
Department of Chemical Engineering, University of Petroleum and Energy Studies, Bidholi via
Premnagar, Dehradun 248007, Uttarakhand, India
43, Judges Bagan, Haridevpur, Kolkata 700082, India. E-mail: sukanchan68@gmail.com,
sukanchan92@gmail.com

Paresh Patel
Department of Pharmaceutical Sciences, Saurashtra University, Rajkot, Gujarat, India

Priya Patel
Department of Pharmaceutical Sciences, Saurashtra University, Rajkot, Gujarat, India

I. Pugazhenthi
P.G. & Research Department of Chemistry, C. Abdul Hakeem College (Autonomous), Melvisharam,
Tamil Nadu 632509, India

M. Nithya Quintoil
Department of Veterinary Public Health, Rajiv Gandhi Institute of Veterinary Education and Research,
Puducherry, India

Prabhat Ranjan
Department of Mechatronics, Manipal University, Jaipur 303007, Rajasthan, India

M. Usha Rani
Department of Physics, School of Advanced Sciences, VIT University, Vellore, India

Mihir Raval
Shivam Institute and Research Centre of Pharmacy, Valasan, Anand, Gujarat, India

S. Wilfred Ruban
Department of Livestock Products Technology, Veterinary College, Hebbal, Bangalore 560024, India

S. Mohammed Safiullah
P.G. & Research Department of Chemistry, C. Abdul Hakeem College (Autonomous), Melvisharam, Tamil Nadu 632509, India

Omar A. El Seoud
Applied Chemistry Department, Sardar Vallabhbhai National Institute of Technology, Ichchhanath, Surat, Gujarat 395007, India

Anamika Singh
Department of Botany, Matreyi College, University of Delhi, Delhi, India

Rajeev Singh
Department of Environment Studies, Satayawati College, University of Delhi, Delhi, India

Ayben Top
Department of Chemical Engineering, İzmir Institute of Technology, Urla, İzmir, Turkey

Sevgi Ulutan
Department of Chemical Engineering, Ege University, Bornova, İzmir, Turkey

Neha Vadgama
Shivam Institute of Pharmacy, Valasan, Gujarat, India

LIST OF ABBREVIATIONS

AF	antiferromagnetic
AOP	advanced oxidation process
ANS	8-anilino-1-napthalenesulfonic acid
BKT	Berezinskii–Kosterlitz–Thouless
BPM	biochemical methane potential
BSA	Bull serum albumen
BSCCO	bismuth strontium calcium copper oxide
CA	cholic acid
CDFT	conceptual density functional theory
CDKIs	cyclin-dependent kinase inhibitors
CDs	cyclodextrins
cmc	critical micelle concentration
CNTs	carbon nanotubes
CPC	cetylpyridinium chloride
DFT	density functional theory
DLS	dynamic light scattering
DSILs	double-salt ionic liquids
DTA	differential thermal analysis
DTDAB	dimethylditetradecylammonium bromide
EC	ethylene carbonate
EPA	Environmental Protection Agency
EPR	enhanced permeation and retention
EU	enzyme units
FA	folic acid
FDA	Food and Drug Administration
FESEM	field emission scanning electron microscope
FRs	folate receptors
FTIR	Fourier transform infrared
HAP	hydroxyapatite
HAV	hepatitis A virus
HBCCO	mercury barium calcium copper oxide
HeTAB	n-hexyltrimethylammonium bromide
HOMO	highest occupied molecular orbital
HPMA	N-(2-hydroxypropyl)methacrylamide

HRT	hydraulic residence time
HTSCs	high-temperature superconductors
ICON	International Council on Nanotechnology
ILs	ionic liquids
JJ	Josephson junction
LUMO	lowest unoccupied molecular orbital
MB	Methylene blue
MBRs	membrane bioreactors
MOS	molecular organic solvent
MRI	magnetic resonance imaging
MWCNT	multiwalled carbon nanotubes
NCPEs	nanocomposite polymer electrolytes
NEs	nanoemulsions
NIOSH	National Institute for Occupational Safety and Health
NIR	near-infrared
NLC	nanostructured lipid carriers
NMF	natural moisturizing factor
NMs	nanomaterials
NPs	nanoparticles
OSHA	Occupational Safety and Health Administration
PAMAM	polyamidoamine
PBMA	poly(butyl methacrylate)
PC	phosphatidylcholine
PEG	polyethylene glycol
PFO	pseudo first-order
PPCP	pharmaceuticals and personal care product
PS	photosensitizers
PSO	pseudo second-order
PTE	periodic table of the elements
PVC	polyvinyl chloride
QDs	quantum dots
RAMs	random access memories
RBC	red blood cells
RO	reverse osmosis
RT	room temperature
RT-PCR	reverse transcriptase polymerase chain reaction
SA	sebacic acid
SEM	scanning electron microscopy
SLNs	solid lipid nanoparticles
SQUID	superconducting quantum interference devices

SSF	simultaneous saccharification and fermentation
SWCNT	single-walled carbon nanotubes
TAAs	tumor-associated antigens
TBCCO	thallium barium calcium copper oxide
TEM	transmission electron microscope
TGA	thermal gravimetric analysis
UPAs	unsaturated polyamides
UPEAs	unsaturated polyesteramides
UV	ultraviolet
VSM	vibrational sample magnetometer
VTF	Vogel–Tammann–Fulcher
XRD	X-ray diffraction
YBCO	yttrium barium copper oxide
ZnONPs	ZnO nanoparticles

PREFACE

This new volume includes a variety of modern research on topics in experimental methods and theoretical implications in physical chemistry. The book emphasizes the fundamental concepts of physical chemistry while also presenting valuable new research developments that emphasize the continuing importance of physical chemistry today.

This volume focuses heavily on new research and helps to fill a need for current information on the science, processes, and applications in the field. It examines the industrial process for emerging materials, determines practical use under a wide range of conditions, and establishes what is needed to produce a new generation of materials.

The book consists of 18 chapters and covers a variety of topics. The chapter authors, affiliated with prestigious scientific institutions from around the world, share their research on new and innovative applications in physical chemistry. The chapters in the volume are divided into several areas, covering

- developments in physical chemistry of modern materials
- polymer science and engineering
- physical chemistry for the life sciences
- nanoscience and nanotechnology
- other selected topics of interest to physical chemists.

This volume will serve as a resource for scientists, researchers, and faculty and advanced students in physical chemistry.

Key features:

- Presents cutting-edge information on emerging physical chemistry innovations
- Provides qualitative and quantitative information on new techniques in physical chemistry
- Covers a variety of new aspects of chemistry and physics, biochemistry development, and technology

PART I

Developments in Physical Chemistry of Modern Materials

CHAPTER 1

AN INVESTIGATION OF KINETIC AND EQUILIBRIUM BEHAVIOR OF ADSORPTION OF METHYLENE BLUE ON A SILICA HYDROGEL

AYBEN TOP[1], YELDA AKDENIZ[2], SEVGI ULUTAN[3], and DEVRIM BALKÖSE[1*]

[1]*Department of Chemical Engineering, İzmir Institute of Technology, Urla, İzmir, Turkey*
[2]*MGM Material Technologies, Urla, İzmir, Turkey*
[3]*Department of Chemical Engineering, Ege University, Bornova, İzmir, Turkey*
Corresponding author. E-mail: devrimbalkose@gmail.com

CONTENTS

ABSTRACT

Methylene blue (MB) adsorption behavior of a silica hydrogel was investigated. Linear and nonlinear forms of Langmuir, Freundlich, and Temkin models were tested to fit equilibrium data. MB adsorption capacity was obtained as ~13 mg/g hydrogel. Nonlinear form of pseudo first order, pseudo second order, and Elovich models agreed with the kinetic data of the hydrogel in slab form with R^2 values greater than 0.99. Although adsorption was quite fast, slow diffusion inside silica hydrogel suggests silica hydrogel in the small pellets or thin film form rather than a slab form seems having more potential in the removal of contaminants.

1.1 INTRODUCTION

Silica hydrogel is a cylindrical fiber network of silica, which contains water entrapped within the three-dimensional disordered silica matrix.[1,2] Major uses of silica hydrogels are oil purification, beer and wine clarification, and protein crystallization. The hydrogels were also proposed as a viscous liquid barrier to stabilize, seal, and treat contaminated soils.[3] The other potential application areas include gel-mediated crystallization of inorganic materials,[4,5] enzyme immobilization,[6] biosensor fabrication,[7] and adsorption.[8]

Silica hydrogels can be prepared by using different precursors including sodium silicate solutions,[2,3,9,10] colloidal silica,[11] a mixture of sodium silicate and colloidal silicate,[12] and tetraethoxy silane,[13] via acid catalyzed hydrolysis reactions. The hydrogels, which contract by the capillary forces created by the surface tension between liquid and vapor in drying at ambient pressure, are transformed into xerogels. If the drying is carried out under supercritical conditions where only vapor phase exists, silica aerogels form.[14] The major advantage of silica hydrogel over silica xerogels and aerogels can be the ease of forming final hydrogel structure by simply casting silica hydrosol into various shapes such as cylindrical pellets, spherical granules, thin films as well as thick sheets.[15] Additionally, after attaining its rigid network form, the silica hydrogel maintains its integrity even immersed in water. Hence, in adsorption applications, the use of silica hydrogels is not likely to be resulted in any turbidity in aqueous phase. However, adsorption properties of silica hydrogels have not been well investigated as silica aerogels and xerogels.

Methylene blue (MB) is a cationic dye frequently used in coloring paper, cotton, and wool fabrics. It has been chosen as a representative water-soluble dye for adsorption experiments. The adsorption of MB on various adsorbents including carbonaceous materials,[16,17] agricultural wastes, industrial solid wastes, biomass, clay minerals, zeolites,[18,19] luffa fibers,[20] halloysite nanotubes,[21] silica aerogels,[13] silica gels,[22,23] and modified silica gels[24] were studied. In this study, it was aimed to investigate MB adsorption behavior of a silica hydrogel prepared via acid hydrolysis of sodium silicate. Adsorption equilibrium and kinetic models were also tested.

1.2 EXPERIMENTAL

1.2.1 PREPARATION AND CHARACTERIZATION OF SILICA HYDROGEL

Silica hydrogel was prepared from a hydrosol precursor obtained by using sodium silicate solution (Aldrich, $d = 1.390$ g/mL having 14% NaOH and 27% SiO_2 by weight) and sulfuric acid (Merck, 98%). In the first step of hydrosol preparation, sodium silicate was diluted (1:1 in volume) with water and, then, added to 1.5 M H_2SO_4 solution using a peristaltic pump at constant flow rate until pH value of the mixture became 1, while stirred at 400 rpm. Next, the sols were cast into glass bottles as 20 mL aliquots and allowed to set for approximately 1 day at room temperature to form a rigid hydrogel. Finally, the hydrogels casted were washed with equal volume of water for 30 min for eight times to remove the by-product, sodium sulfate, and unreacted sulfuric acid. The washing process was monitored by pH and conductivity measurements using a pH meter (EMAF EM 78 pH) and a conductivity meter (Labconco), respectively. The amount of sulfuric acid and sodium sulfate leached from the silica hydrogel slab were calculated from the measured pH and conductivity of the solutions. Water content of the purified hydrogel was determined by Shimadzu TGA-50 model thermal gravimetric analysis (TGA) instrument. TGA analysis was performed between 25 and 400°C temperature range and at 10°C/min heating rate. Fourier transform infrared (FTIR) spectrum was taken after drying the silica hydrogel at 100°C using KBr disc technique. The spectrum was obtained between 400 and 4000 cm^{-1} wavelength range by employing Shimadzu 8601PC model FTIR spectrophotometer.

1.2.2 ADSORPTION EQUILIBRIUM AND KINETICS

In all adsorption experiments, freshly prepared MB (Aldrich) solutions were employed. Adsorption equilibrium experiments were carried out by contacting different amounts of ground silica hydrogel (0.03, 0.06, 0.11, 0.31, and 0.54 g) having approximately 1–3 mm particle size with 50 mL of 20 ppm MB solutions at 25°C for 1 week by shaking at 100 rpm. After separation of the aqueous phase from silica hydrogel particles by centrifugation (Rotofix 32 Hettich) at 4000 rpm, MB concentrations were determined using the Jasco 700 UV/visible spectrophotometer. Adsorption kinetic experiments at 25°C were conducted using intact hydrogels casted in glass bottles. A volume of 70 mL of MB solution (2 or 5 ppm) was added over the hydrogel (~20 mL) and stirred at 600 rpm rate with a magnetic bar placed over each silica hydrogel layer. UV–visible absorbance values of MB solutions were measured in situ by using Avantes model fiber-optic spectrophotometer. MB concentrations were determined using a calibration curve obtained at 664 nm wavelength. The absorbance-versus-concentration calibrations curves were linear up to 20 and 5 ppm in Jasco UV/visible spectrophotometer and fiber-optic spectrophotometer, respectively. Thus, the concentration of MB solutions for in situ measurements of adsorption kinetics was chosen below 5 ppm. In the analysis of the equilibrium and kinetic data, model fitting parameters and correlation coefficient (R^2) values were obtained using SigmaPlot software. Root mean squared error (RMSE)[25] and Δq (%)[26] values were calculated using the following respective equations:

$$\mathrm{RMSE}\sqrt{\frac{\sum (q_{\exp} - q_{\mathrm{cal}})^2}{n}} \tag{1.1}$$

$$\Delta q(\%) = 100 \times \sqrt{\frac{\sum \left[\left(q_{\exp} - q_{\mathrm{cal}} \right) \Big/ q_{\exp} \right]^2}{(n-1)}} \tag{1.2}$$

where $q_{\exp}$ and q_{cal} are experimental and predicted values of amount of MB sorbed onto the silica hydrogel and n is the number of experimental points. The photographs of the MB solutions over silica hydrogel was taken by using Nikon, Coolpix 995 digital camera at nearly 1 week period up to 5 weeks to observe the diffusion of the blue-colored boundary in silica hydrogel.

1.3   RESULTS AND DISCUSSION

1.3.1   CHARACTERIZATION OF SILICA HYDROGEL

Silica hydrogel was prepared at pH 1 to obtain uniform and transparent silica hydrogels. The silica particles formed had positive charges since the pH was lower than the isoelectic point 1.3.[2] Thus, the silica hydrosols obtained had long enough gelation times to be transferred to containers with different geometries. The silica gel obtained by drying the silica hydrogel that was prepared under the same conditions as in the present study had 760 m^2/g surface area, 0.31 cm^3/g total pore volume, and average pore diameter 1.6 nm as characterized by argon gas adsorption at $-196°C$.[27]

The leached amount of the sulfuric acid and sodium sulfate during washing process were used to determine composition of the hydrogel after washing for 8 times. pH and Na_2SO_4 concentration of the washing solutions are shown in Figure 1.1. Slow diffusion of protons and Na_2SO_4 in the silica hydrogel phase necessitated the repeated washing process. As seen in Figure 1.1, pH of the first washing solution observed as 1.82, increased to 2.2 at the end of the last washing process. Na_2SO_4 concentration in washing solution decreased with repeated washing and obtained as 0.025 M in the 8th washing. The composition of silica hydrogel was found as $3.2 \pm 0.3\%$ Na_2SO_4, $1.7 \pm 0.1\%$ H_2SO_4, 7.95% SiO_2, and $87.1 \pm 0.3\%$ H_2O from the material balance results of each step.

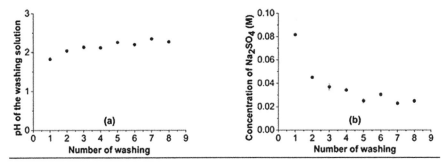

FIGURE 1.1   (a) pH and (b) concentration of sodium sulfate of the washing solutions at the end of each washing step.

FTIR spectroscopy was used to assess purity of the silica hydrogel. FTIR spectrum of dried silica hydrogel is given in Figure 1.2. The bands observed at around 470 and 810 cm^{-1} are assigned to Si–O–Si bending and symmetric stretching vibrations, respectively.[28,29] Other vibration bands include Si–O stretching mode at 960 cm⁻¹, asymmetric Si–O–Si stretching mode TGA was used to determine water content of the silica hydrogel. In the thermogravimetric curve of the hydrogel, constant mass of the hydrogel was reached at about 250°C with 90.6% at 1080 cm⁻¹, hydrogen bonded OH stretching vibration of SiOH groups and adsorbed water at 3440 cm⁻¹.[29]

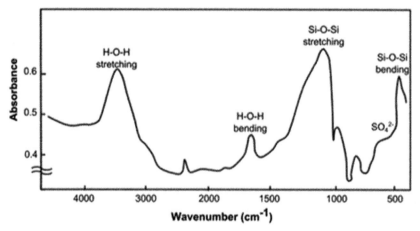

FIGURE 1.2   FTIR spectrum of the silica gel dried at 100°C.

The band at 1645 cm⁻¹ is due to the overlapping of stretching vibration of Si–O and H₂O bending vibrations.[30] Thus, the bands obtained related to Si–O–Si vibrational modes clearly show the formation of silica network structure.[31] However, the shoulder observed at 620 cm⁻¹ indicates the presence of sulfate ions in the silica hydrogel. The composition determined by the conductivity measurements also confirmed the presence of sulfate ions in silica hydrogel mass loss that corresponds to dehydration of water (Fig. 1.3). This result indicated that the ratio of water to silica in the hydrogel is approximately 9 mL:1 g, which is consistent with the ratio obtained from the estimated chemical composition of the silica hydrogel (8.7 mL:1 g). It is possible to change water sorption capacity of silica hydrogels by changing the surface properties of silica or controlling the addition amount of water during the synthesis. For example, water sorption capacity values were measured as ~2–2.3 and ~0.12–0.13 mL water/g silica for hydrophilic and hydrophobic silica xerogels, respectively.[31] In another study, silica hydrogels with water sorption capacity spanning from 1.8 to 5.4 mL water/g silica were synthesized by adjusting feed flowrates during the synthesis.[32]

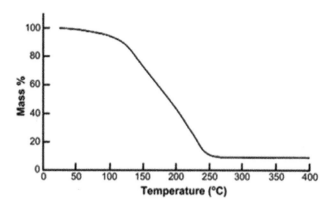

FIGURE 1.3 TGA curve of the silica hydrogel.

1.3.2 ADSORPTION EQUILIBRIUM

Silica hydrogel was ground to 1–3 mm particle size to ensure the attainment of equilibrium in adsorption isotherm experiments. The adsorption equilibrium data were fitted to linear and nonlinear forms of Langmuir, Freundlich, and Temkin models as shown in Table 1.1. In these equations, C is MB concentration in solution; q is MB concentration in the hydrogel; q_m is the maximum amount of MB concentration in the hydrogel obtained from Langmuir model; K_L, K_F, and K_T are Langmuir, Freundlich, and Temkin equilibrium constants, respectively. n is the other Freundich constant indicating heterogeneity of the adsorption, and B_T is the other Temkin constant related to heat of adsorption. These three popular isotherms differ based on the assumptions used to derive the equilibrium model. Langmuir isotherm assumes monolayer adsorption with heat of adsorption independent of surface coverage. Conversely, heat of adsorption is assumed to decrease linearly and logarithmically for the respective Temkin and Freundlich models with increasing surface coverage.[33,34] Therefore, the fitting of experimental

TABLE 1.1 Adsorption Isotherm Models.

Isotherm	Nonlinear form	Linear form
Langmuir	$q = \dfrac{q_m K_L C}{1 + K_L C}$	$\dfrac{C}{q} = \dfrac{1}{q_m K_L} + \dfrac{C}{q_m}$
Freundlich	$q = K_F C^n$	$lnq = \ln K_F + nlnC$
Temkin	$q = B_T \ln (K_T C)$	$q = B_T \ln K_T + B_T lnC$

data to these models gives valuable information about the adsorbent–adsorbate interactions. Adsorption isotherm models fitting parameters along with correlation coefficient (R^2), RMSE, and normalized standard deviation (Δq (%)) values for the MB–silica hydrogel system are given in Table 1.2. Linear plots of Langmuir, Freundich and Temkin models are shown in Figure 1.4. Comparison of the plots of the experimental isotherm with the predicted isotherms obtained from the linear and nonlinear model fitting parameters are given in Figure 1.5a,b, respectively. Compared to the nonlinear analysis, different fitting parameters were estimated for linear Langmuir and Freundlich models. R^2 and RMSE values suggested that linearization lowered fit quality. Δq (%) values, on the other hand, were found to be quite sensitive to the agreement of the first portion of the experimental and estimated data points. Thus, they did not correlate with the R^2 and RMSE values. However,

TABLE 1.2   Adsorption Isotherm Model-Fitting Parameters.

Model/parameter	Linear model prediction	Nonlinear model prediction
Langmuir		
q_m (mg/g)	13.6	12.8
K_L (L/mg)	0.378	0.491
R^2	0.960	0.966
RMSE	0.902	0.839
Δq (%)	18.40	23.46
Freundlich		
n	0.504	0.400
K_F (mg/g)(L/mg)$^{1/n}$	3.571	4.258
R^2	0.865	0.955
RMSE	1.211	0.970
Δq (%)	27.96	41.78
Temkin		
B_T	2.783	2.783
K_T (L/mg)	4.695	4.695
R^2	0.957	0.957
RMSE	0.727	0.727
Δq (%)	19.21	19.21

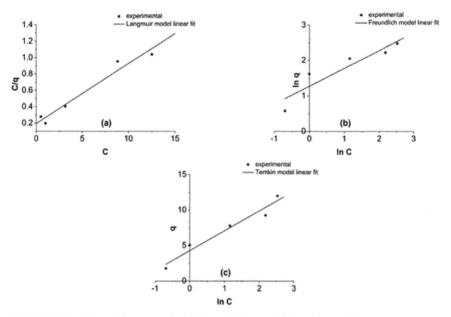

FIGURE 1.4   Linear (a) Langmuir, (b) Freundlich, and (c) Temkin model plots.

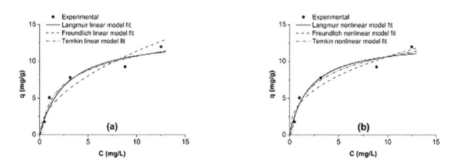

FIGURE 1.5   Comparison of the adsorption isotherms obtained from the (a) linear and (b) nonlinear isotherm model fit parameters.

for both linear and nonlinear Temkin models, same fitting parameters were obtained. In the case of biosorption of MB on an algae, no significant effect of linearization on fitting parameters was observed for Temkin model but for Langmuir and Freundlich model fitting parameters changed upon linearization similar to the results of the current study. On the other hand, in contrast to our study, R^2 values were observed to increase at certain experimental conditions for Langmuir model when linear form of the equation was used.

Interestingly, though different model parameters were estimated similar R^2 values were obtained both in the linear and nonlinear form of Freundlich model.[35] An increase in R^2 values were also observed when linear forms of Langmuir and Freundlich model were used in the case of phosphate adsorption on mesoporous MCM-41 adsorbents.[36] However, there are also reports in which improvement of R^2 values and, hence, better fit quality was obtained upon the use of nonlinear curve fitting.[37,38] The discrepancy of the results obtained for the different adsorbent–adsorbate systems indicated that linearization of an isotherm model changes the error distributions as a result higher or lower R^2 values can be possible which may lead to improper interpretation of the sorption data. Thus, to avoid these inevitable misleading outcomes of linearization, nonlinear form of the isotherms should be used in the assessment of equilibrium data.[39,40]

Our results revealed that none of the isotherm models under investigation agreed with the experimental data completely ($R^2 > 0.99$) most probably due to the complexity of the adsorption. Indeed, isotherm models showed no significant superiority to each other but Langmuir and Freundlich models fitted the experimental data fairly at certain surface coverage regions only. Comparison of the RMSE and Δq (%) values nonlinear model fitting of the three isotherms, on the other hand, suggested modest agreement of Temkin model with the experimental data due to its more evenly distributed errors. For simplicity, by comparing R^2 values obtained in the nonlinear curve fitting of the isotherm models, the adsorption equilibrium of MB on to silica hydrogel can be described more likely by Langmuir model due to its superior fitting ability at low coverage region. However, at higher coverage values, Freundich model better matched to the experimental data which suggest that certain degree of surface heterogeneity is possible due to the different chemical functionalities at the silica surface.[41] Value of the exponent of Freundlich isotherm, n, obtained between 0 and 1 ($\sim$0.4) also indicated heterogeneous adsorption. Adsorption equilibrium of malachite green on silica hydrogels prepared at different pH values 4.5, 7, and 9.5 were fitted to Freundlich isotherm with respective n values between 0.78, 0.88, and 0.88. It was suggested that the silica hydrogel prepared at acidic pH have higher surface heterogeneity as n value is closer to 0.[12,42] In the current study, the silica hydrogel was prepared at more acidic conditions. Thus, lower n value or higher surface heterogeneity obtained is consistent with the study of Perullini et al.[12] Silica in different aerogel forms also exhibited surface heterogeneity for MB sorption with Freundlich exponent between 0.4 and 0.7. Similar to the current study, these aerogels followed Langmuir isotherm having slightly higher R^2 values (0.972–0.986) than Freundlich model.[13]

From Langmuir model analysis, maximum MB sorption capacity (q_m) of the silica hydrogel was estimated as ~13 mg/g which corresponds to ~130 mg/g dry silica. This value is comparable to MB sorption capacity of silica aerogels (110–220 mg/g),[13] *p-tert*-butyl-calix(*n*)arene-modified silica gels (~20–190 mg/g),[24] and citric acid-modified silica gel (~125 mg/g).[43] Another Langmuir constant, K_L, is the adsorption equilibrium constant, simply the ratio between the rates of adsorption and desorption. Thus, it is related to free energy of adsorption and higher value of K_L suggests higher degree of attractive interactions between the sorbent and sorbate. Compared to unmodified forms of silica, K_L values were observed to increase upon modification with citric acid,[43] and amine functionalization[44] for MB, and heavy metal sorption, respectively. K_L value of the silica hydrogel–MB pair obtained in the current study (0.3) lies in the range of those obtained for MB sorption on unmodified and modified silica gels/aerogels ($K_L \approx 0.05$–0.7).[13,24,43] Considering both K_L and q_m values of the silica hydrogel is in the same order of magnitude as its gel/aerogel counterparts, silica hydrogels can be proposed to be potential adsorbent in cationic dye removal processes.

1.3.3 MB ADSORPTION KINETICS

Adsorption kinetic measurements were carried out using the system given in Figure 1.6 at two different initial concentrations of MB. Basically, the MB solution over the rigid transparent silica hydrogel layer was kept under continuous stirring by means of a magnet to make sure that there was no resistance to transfer of MB in solution. The concentration of solution was monitored immersing the fiber-optic spectrophotometer's probe in the solution (Fig. 1.6).

The transfer of MB from solution to silica hydrogel in the short time period at the interface was modeled using linear and nonlinear forms of pseudo first-order (PFO), pseudo second-order (PSO), Elovich kinetic models, and intraparticle solid diffusion model equations in Table 1.3. In these equations, q is the concentration of MB in silica hydrogel at time t, q_e is the concentration of MB in silica hydrogel at equilibrium, k_1 and k_2 are the rate constants of the PFO and PSO rate expressions, respectively, α and β are constants of Elovich model, and k_p is the intraparticle diffusion rate constant. Linear and nonlinear plots of these models are given in Figures 1.7 and 1.8. Model parameters and R^2, RMSE, and Δq (%) values are summarized in Table 1.4. In the linear fittings of PFO model, q_e values which were calculated by assuming all MB in solution was adsorbed by silica hydrogel were

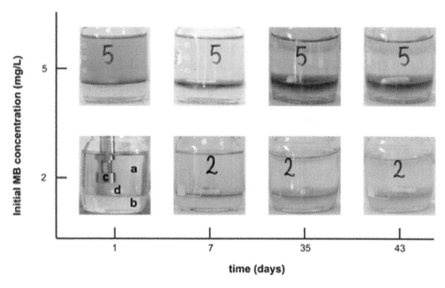

FIGURE 1.6 System of methylene blue solution over the silica hydrogel and diminishing ethylene blue concentration with time for 2 and 5 mg/L initial concentrations, (a) aqueous phase, (b) silica hydrogel phase, (c) fiber optic spectrometer probe, and (d) magnetic bar.

TABLE 1.3 Adsorption Kinetic Models.

Kinetic model	Nonlinear form	Linear form
Pseudo first order	$q = q_e\left(1 - e^{-k_1 t}\right)$	$\ln\left[\dfrac{q_e - q}{q_e}\right] = -k_1 t$
Pseudo second order	$q = \dfrac{k_2 q_e^2 t}{1 + k_2 q_e t}$	$\dfrac{t}{q} = \dfrac{1}{k_2 q_e^2} + \dfrac{1}{q_e} t$
Elovich	$q = \dfrac{1}{\beta} \ln\left(1 + \alpha\beta t\right)$	$q = \dfrac{1}{\beta} \ln\left(\alpha\beta\right) + \dfrac{1}{\beta} \ln\left(t\right)$ (if $\alpha\beta t \gg 1$)
Intraparticle	$q = k_p t^{0.5}$	$q = k_p \tau$ (where $\tau = t^{0.5}$)

used. Indeed, essentially clear solution was obtained over the hydrogel after more than 40 days confirming this assumption (Fig. 1.6). For the kinetic experiments at $C_i = 2$ mg/L, the PFO fit spanned linear within the first 70 min, whereas at $C_i = 5$ mg/L, this initial linear region only corresponds to the first 30 min. It was reported that in the majority of the kinetic studies, the PFO model fitted to the experimental data for the initial 20–30 min.[45] Conversely, for the PSO linear fits, initial regions of the plots deviated from the linearity for both initial concentrations. The linear region started after ~30 and ~10 min for C_i = 2 mg/L and C_i = 5 mg/L, respectively. Similarly, initial experimental data points taken during the first 10 and 5 min periods for C_i = 2 mg/L and C_i = 5 mg/L, respectively, did not fit to the linearized Elovich model. The initial deviations may be due to the fact that the multiplication of α and parameters are not high enough to satisfy $\alpha\beta t \gg 1$ criteria used to derive the linear form of the model.[46,47] Linearized intraparticle diffusion model fits failed to pass through the origin and presented a few different linear regions suggesting intraparticle diffusion is not the sole rate-limiting step.[48] Linear PFO model gave the highest R^2 value (0.971) for C_i = 2 mg/L, whereas this model gave the poorest fit at C_i = 5 mg/L. Similar R^2 values (0.967–0.970) were obtained for PSO, Elovich, and intraparticle linear models at C_i = 5 mg/L. Thus, linear analyses of the models indicated none of these models could describe the kinetics of the MB sorption onto silica hydrogel thoroughly. Based on the linear ranges of the models a composite model composed of PFO model and PSO/Elovich model corresponding to the initial and later stages of the sorption, respectively, can be suggested. The kinetic study of MB sorption on silica aerogels showed that experimental data can be fitted to both PSO and PFO model depending on the pore structure of the silica.[13] MB sorption kinetic data of the surface-modified silica gels and PAN–silica composite were reported to follow PFO and PSO models, respectively.[24,49] In the case of silica in monolith gel forms, in general, PSO was more likely to describe MB sorption kinetics rather than PFO model.[8] Thus, MB sorption kinetics of silica-based materials has a strong dependence on their surface properties, composition, and physical form. Likewise adsorption equilibrium models, the use of nonlinear curve fitting improved fitting quality of the PFO, PSO, and Elovich models with R^2 values greater than 0.99. RMSE values of them were also obtained in agreement with each other. However, PFO model at C_i = 5 mg/L presented the lowest Δq (%) value. This noticeable result is due to the smaller deviations of PFO model in the initial portion of the kinetic data as larger errors of PSO and Elovich model predictions in the in the early data points were more severely punished in the calculation of Δq (%). Thus, these three models were found to fit the entire kinetic data by arising questions overshadowing their physical interpretation as they became indistinguishable correlation functions. In the sorption kinetics analyses of MB onto activated carbon and three different fungi, *Rhizopus* sp. R-18, *Penicillium candidum*, and *Penicillium chrysogenum*, it was also indicated that the use of nonlinear models resulted in superior agreement with the experimental data compared to their linearized forms.[26,50,51]

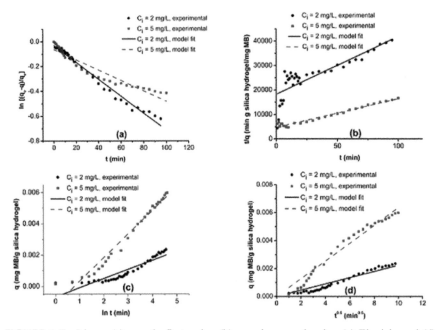

FIGURE 1.7   Linear (a) pseudo first-order, (b) pseudo second-order, (c) Elovich, and (d) intraparticle model plots.

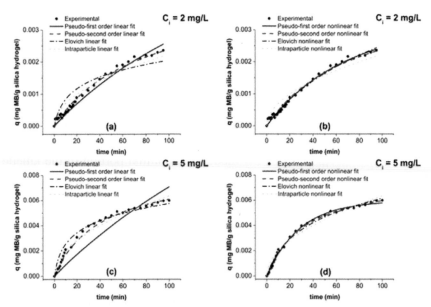

FIGURE 1.8   Comparison of the linear kinetic models (a) at $C_i = 2$ mg/L, (c) at $C_i = 5$ mg/L and nonlinear kinetic models (b) for $C_i = 2$ mg/L, and (d) at $C_i = 5$ mg/L.

TABLE 1.4 Kinetic Model-Fitting Parameters.

Model parameters	$C_i = 2$ mg/L		$C_i = 5$ mg/L	
	Linear fit	Nonlinear fit	Linear fit	Nonlinear fit
First-order model				
k_1 (min^{-1})	0.0073	0.0157	0.0051	0.0352
$q_e \times 10^3$ (mg/g)		3.1		6.0
R^2	0.971	0.991	0.829	0.996
RMSE	1.23×10^{-4}	0.69×10^{-5}	8.32×10^{-4}	1.32×10^{-4}
Δq (%)	26.25	21.43	35.69	12.17
Second-order model				
k_2 (L/mg/min)	3.078	2.082	2.839	3.943
$q_e \times 10^3$ (mg/g)	4.2	4.9	8.5	7.9
R^2	0.716	0.991	0.968	0.996
RMSE	0.81×10^{-5}	0.72×10^{-5}	1.78×10^{-4}	1.24×10^{-4}
Δq (%)	21.85	21.27	12.21	18.37
Elovich model				
$\alpha \times 10^4$ (mg/g/min)	1.85	0.52	6.74	2.95
β (g/mg)	1701	526	667	423
R^2	0.838	0.990	0.967	0.992
RMSE	3.04×10^{-4}	0.76×10^{-5}	4.71×10^{-4}	1.79×10^{-4}
Δq (%)	64.09	21.10	80.88	26.09
Intraparticle model				
$k_p \times 10^4$, (mg/g/min$^{0.5}$)	2.24	2.24	6.36	6.36
R^2	0.923	0.923	0.970	0.970
RMSE	2.07×10^{-4}	2.07×10^{-4}	3.55×10^{-4}	3.55×10^{-4}
Δq (%)	46.69	46.69	76.44	76.44

The difference in the fitting quality of linear and nonlinear models can be attributed to change in the error distribution during linear transformation.[50] Intraparticle diffusion model, on the other hand, presented the same model parameters for both linear and nonlinear fitting as linearization is simply the transformation of independent variable with a known exponent into another independent parameter in linear form without distorting the original expression due to the simplicity of the model.

MB sorption behavior of the hydrogel for the extended time interval can be tracked using the pictures taken within several weeks. As given in Figure 1.6, when MB solution was placed over the silica hydrogel layer,

MB was first adsorbed at the interface of solution and silica hydrogel layer. The adsorbed MB at the interface thereafter diffused into silica hydrogel with time forming a blue-colored region with the highest intensity at the interface. The steps of the mass transfer of MB from the aqueous phase to silica hydrogel and in silica hydrogel and their boundary conditions are as the followings:

1) MB solution placed over the silica hydrogel layer, $t = 0$, $C = C_i$, $q = 0$, $l = 0$;
2) MB is first adsorbed at the interface of water and silica hydrogel, $t = t$, $C = C$, $q = q$, $l = 0$ in a short period measured in hours (fast);
3) the adsorbed MB at the interface diffuses into silica hydrogel with time, $t = t$, $C = C$, $q = q$ for $0 \leq z \leq 1$ (slow); and
4) the equilibrium is attained in solution and silica hydrogel phases in a long period measured in weeks, $t = \infty$, $C = C_e$, $q = q_e$, for $0 \leq z \leq l_e$ (slow).

where t is time; z is the distance coordinate from the interface, l is the distance of blue color moved from the interface of silica hydrogel; C and q is the MB concentration in the aqueous phase and in silica hydrogel, respectively. The subscripts i and e refer to initial and equilibrium values, respectively. Figure 1.6 indicated that the transport of MB into silica hydrogel is a complicated process since both the solution concentration was decreasing with time and the adsorbed MB was diffusing in the silica hydrogel simultaneously.

The dye solutions became clear after weeks due to the resistance of the dye transport inside the silica hydrogel in the slab form. Conductivity measurements showed that sulfate ions remaining in silica hydrogel diffuse slowly from the silica hydrogel to the aqueous phase. The small amount of salt can reduce the diffusivity of MB in water and hence its sorption. It was reported that adsorption of MB on zeolites was reduced maximum 10% in the presence of NaCl and $CaCl_2$ even at concentrations as high as 0.2 M.[52] Since continuous stirring of the aqueous phase was provided during MB adsorption and amount of released sulfate ions is quite low, the presence of these ions seems negligible effect on adsorption. It is known that solution pH dictates the charge of the dye and surface charge of the adsorbent if they have ionizable groups. As a result of the nature of these charges electrostatic attractions and repulsions, dispersion forces, and H-bonding interactions may compete. pH of the solution was measured to decrease from ~5 to ~2 in the course of the adsorption most probably due to the proton release from the hydrogel. pK_a of MB is around 3.8.[53] Apparently, initially positively charged MB has become neutral and initially negatively charged silica has proceeded to onset of neutralization at the later stage of sorption. Thus, initial fast sorption rate can also be as a consequence of strong electrostatic attractions between the dye and the silica surface in addition to the availability of the sorption sites. Inside the hydrogel, silica surface is positively charged with associated sulfate and sodium ions and protons due to it acidic preparation conditions and MB is neutral. It is likely that electrostatic attractions between silica and MB are no longer available inside the hydrogel but replaced by some other forces. Hence, these forces along with the tightly packed silica network structure may contribute the slow diffusion of the MB inside the hydrogel.

1.4 CONCLUSIONS

MB adsorption on a silica hydrogel was investigated. Linear and nonlinear forms of the equilibrium and kinetic models were tested. In the case of adsorption equilibrium, the use of nonlinear models improved fit quality of Langmuir and Freundlich models, though none of the models could afford to agree with the experimental data for the entire coverage range. The value of Freundich exponent suggested surface heterogeneity of the silica hydrogel in consistent with the similar hydrogels. Adsorption capacity based on dry silica and Langmuir interaction parameter of the hydrogel was found to be in the same order of silica gels and aerogels.

PSO, PFO, Elovich, and intraparticle models were applied to short-term kinetics. Linear models failed to fit the entire data. Nonlinear PSO, PFO, and Elovich models were found to describe the experimental data thoroughly with R^2 values greater than 0.99. These outstanding agreements of between the models and the experimental data, however, have overshadowed the physical interpretation of these models. The hydrogel in the slab form was observed to attain equilibrium quite sluggishly, due to the slow diffusion of MB through the hydrogel. The resistance of MB transport inside the silica hydrogel can be advantageous in the development of a single-use storage time indicators in such areas as food and paint packaging's where the blue-colored front in silica hydrogel would indicate the time. For the adsorbent applications, on the other hand, small pellet or thin film form of the hydrogels should be used to maximize the exposure area as well as to provide efficient use of the material. Additionally, dyed silica hydrogel can find some application areas such that its utilization as pigment in plastics industry after grinding down to an appropriate size and drying.

KEYWORDS

- silica hydrogel
- xerogels
- methylene blue
- diffusion
- adsorption

REFERENCES

1. Cupane, A.; Levantino, M.; Santangelo, M. G. Near-Infrared Spectra of Water Confined in Silica Hydrogels in the Temperature Interval 365-5 K. *J. Phys. Chem. B.* **2002,** *106* (43), 11323–11328.

2. Iler, R. K. *The Chemistry of Silica: Solubility, Polymerization, Colloid and Surface Properties and Biochemistry*; John Wiley & Sons, Inc.: New York, 1979.

3. Kaczmarek, M.; Kazimierska-Drobny, K. Estimation–Identification Problem for Diffusive Transport in Porous Materials Based on Single Reservoir Test: Results for Silica Hydrogel. *J. Colloid Interface Sci.* **2007,** *311* (1), 262–275.

4. Asenath-Smith, E.; Hovden, R.; Kourkoutis, L. F.; Estroff, L. A. Hierarchically Structured Hematite Architectures Achieved by Growth in a Silica Hydrogel. *J. Am. Chem. Soc.* **2015,** *137* (15), 5184–5192.

5. Yokoi, T.; Kawashita, M.; Kawachi, G.; Kikuta, K.; Ohtsuki, C. Synthesis of Calcium Phosphate Crystals in a Silica Hydrogel Containing Phosphate Ions. *J. Mater. Res.* **2009,** *24*, 2154–2160.

6. Dong, W.; He, H.; Gong, J.; Yang, V. C. Immobilization of Penicillin G Acylase onto Amino-Modified Silica Hydrogel. *Front. Chem. Chin.* **2010,** *4* (1), 87–90.

7. Ferro, Y.; Perullini, M.; Jobbagy, M.; Bilmes, S. A.; Durrieu, C. Development of a Biosensor for Environmental Monitoring Based on Microalgae Immobilized in Silica Hydrogels. *Sensors* **2012,** *12* (12), 16879–16891.

8. Liu, H.; You, L.; Ye, X.; Li, W.; Wu, Z. Adsorption Kinetics of an Organic Dye by Wet Hybrid Gel Monoliths. *J. Sol–Gel Sci. Tech.* **2008,** *45* (3), 279–290.

9. Balköse, D. Effect of Preparation pH on Properties of Silica Gel. *J. Chem. Technol. Biotechnol.* **1990,** *49* (2), 165–171.

10. Titulaer, M.; Van Miltenburg, J.; Jansen, J.; Geus, J. Thermoporometry Applied to Hydrothermally Aged Silica Hydrogels. *Recl. Trav. Chim. Pays-Bas.* **1995,** *114* (8), 361–370.

11. Tantemsapya, N.; Meegoda, J. N. Estimation of Diffusion Coefficient of Chromium in Colloidal Silica Using Digital Photography. *Environ. Sci. Technol.* **2004,** *38* (14), 3950–3957.

12. Perullini, M.; Jobbágy, M.; Japas, M. L.; Bilmes, S. A. New Method for the Simultaneous Determination of Diffusion and Adsorption of Dyes in Silica Hydrogels. *J. Colloid Interface Sci.* **2014,** *425*, 91–95.

13. Liu, G.; Yang, R.; Li, M. Liquid Adsorption of Basic Dye Using Silica Aerogels with Different Textural Properties. *J. Non-Cryst. Solids* **2010**, *356* (4), 250–257.

14. Dorcheh, A. S.; Abbasi, M. Silica Aerogel: Synthesis, Properties and Characterization. *J. Mater. Process. Technol.* **2008**, *199* (1), 10–26.

15. Krupa, I.; Nedelčev, T.; Chorvát, D.; Račko, D.; Lacík, I. Glucose Diffusivity and Porosity in Silica Hydrogel Based on Organofunctional Silanes. *Eur. Polym. J.* **2011**, *47* (7), 1477–1484.

16. Li, Y.; Du, Q.; Liu, T.; Peng, X.; Wang, J.; Sun, J.; Wang, Y.; Wu, S.; Wang, Z.; Xia, Y. Comparative Study of Methylene Blue Dye Adsorption onto Activated Carbon, Graphene Oxide, and Carbon Nanotubes. *Chem. Eng. Res. Des.* **2013**, *91* (2), 361–368.

17. Vargas, A. M.; Cazetta, A. L.; Kunita, M. H.; Silva, T. L.; Almeida, V. C. Adsorption of Methylene Blue on Activated Carbon Produced from Flamboyant Pods (*Delonix regia*): Study of Adsorption Isotherms and Kinetic Models. *Chem. Eng. J.* **2011**, *168* (2), 722–730.

18. Rafatullah, M.; Sulaiman, O.; Hashim, R.; Ahmad, A. Adsorption of Methylene Blue on Low-Cost Adsorbents: A Review. *J. Hazard. Mater.* **2010**, *177* (1), 70–80.

19. Acemioglu, B.; Ertas, M.; Alma, M. H.; Usta, M. Investigation of the Adsorption Kinetics of Methylene Blue onto Cotton Wastes. *Turk. J. Chem.* **2014**, *38*, 454–469.

20. Demir, H.; Top, A.; Balköse, D.; Ülkü, S. Dye Adsorption Behavior of Luffa Cylindrica Fibers. *J. Hazard. Mater.* **2008**, *153* (1), 389–394.

21. Zhao, M.; Liu, P. Adsorption Behavior of Methylene Blue on Halloysite Nanotubes. *Microporous Mesoporous Mater.* **2008**, *112* (1), 419–424.

22. Gaikwad, R.; Misal, S. Sorption Studies of Methylene Blue on Silica Gel. *Int. J. Chem. Eng. Appl.* **2010**, *1* (4), 342–345.

23. Yanishpolskii, V.; Skubiszewska-Zieba, J.; Leboda, R.; Tertykh, V.; Klischar, I. Methylene Blue Sorption Equilibria on Hydroxylated Silica Surfaces as well as on Carbon–Silica Adsorbents (Carbosils). *Adsorpt. Sci. Technol.* **2000**, *18* (2), 83–95.

24. Chen, M.; Chen, Y.; Diao, G. Adsorption Kinetics and Thermodynamics of Methylene Blue onto *p-tert*-Butyl-calix (4, 6, 8) Arene-Bonded Silica Gel. *J. Chem. Eng. Data* **2010**, *55* (11), 5109–5116.

25. Kothawala, D.; Moore, T.; Hendershot, W. Adsorption of Dissolved Organic Carbon to Mineral Soils: A Comparison of Four Isotherm Approaches. *Geoderma* **2008**, *148* (1), 43–50.

26. Lin, J.; Wang, L. Comparison between Linear and Non-linear Forms of Pseudo-first-order and Pseudo-second-order Adsorption Kinetic Models for the Removal of Methylene Blue by Activated Carbon. *Front. Environ. Sci. Eng. Chin.* **2009**, *3* (3), 320–324.

27. Ülkü, S.; Balköse, D.; Baltacıoğlu, H. Effect of Preparation pH on Pore Structure of Silica Gels. *Colloid Polym. Sci.* **1993**, *271* (7), 709–713.

28. Vijayalakshmi, U.; Balamurugan, A.; Rajeswari, S. Synthesis and Characterization of Porous Silica Gels for Biomedical Applications. *Trends Biomater. Artif. Organs* **2005**, *18* (2), 101–105.

29. Fidalgo, A.; Ilharco, L. M. The Influence of the Wet Gels Processing on the Structure and Properties of Silica Xerogels. *Microporous Mesoporous Mater.* **2005**, *84* (1), 229–235.

30. Chukin, G.; Apretova, A. Silica Gel and Aerosil IR Spectra and Structure. *J. Appl. Spectrosc.* **1989**, *50* (4), 418–422.

31. Sarawade, P. B.; Kim, J.-K.; Hilonga, A.; Quang, D. V.; Kim, H. T. Synthesis of Hydrophilic and Hydrophobic Xerogels with Superior Properties Using Sodium Silicate. *Microporous Mesoporous Mater.* **2011,** *139* (1), 138–147.

32. Tong, C. L.; Stroeher, U. H.; Brown, M. H.; Raston, C. L. Continuous Flow Vortex Fluidic Synthesis of Silica Xerogel as a Delivery Vehicle for Curcumin. *RSC Adv.* **2015,** *5* (11), 7953–7958.

33. Ho, Y.; Porter, J.; McKay, G. Equilibrium Isotherm Studies for the Sorption of Divalent Metal Ions onto Peat: Copper, Nickel and Lead Single Component Systems. *Water, Air, Soil Pollut.* **2002,** *141* (1–4), 1–33.

34. Elmorsi, T. M.; Mohamed, Z. H.; Shopak, W.; Ismaiel, A. M. Kinetic and Equilibrium Isotherms Studies of Adsorption of Pb(II) from Water onto Natural Adsorbent. *J. Environ. Protect.* **2014,** *5* (17), 1667–1681.

35. Hammud, H. H.; Fayoumi, L.; Holail, H.; Mostafa, E.-S. M. Biosorption Studies of Methylene Blue by Mediterranean algae Carolina and its Chemically Modified Forms. Linear and Nonlinear Models' Prediction Based on Statistical Error Calculation. *Int. J. Chem.* **2011,** *3* (4), 147.

36. Chen, X. Modeling of Experimental Adsorption Isotherm Data. *Information* **2015,** *6* (1), 14–22.

37. Armagan, B.; Toprak, F. Optimum Isotherm Parameters for Reactive Azo Dye onto Pistachio Nut Shells: Comparison of Linear and Non-linear Methods. *Pol. J. Environ. Stud.* **2013,** *22* (4), 1007–1011.

38. Subramanyam, B.; Das, A. Linearised and Non-linearised Isotherm Models Optimization Analysis by Error Functions and Statistical Means. *J. Environ. Health Sci. Eng.* **2014,** *12* (1), 92.

39. Kumar, K. V.; Porkodi, K.; Rocha, F. Isotherms and Thermodynamics by Linear and Non-linear Regression Analysis for the Sorption of Methylene Blue onto Activated Carbon: Comparison of Various Error Functions. *J. Hazard. Mater.* **2008,** *151* (2), 794–804.

40. Foo, K.; Hameed, B. Insights into the Modeling of Adsorption Isotherm Systems. *Chem. Eng. J.* **2010,** *156* (1), 2–10.

41. Fubini, B.; Bolis, V.; Cavenago, A.; Garrone, E.; Ugliengo, P. Structural and Induced Heterogeneity at the Surface of Some Silica Polymorphs from the Enthalpy of Adsorption of Various Molecules. *Langmuir* **1993,** *9* (10), 2712–2720.

42. Haghseresht, F.; Lu, G. Adsorption Characteristics of Phenolic Compounds onto Coal-Reject-Derived Adsorbents. *Energy Fuels* **1998,** *12* (6), 1100–1107.

43. Kushwaha, A. K.; Gupta, N.; Chattopadhyaya, M. Enhanced Adsorption of Methylene Blue on Modified Silica Gel: Equilibrium, Kinetic, and Thermodynamic Studies. *Desalinat. Water Treat.* **2014,** *52* (22–24), 4527–4537.

44. Kushwaha, A. K.; Chattopadhyaya, M. Surface Modification of Silica Gel for Adsorptive Removal of Ni^{2+} and Cd^{2+} from Water. *Desalinat. Water Treat.* **2015,** *54* (6), 1642–1650.

45. Ho, Y.; McKay, G. A Comparison of Chemisorption Kinetic Models Applied to Pollutant Removal on Various Sorbents. *Process Saf. Environ. Prot.* **1998,** *76* (4), 332–340.

46. Wu, F.-C; Tseng, R.-L.; Juang, R.-S. Characteristics of Elovich Equation Used for the Analysis of Adsorption Kinetics in Dye–Chitosan Systems. *Chem. Eng. J.* **2009,** *150* (2), 366–373.

47. Sparks, D.; Jardine, P. Comparison of Kinetic Equations to Describe Potassium-Calcium Exchange in Pure and in Mixed Systems. *Soil Sci.* **1984,** *138* (2), 115–122.

48. Zheng, H.; Liu, D.; Zheng, Y.; Liang, S.; Liu, Z. Sorption Isotherm and Kinetic Modeling of Aniline on Cr-Bentonite. *J. Hazard. Mater.* **2009,** *167* (1), 141–147.

49. Ayad, M. M.; El-Nasr, A. A.; Stejskal, J. Kinetics and Isotherm Studies of Methylene Blue Adsorption onto Polyaniline Nanotubes Base/Silica Composite. *J. Ind. Eng. Chem.* **2012,** *18* (6), 1964–1969.

50. Kumar, K. V. Linear and Non-linear Regression Analysis for the Sorption Kinetics of Methylene Blue onto Activated Carbon. *J. Hazard. Mater.* **2006,** *137* (3), 1538–1544.

51. Vrtoch, L.; Augustin, J. Linear and Non-linear Regression Analysis for the Biosorption Kinetics of Methylene Blue. *Nova Biotechnol. Chim.* **2009,** *9*, 199–204.

52. Han, R.; Zhang, J.; Han, P.; Wang, Y.; Zhao, Z.; Tang, M. Study of Equilibrium, Kinetic and Thermodynamic Parameters about Methylene Blue Adsorption onto Natural Zeolite. *Chem. Eng. J.* **2009,** *145* (3), 496–504.

53. Kim, J. R.; Santiano, B.; Kim, H.; Kan, E. Heterogeneous Oxidation of Methylene Blue with Surface-Modified Iron-Amended Activated Carbon. *Am. J. Anal. Chem.* **2013,** *4*, 115–122.

CHAPTER 2

ADVANCEMENTS IN BORIC ACID PRODUCTION FROM BORON MINERALS

MEHMET GÖNEN*

Department of Chemical Engineering, Engineering Faculty, Süleyman Demirel University, Batı Yerleşkesi, Isparta 32260, Turkey

E-mail: mehmetgonen@sdu.edu.tr

CONTENTS

ABSTRACT

Following is an overview of boric acid production by environmental considerations from different boron sources, along with its characterization methods. Boric acid is an important primary compound which is currently produced from the reaction of sulfuric acid and colemanite mineral in Turkey. The main problem in boric acid production is the formation of wastewater streams bearing boron above the accepted standard limits. Especially, the gypsum with high boron content creates storage problems. As sulfuric acid dissolves almost everything in mineral, removal of impurities from boric acid consumes much water in separation step. The use of CO_2 in boric acid production would be an alternative method in which both CO_2 would be mineralized and boric acid would be extracted from mineral. The review provides an evaluation of current techniques for improving the process and to promote the cleaner production in the boron industry.

2.1 INTRODUCTION

Boron element having an atomic number of 5 is a metalloid and found between metals and nonmetallic elements in the periodic system. The boron element combines with oxygen in a trigonal and tetrahedral coordination and those anions react with calcium, sodium, magnesium, or a combination of those alkaline and alkaline earth metals to form natural boron minerals such as colemanite, tincal, and ulexite in nature.[47,55] Although there are 200 boron compounds defined in the world, only calcium and sodium borates shown in Table 2.1 have commercial value.

TABLE 2.1   Important Borate Minerals.

Mineral	Formula	B_2O_3 (%)	H_2O (%)
Colemanite	$Ca_2B_6O_{11} \cdot 5H_2O$	50.8	21.9
Ulexite	$NaCaB_5O_9 \cdot 8H_2O$	43.0	35.6
Kernite	$Na_2B_4O_7 \cdot 4H_2O$	50.9	26.4
Tincalconite	$Na_2B_4O_7 \cdot 5H_2O$	47.8	30.9
Tincal (borax)	$Na_2B_4O_7 \cdot 10H_2O$	36.5	47.2
Pandermite	$Ca_4B_{10}O_{19} \cdot 7H_2O$	49.8	18.1

Boric acid (H_3BO_3) is an important boron compound which is utilized in many applications directly as wood preservative,[16] insecticides, and fungicides.[47] It is also used in the production of other chemicals such as zinc borates,[52] borate esters,[30] and boron nitrides.[13] Turkey has the largest boron reserves in the world and its boron reserve was reported as 883 Mt based on B_2O_3 wt% content given by Eti Maden Inc. in 2008.[47] According to the reserves explored recently, Turkey has 70% of the world boron reserves and they are mainly located in the western Anatolia. Eti Maden Inc., which is a state-owned chemical and mining company, is responsible for the production of boron products in Turkey. Its factories are located in western Anatolia (Bandirma, Bigadiç, Emet, and Kirka). Plants are designed according to the mineral present in the region; for example, Kirka and Emet plants produce borax and boric acid, respectively.

Boric acid can be produced either from solid ores of colemanite, kernite, tincal, ulexite, or from brines having dissolved boron compounds.[7] It is usually produced from the heterogeneous reaction of colemanite mineral and sulfuric acid in Turkey.[10] The challenges in boric acid production are discussed based on the recent studies. The waste formed in the boron industry of Turkey is a key environmental concern as it produces a large amount of wastewater and solid waste discharges.[8] To eliminate or decrease the amount of those wastes, various acids,[15,19,40] carbon dioxide gas,[9,23] and SO_2 gases[18] have been investigated in boric acid production.

In this study, we have attempted to summarize the studies related to boric acid production, its utilization in the synthesis of other boron compounds and discuss the parameters involved in the boric acid production process, considering the environmental impacts. A special emphasis was given to the waste generation from the colemanite–sulfuric acid process. Characterization methods of those boron products and streams having boric acid have been summarized.

2.1.1 PHYSICAL AND CHEMICAL PROPERTIES

Boric acid is a white crystalline powder with a triclinic structure. It is soluble in water and alcohols as it has hydroxyl groups in its structure. Its solubility is 4.7 wt% at 20°C in water.[45] Its solubility increases with the rise in temperature. Boric acid dissolves in water accepting the hydroxyl group of water to form a borate anion and a proton as shown in eq. (2.1) commonly known to be Lewis acid. An equilibrium constant for the reaction in eq. (2.1) is given as 5.80×10^{-10} at 25°C.[49] At 0.12-M boric acid concentration, the pH was measured as 5.01; when the concentration was increased to 0.75 M, the pH value reached to 3.69.[17] There is a deviation between the measured pH and the calculated pH from the equilibrium constant as boric acid molecules give a condensation reaction with increased concentration.[17] Polymeric borate anions ($B_3O_3(OH)_4^-$, $B_3O_3(OH)_5^{2-}$, $B_6O_6(OH)_4^-$, $B_4O_5(OH)_4^{2-}$, $B_3O_4(OH)_3^{2-}$) are formed depending on the boric acid concentration in the solution.[50,55]

$$B(OH)_{3(s)} + H_2O_{(l)} \leftrightarrow B(OH)_{4\,(aq)}^- + H_{(aq)}^+ \qquad (2.1)$$

The solubility and the pH of boric acid are substantially important parameters during its production. For instance, the solubility of boric acid should be taken into account during the reaction, filtration, and crystallization steps. The amount of boric acid could be kept within solubility range during the reaction and filtration by changing the solid–liquid ratio of feed streams, varying the temperature. A filter cake is usually the desired product in most of filtration processes. However, in boric acid production from insoluble borate salts such as colemanite, by-product, calcium sulfate, precipitates, and forms the filter cake, the boric acid being in the filtrate is crystallized by saturating the solution and lowering the solution temperature. The parameters such as temperature, feed rate, and water flow rate need to be optimized in boric acid separation.

2.2   BORIC ACID PRODUCTION

Boric acid is mainly produced from sodium and calcium borates in America and Europe, respectively, depending on the abundance of minerals. Boric acid is also found as a sassolite mineral in nature. There are many studies in literature for boric acid production from different boron minerals using different acids as given in Table 2.2. The production process is mainly tailored according to the type of boron minerals, its state (solid or dissolved in aqueous phase), and the product separation from the reaction mixture.[7] There are two main approaches for processing of borate salts. In the first one, borate minerals are excavated in an open-pit mine and further processed for separation and size reduction before reaching the plant. This approach is suitable if borate deposits are on the surface or in an accessible depth. In the second approach, if borate minerals are soluble in water or well below the ground surface in which open-pit mining is not feasible, in situ borate leaching is applied as performed by Fort Cady Minerals Corp.[27] Boric acid production process consists of a reactor, filtration, crystallization, and drying units.[21] It is usually designed as a continuous process by utilizing several reactors in cascade mode. Mineral processing operations, for example, excavation, size reduction, and mineral purification are out of the scope of this review. Thus, minerals having a certain particle size, less than 150 µm, are used in the reaction. However, impurities, for example, arsenic compounds like realgar (As_2S_2) and orpiment (As_2S_3) being in mineral between 0.03 and 4 wt%[12] must be separated to decrease separation cost of the product and acid consumption in the process.[14]

2.2.1 BORIC ACID FROM SODIUM BORATES

Boric acid is produced from sodium borates especially in the United States as there is an abundant kernite ($Na_2B_4O_7 \cdot 4H_2O$) reserves.[49] Since sodium borates are soluble in water, they might be present as dissolved salt in lakes. Searles Valley minerals produce boron compounds from the brines in Trona, California. If the boron mineral includes some impurities, it is initially treated to remove those insoluble impurities such as clay and carbonates by using coagulants.[54] After that separation, either soluble boron mineral is crystallized to produce borax, or it is concentrated and used as a raw material in boric acid production. Being known that the other acids (nitric and hydrochloric) proposed for boric acid production are more expensive than the sulfuric acid, they cannot be used in a commercial scale. Organic acids such as oxalic acid,[2] acetic acid,[19] and citric acid[20] were investigated for boric acid production from sodium borates as given in Table 2.2.

Tolun and his coworkers studied the electrolysis of borax solution to produce boric acid and sodium hydroxide. A mercury cathode and a cation exchange membrane were utilized in the system. It was reported that the working temperature must be between 60°C and 100°C and the working voltage has to be between 3.5 and 5 V.[62] However, this process was never commercialized in industry due to the high energy prices. Since borax is soluble in water, the reaction given in eq. (2.2) is a homogeneous reaction in which the value of x (5 and 10) represents the sodium borate pentahydrate and decahydrate, respectively.

$$Na_2B_4O_7 \cdot xH_2O + H_2SO_4 \rightarrow 4B(OH)_3 + Na_2SO_4 + (x-5)H_2O \qquad (2.2)$$

The separation of boric acid and sodium sulfate is performed based on their solubility at different temperatures. The production of boric acid from sodium tetraborate was studied by Emil in 1934 by using 66° Be of sulfuric acid at a temperature of between 90°C and 100°C.[21] Sodium sulfate, having lower solubility at higher temperatures, was filtered from

TABLE 2.2 Previous Studies on the Reactions of Different Boron Mineral with Different Acids.

Boron mineral	Proton source	Conditions	By-products	References
Colemanite	Sulfuric acid and hydrochloric acid	H_2SO_4 (0.5–2 M), HCl (1 and 2 M), $T = 35°C$	Calcium sulfate and gypsum	[24]
Colemanite	Phosphoric acid	H_3PO_4 (1.43–19.52 wt.%), $T = 2.5–35°C$	Calcium hydrogen phosphate, calcium phosphate	[60]
Colemanite	Phosphoric acid	H_3PO_4 (2.5–2.9 M), $T = 94°C$	Calcium phosphate	[69]
Colemanite	Nitric acid	HNO_3 (2.0–2.4 M), $T = 94°C$	Calcium nitrate	[68]
Colemanite	Acetic acid	CH_3COOH (1.7–10.3 M) $T = 10.6–50°C$	Calcium sulfate and calcium acetate	[6,42]
Colemanite	Propionic acid and sulfuric acid	H_2SO_4 (30–90%), $C_3H_6O_2$ (10–70 wt%), $T = 90°C$	Calcium sulfate, calcium propionate	[10]
Colemanite	Propionic acid	$C_3H_6O_2$ (10–16 M), $T = 288–308K$	Calcium propionate	[70]
Colemanite	Citric acid	$C_6H_8O_7$ (0.025–0.3 M), $T = 25–80°C$	Calcium citrate	[11]
Colemanite	Oxalic acid	$H_2C_2O_4$ (0.05–0.75 M) $T = 30–60°C$	Calcium oxalate	[4]
Colemanite	Carbon dioxide	$P = 7.3–9.1$ MPa, $T = 35–60°C$	Calcite	[9]
Tincal	Oxalic acid	$C_2H_2O_4$ (1–10% by w/v) $T = 20–50°C$	Sodium oxalate	[2]
Tincal	Phosphoric acid	H_3PO_4 (0.8–25 M), $T = 10–50°C$	Sodium phosphate	[1]
Tincal (Borax)	Sulfuric acid	(95.1–98%), $T = 98°C$	Sodium sulfate	[37]
Ulexite	Phosphoric acid	H_3PO_4 (0.25–2.5 M) $T = 273–303$ K	Calcium phosphate and sodium phosphate	[15]
Ulexite	Acetic acid	CH_3COOH (0.025–0.50 M)	Calcium acetate and sodium acetate	[19]
Ulexite	Sulfuric acid	H_2SO_4 (0.23–4.9 M)	Calcium sulfate and sodium sulfate	[63]
Ulexite	Oxalic acid	$C_2H_2O_4$ (0.05–0.5 M) $T = 30–60°C$	Calcium oxalate	[3]

the reaction liquor at the reaction temperature and boric acid was crystallized and filtered at 30–35°C. The filtrate stream saturated with boric acid was recycled to the reactor.[21] Boric acid production was investigated from alkaline metal borate and concentrated sulfuric acid by using superheated steam at different temperatures (100–400°C). The alkaline metal to sulfur mole ratio was given as 1.9 to 2.1 and the particle size of metal borate was pointed between 100 and 200 mesh screen sizes. It was reported that steam separated the boric acid from the reaction zone as it volatilizes at the temperature of operation.[53] The utilization of CO_2 in borax solution to produce boric acid was suggested by Nelson in 1962. He suggested that by varying the temperature (26–46°C) and CO_2 pressure (48–615 psi), sodium bicarbonate and boric acid could be crystallized successively from the solution. For example, at 46°C, sodium bicarbonate was formed by introducing CO_2 into borax solution at 615 psi, and sodium bicarbonate was filtrated and washed then the temperature of solution was lowered to 32°C to crystallize boric acid at 48 psi CO_2 pressure.[38] CO_2 gas was used to produce the mixture of boric carbonate and sodium bicarbonate from borax solutions.[41] The experiments were carried out in three different systems: an autoclave, a reactor with jacket, and a gas absorption column at different conditions. The reaction between sodium borate solution and CO_2 gas is shown in eq. (2.3).

$$Na_2B_4O_7 \cdot nH_2O + (7-n)\,H_2O + 2CO_2 \rightarrow 2NaHCO_3 + 4H_3BO_3 \qquad (2.3)$$

The maximum $NaHCO_3$ value (79.4%) was obtained in the autoclave system at the conditions of 600 psi CO_2 pressure, a temperature of 65°C, and for a 120-min reaction time. $NaHCO_3/B_2O_3$ ratio could be adjusted depending on where the product would be used. It was pointed out that mixture of boric acid and sodium bicarbonate could be separated by extracting the boric acid with methyl alcohol. After the filtration of the remaining mixture, sodium bicarbonate with the purity (98–99 wt%) and 87% efficiency was obtained.[41] It is seen that no separation step was required for the final products' mixture, similarly used in ceramic and glass industries.

The electrochemical production of boric acid from borax pentahydrate $(Na_2B_4O_7 \cdot 5H_2O)$ was studied by Elbeyli and coworkers.[67] The electrolysis cell has two compartments separated by Nafion®324 membrane. The electrolysis was performed at 85°C by feeding saturated borax solution to the anode compartment and distilled water to the cathode compartment. Different current densities (1.2, 1.8, and 2.9 kA m^{-2}) were applied by DC power source. The optimum conditions were reported as 1.8 kA m^{-2} current density and 20% NaOH concentration. The specific energy consumption was reported as 2000 kW t^{-1} of boric acid.[67] The performance and service life of the membrane used in this method depends intensively purity of anode electrolyte (borax solution). The presence of other cations, such as Mg and Ca, may cause the fouling of pores. Thus, pretreatment of the feed solution by ion exchange resins was required and all of them increase the overall cost of the process.

2.2.2   BORIC ACID FROM CALCIUM BORATES

Boric acid is commercially produced from the reaction of colemanite and sulfuric acid at 88–92°C under atmospheric pressure in Turkey as shown in eq. (2.4).[10,32] Boric acid dissolves in aqueous phase as a product and gypsum $(2CaSO_4 \cdot 2H_2O)$ is formed as a solid by-product in the reaction mixture. The reaction could also be carried out at higher temperatures (98–102°C) to increase the reaction rate, but calcium sulfate hemihydrate $(2CaSO_4 \cdot 1/2H_2O)$ is formed as a by-product, which causes an increase in the amount of waste water discharged from process.[10]

$$Ca_2B_6O_{11} \cdot 5H_2O_{(s)} + 2H_2SO_{4(l)} + 6H_2O_{(l)} \rightarrow 2CaSO_4 \cdot 2H_2O_{(s)} \\ + 6H_3BO_{3(aq)} \tag{2.4}$$

The heat of reaction in eq. (2.4) was calculated as -329.46 kJ mol^{-1} by using the molar heat of formation of reactants and products. The reaction between colemanite and sulfuric acid is highly exothermic ($\Delta H < 0$). The continuous boric acid production process designed by SuperPro Designer (Version 9.0) consists of a reactor, filtration-washing unit (belt filtration), crystallization, and decantation and drying units as shown in Figure 2.1. The reaction is carried out in several reactors working in staggered mode with the total residence time of 3–3.5 h.[34] Calcium sulfate (gypsum) and unreacted inert solids in the colemanite mineral are separated in a belt filter and the filtrate has boric acid 15–18 wt% depending on stream temperature. However, the filter cake is further washed by hot water to recover the boric acid retained, and the filter cake is discharged into a waste water dam in the form of slurry, containing some boric acid. Since boron level in water is critically important for plants and living organisms, that slurry should not be discharged into rivers or lakes.[44] The boron industry in Turkey is producing a significant amount of wastewater related to its production rate.

In the crystallization unit, the boric acid solution is concentrated by evaporating water under vacuum, and after reaching the desired saturation level, the solution is cooled to 35°C to crystallize the boric acid.[34] A centrifuge separates the boric acid crystals and the liquid stream is recycled to the reactor. The wet boric acid crystals from the centrifuge are washed and filtered to remove $MgSO_4$ precipitated as an impurity. Then, some part of the filtrate is recycled to the reactor and the rest is discharged to prevent Mg accumulation in the process.[10] The boric acid having 7–8 wt% water is dried in a fluidized bed dryer by using hot air. There are several issues in this drying operation: dehydration of boric acid, fluidization of smaller boric acid particles, and its sublimation. The temperature of air stream is important as boric acid dehydrates at higher temperatures such as 100 and 130°C. It was pointed that drying at 343 K cannot decompose or cause sublimation of boric acid.[43] Thus, temperature in boric acid drying must be controlled or an outlet gas stream must be scrubbed by water for recovering either fluidized boric acid particle or the sublimed product.

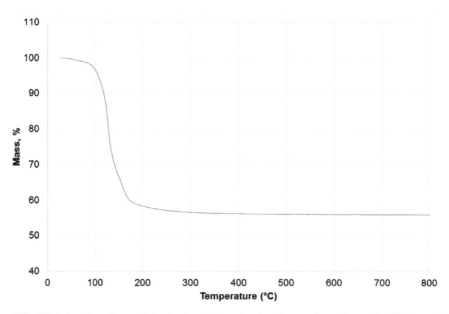

FIGURE 2.1 Flowsheet of the boric acid production from colemanite and sulfuric acid process (designed by SuperPro Designer, Version 9.0).

Instead of sulfuric acid, carbonic acid could be utilized in boric acid production from colemanite mineral. Carbon dioxide gas dissolves in water by forming carbonic acid according to eq. (2.5).[46] The consumption of atmospheric CO_2 by oceans occurs based on CO_2 dissolution and reaction with cations in water as pointed out in literature.[51] However, the equilibrium between CO_2 emission rate and CO_2 consumption either by plants or oceans has been destroyed. We are producing much more CO_2 than the one the ecosystem can consume. As a consequence, CO_2 concentration in the atmosphere has reached 400 ppm.[57] The formed carbonic acid produces a proton (eq. (2.6)) to become utilized in acid leaching of the mineral. The formed bicarbonate anion further dissociates into a carbonate ion and H^+ as shown in eq. (2.7). When CO_2 dissolves in water according to eq. (2.5), the carbonic acid formed attacks the mineral structure. Ca^{2+} cations from the mineral and carbonate anions from CO_2 react to produce calcium carbonate. CO_2 could be utilized for the colemanite mineral, which can, in turn, be good source for CO_2 mineralization.

$$CO_{2(g)} + H_2O_{(l)} \leftrightarrow H_2CO_{3(aq)} \qquad (2.5)$$

$$H_2CO_{3(aq)} \leftrightarrow HCO_3^-{}_{(aq)} + H^+{}_{(aq)} \qquad (2.6)$$

$$HCO_3^-{}_{(aq)} \leftrightarrow CO_3^{2-}{}_{(aq)} + H^+{}_{(aq)} \qquad (2.7)$$

Gülensoy and Kocakerim[23] investigated the formation of calcite in colemanite minerals by passing CO_2 gas through a suspension of colemanite and water. About 24.0% boric acid extraction efficiency was obtained by dissolution of colemanite mineral at room temperature. When the temperature was raised to 50°C, the dissolution efficiency of colemanite increased to 58.9% for 1 h of mixing. The extraction efficiency of boric acid from calcined colemanite mineral increased parallel to the calcination temperature. Although calcination increases extraction efficiency, it consumes a large amount of energy, and thus, it is not feasible to use this at industrial scale. A boric acid extraction efficiency of 96.3% was obtained from colemanite mineral which was calcined at 400°C.[23]

Ata and coworkers studied the extraction of boric acid from colemanite mineral suspension saturated with CO_2 utilizing the Taguchi method to determine the optimum conditions. A boric acid extraction efficiency of 54% was found at 70°C, and a gas flow rate of 711 mL min^{-1} was achieved under atmospheric pressure for the reaction time of 90 min. After increasing the CO_2 pressure to 2.7 atm in a closed reactor, 75% of the boric acid extraction efficiency was obtained at reaction temperature (70°C), and for the reaction time of 120 min. The boric acid extraction efficiency from calcined colemanite mineral was determined as 99.6%.[5] The 75% boric acid extraction efficiency could be enhanced by increasing the CO_2 pressure in the reactor as CO_2 solubility increases with increased pressure. On the other hand, the calcination of minerals does not apply for a for large scale production as it consumes much energy.

The overall reaction between solid colemanite and carbonic acid dissolved in aqueous phase is shown in eq. (2.8). The intermediate steps of colemanite and CO_2 reaction can be found in our previous study.[9] The heat of reaction in eq. (2.8) is theoretically calculated as -107.76 kJ mol^{-1} by using the molar formation energies of reactant and products. The heat of reaction in eq. (2.4) is approximately three times greater than the one calculated for eq. (2.8).

$$Ca_2B_6O_{11} \cdot 5H_2O_{(s)} + 2CO_{2(g)} + 4H_2O_{(l)} \rightarrow 2CaCO_{3(s)} + 6H_3BO_{3(s)} \qquad (2.8)$$

The extraction of boric acid from a colemanite mineral was investigated by supercritical carbon dioxide in the aqueous phase.[9] Increasing CO_2 pressure beyond the critical points of CO_2 has an accelerated reaction rate that induces a higher conversion. A boric acid extraction of 72.7% was obtained at 35°C and 7.7 MPa CO_2 pressure for 2 h of reaction time. Increasing the CO_2 pressure to 9.0 MPa at a 60°C reaction temperature, brought the rise in conversion of 96.9% for a 2-h reaction time. In our recent study, the effect of pressure between 4 and 12 MPa was investigated at 50°C and for 30 min of reaction time. It was found that up to the critical pressure of CO_2, there is significant enhancement in conversion. On the other hand, reaction carried out at 70°C at 6 MPa pressure for 60 min resulted in a 99% conversion. As CO_2 has critical points ($T_c = 31.1$°C and $P_c = 73$ atm), the reaction carried out in the supercritical region of CO_2 has enhanced reaction rate by decreasing mass transfer limitations. There are two factors (gas solubility and reaction rate) competing each other with rising temperature. When temperature is raised, the gas solubility decreases and reaction rate increases according to the Arrhenius equation. In that study, it was shown that increasing reaction temperature improved reaction rate more. Boric acid was extracted and CO_2 was mineralized to calcium carbonate, which is a thermodynamically stable form.[22] The utilization of CO_2 from the power station can help to decrease the CO_2 emissions. On the other hand, the separation of boric acid from reaction mixture would be more practical based on the solubility of the by-products.

The purity of the product is very important, since it is related to its properties and its price. In current boric acid productions, $CaSO_4 \cdot 2H_2O$ and $MgSO_4$ are formed as by-products from the reaction between colemanite and sulfuric acid. Since natural colemanite ore contains dolomite $(CaMg(CO_3)_2)$ and clay minerals as an impurity and if colemanite ore bears sodium magnesium borates, $MgSO_4$ is formed as by-product which has higher solubility in water (35.1 g 100 mL^{-1} at 20°C).[35] Because of this higher solubility, SO_4^{2-} ions are transported by a mother liquor solution to the final crystallization step where boric acid is obtained. To decrease the dissolution of dolomite and clay minerals in colemanite ore, propionic acid was used in the dissolution of colemanite ore.[10] The use of propionic acid decreased Mg concentration to half of the value in the case of H_2SO_4. When CO_2 is used in the dissolution of the colemanite ore, one of the benefits would be a decrease for impurities related with Mg since $MgCO_3$ would be produced as by-product, which has a lower solubility value (0.18 g 100 mL^{-1} H_2O at 25°C)[35] than the solubility of $MgSO_4$ (35.7 g 100 mL^{-1} H_2O at 25°C). To produce highly pure boric acid in the current process, final wet solid boric acid is washed by purified water.

2.2.3   THE USE OF ULTRASOUND IN BORIC ACID PRODUCTION

The use of ultrasound is considered to be an effective agitation method for enhancing the rate of reaction, mass transfer, and increasing the yield and selectivity for both homogeneous and heterogeneous reactions.[25,61] Ultrasound applies the transmission of waves that compress and stretch the molecular spacing of a physical medium. As the ultrasound crosses the medium, the average distance between the molecules vary as they oscillate about their mean position.[48] Cavitation bubbles are caused by large negative pressure of ultrasonic waves when crossing the liquid, the distance between the molecules is large enough that it exceeds the minimum distance that the liquid requires to stay intact. When this liquid breaks down, they create voids that are classified as cavitation bubbles.[26] Microstreaming is a phenomenon used in ultrasound and is caused by cavitation, which are the asymmetric or symmetric implosive collapse of microbubbles. The liquid surrounding the solid help create shock waves to cause a microscopic turbulence and sometimes the thinning of the solid–liquid film. Microstreaming relates to the increase of the mass transfer coefficient which also affects the dissolution and decrease in particle size.[26] Cavitation from ultrasonic energy has physical and chemical effects on both reactants and products. The utilization of ultrasound leads to change in properties of crystals, a decrease in the induction time of nucleation,[36] narrow the particle size distribution,[59] and increase the nucleation rate.[36] Those changes directly relate to the mass transfer coefficient and the decrease in particle size through the ionic lattice or arrangement of the product layer.[26] Solid–liquid reactions leading to crystallization of the product are significantly important in the production of industrial chemicals. As summarized above, ultrasound not only decreases the particle size of solid reactant but also removes the precipitated product or byproduct layer on the solid reactant.

In the production of boric acid from colemanite mineral in the aqueous phase, the reaction mixture is heterogeneous, as colemanite is insoluble in water. Such solid–liquid reactions were studied utilizing ultrasonic energy for decreasing the particle size of colemanite,[58] increasing the dissolution rate of colemanite in H_2SO_4 solution.[40] The improvement of mass transfer rate and the decrease in particle size influence the rate of reaction in a heterogeneous system where ultrasonic energy used.[40] Although ultrasound was used externally in heterogeneous solid–liquid reaction, the properties of crystallization, the dissolution rate of colemanite and precipitation rate of the gypsum were enhanced by decreasing thickness of layer formed on reactant and decreasing mass transfer resistance on solid–liquid interfaces in the suspension.[58] One of the main reasons why the use of ultrasound is being associated with the efficiency in the production of boric acid is due to the very small particle sizes of colemanite that it has to be crushed. By using an ultrasound, it allows more flexibility to the increase in particle size being that one of the mechanical effects of an ultrasound allow the nonreactive coatings from the surfaces of the solids to dissolve.[58]

2.3 CHARACTERIZATION EXPERIMENTS

CO_2 was utilized as a reactant in the production of boric acid from colemanite mineral. Reaction was carried out in an aqueous phase batch reactor. The effects of pressure, temperature, reaction time, particle size were investigated in our previous studies.[9,22] The solid phases formed at the end of colemanite–CO_2 reaction at different conditions were separated from aqueous phase by vacuum filtration. Filter cake was washed by deionized water and filtered two times. Finally, filter cake was dried in an air circulation oven at 105°C until reaching to the constant weight. The filtrate was evaporated to crystallize dissolved boric acid. Solid products formed at the end reaction as filter cake and powder crystallized from aqueous phase were examined by

X-ray diffraction, Fourier transform infrared spectroscopy (FTIR), thermal gravimetric analysis (TGA). X-ray diffractometer (Philips Xpert-Pro) was used to analyze the crystal structures of the colemanite and the formed solid products with Cu $K\alpha$ radiation at 45 kV and 40 mA. The registrations were performed in the 2θ range of 5–80°. The transmission spectra of KBr pellets prepared by mixing 4.0 mg of sample and 196 mg of KBr in an agate mortar and pressing the mixture under 8 t were obtained using FTIR spectrophotometer (Perkin Elmer Spectrum BX). TGAs were performed in Perkin Elmer-Diamond TG/DTA. Powder samples (10–15 mg) were loaded into an alumina pan and heated from 30°C to 800°C at 10 C min^{-1} under N_2 flow (40 mL min^{-1}).

2.4 BORIC ACID CHARACTERIZATION

Boric acid has various uses either in consumer goods or in industrial products. It is mainly utilized as a raw material in the production of synthetic metal borates, for example, zinc borates, lithium borates, copper borates, and calcium borates; boron esters, boron nitride, etc. In most of its application its characterization is extremely important. The basic analysis for the determination of boric acid in aqueous solution is the analytical titration. The amount of boric acid dissolved in water can be determined by a conventional acid base titration in the presence of phenolphthalein and mannitol as a chelating agent.[29] The consumption of $B(OH)_4^-$ in the aqueous phase by the formation of the complex with alcohol accelerates the release of proton which is detected in titration. The other characterization methods can be summarized as infrared spectroscopy, X-ray powder diffraction, and thermal analysis. Each of those methods will be discussed below from the studies in the literature.

2.4.1 FOURIER TRANSFORM INFRARED SPECTROSCOPY

Vibrational spectroscopy is a traditional method for characterization and identification of boron compounds. As boron atom can coordinate in trigonal and tetragonal forms with oxygen, there are many structures possible in boron compounds. Borates may exist either in monomeric or polymeric forms. Polymerization occurs by elimination of water from two hydrated borate molecules forming chains, sheets, or networks.[31] FTIR spectra of powder boric acid can be recorded by using either KBr disc method or attenuated total reflectance (ATR) technique. In KBr disc method, a small amount of sample is mixed with highly pure KBr and pressed to form a pellet. In ATR technique, a small amount of sample is placed on sample holder of ATR kit and analysis is performed. FTIR spectrum of boric acid obtained from colemanite mineral by CO_2 leaching is shown in Figure 2.2. The broad band observed between 2700 and 3500 cm^{-1} in Figure 2.2 is related to the hydroxyl groups in boric acid structure. The weak band at 856 cm^{-1} and strong band at 1435 cm^{-1} observed are caused by symmetric stretching vibration of B–O and by asymmetric stretching vibration of B–O in the BO_3 structure, respectively. The band observed at 1190 cm^{-1} belongs to in-plane bending vibration of the B–O–H structure. Asymmetric and symmetric stretching vibrations of B–O bonding in tetrahedral structure (BO_4) are seen at 1114 and 691 cm^{-1}.

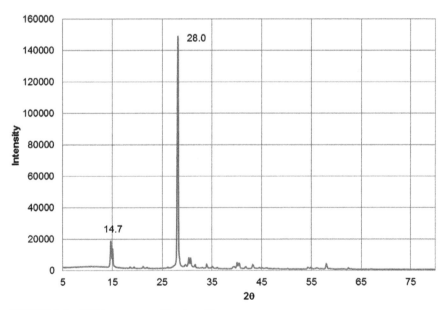

FIGURE 2.2   FTIR spectrum of boric acid.

2.4.2   X-RAY POWDER DIFFRACTION

The peaks at the 2θ value of 14.7° and 28.0° represent the characteristic peak of boric acid as inferred from its crystallographic data (JCPDS 30-0199). The maximum peak at the 2θ value of 28.0° on Figure 2.3 is caused by the reflection from the (0 0 2) plane.

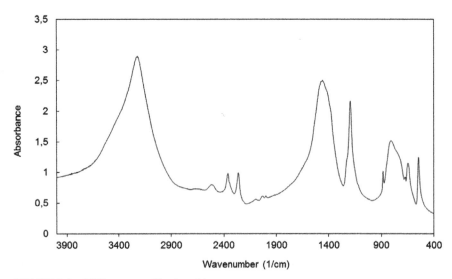

FIGURE 2.3   XRD pattern of boric acid.

2.4.3   THERMAL GRAVIMETRIC ANALYSIS

The boric acid powder obtained from colemanite–CO_2 reaction has a mass loss of 43.6 wt% in the temperature range of 60–200°C (Fig. 2.4). This thermal behavior is in good agreement with the theoretical mass loss of 43.8 wt%. When boric acid is heated slowly hydroxyl groups in the structure polymerize and give a condensation reaction to form water. The release of water from the bulk of boric acid resulted in a mass loss. There are also phase transitions in boric acid depending on temperature during heating process.

2.5   ENVIRONMENTAL CONCERNS IN BORIC ACID PRODUCTION

Boron is an inorganic element which is found in nature as compounds where it combines to oxygen and alkaline and alkaline earth metals (Na, Ca, and Mg). Boron could be in low concentrations (<0.1 mg L^{-1}) in low-saline ground water and rivers[66] and in high concentrations in saline water (e.g., 4.7 mg L^{-1} for sea water). Boron is an essential element for plants; however, it can be toxic beyond a certain concentration (1–2 mg L^{-1}).[28,56] Thus, its concentration in soil and irrigation water is an important parameter for efficient production and must be monitored, especially in boron deposition areas. The concentration of boron in water is also important for human

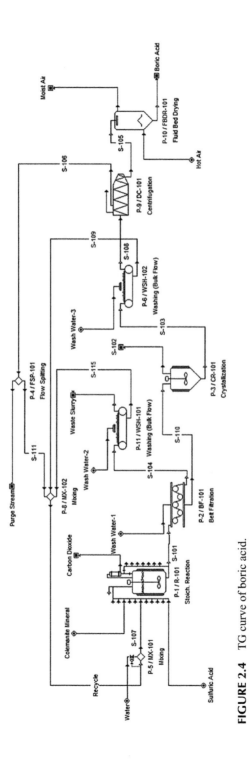

FIGURE 2.4 TG curve of boric acid.

beings. The directive (98/93/EC) in European Union (EU) which is responsible for regulating the quality of water for human consumption points out the present boron level concentration as 1 mg L^{-1} which was greater than the value of 0.5 mg L^{-1} reported by World Health Organization (WHO).[65] But recently, the critical value of boron in water was changed to 2.4 mg L^{-1} by WHO in 2011. As the some of the European member states in the Mediterranean basin (e.g., Cyprus, Greece, Italy, and Spain) and associate member, Turkey have higher boron concentrations in their ground and surface water bodies, it is being increasingly difficult for them to provide the value reported by EU.[33,64]

The possible causes of high boron concentrations in water might be from anthropogenic sources, such as waste water release from boron mineral processing plants, or the use of boron chemicals in other formulations (detergents, fertilizers) or natural sources, for example, dissolution of minerals in water.[65] As some of the boron minerals such as tincal and sassolite are soluble in water, boron concentrations might be higher than the toxic level in ground water and surface fresh waters in regions where those minerals are abundant. For instance, the Searles Lake in California bears dissolved boron mineral from which boron compound is separated. On the other hand, excavation and processing of other insoluble boron minerals may cause boron pollution in the environment. Another example is the Simav River running in the vicinity of the boron deposition area in Turkey reported as polluted by boron compounds due to the boron facilities in the region. The boron concentration of the Simav River was 0–0.5 mg L^{-1} before it enters the boron deposition area, but after the mixing of waste streams from those facilities, boron concentration was reported as 4–7 mg L^{-1}.[39] Those values are above the limits determined for humans and plants.

2.6 CONCLUSION

Boric acid is produced mainly from colemanite and ulexite in the world depending on the availability of the raw mineral. It is an important boron compound which is utilized in the production of other commercial boron compounds or formulated directly in consumer end products. In this review, the conventional boric acid production process was investigated based on arisen environmental issues and environmentally friendly processes in which CO_2 utilized are discussed. The main problem in boric acid production is the formation of wastewater streams bearing boron above the accepted standard limits. Especially, the gypsum with high boron content creates storage problems. As sulfuric acid dissolves almost everything in mineral, removal of impurities from boric acid requires much water in separation step. The boron concentration in waste streams of production facilities must be controlled and discharged based on acceptable limits. The use of carbon dioxide in the boron industry will provide a more sustainable and efficient process to help reduce CO_2 emissions from the boric acid production process. Infrared spectroscopy, X-ray diffraction, and thermal analysis are important techniques to determine qualitatively not only boric acid but also other boron compounds. The amount of boric acid in an aqueous phase can be found precisely by using analytical titration.

ACKNOWLEDGMENTS

The Scientific and Technical Research Council of Turkey (TUBİTAK) is greatly acknowledged for supporting the project 111M639 in which boric acid characterization was performed.

KEYWORDS

- boric acid
- colemanite
- boron mineral
- environmental pollution
- carbon dioxide

REFERENCES

1. Abali, Y.; Bayca, S. U.; Guler, A. E. The Dissolution Kinetics of Tincal in Phosphoric Acid Solutions. *Int. J. Chem. Reactor Eng.* **2007,** *5*, A115.
2. Abali, Y.; Bayca, S. U.; Mistincik, E. Kinetics of Oxalic Acid Leaching of Tincal. *Chem. Eng. J.* **2006,** *123* (1–2), 25–30.
3. Alkan, M.; Doğan, M.; Namli, H. Dissolution Kinetics and Mechanism of Ulexite in Oxalic Acid Solutions. *Ind. Eng. Chem. Res.* **2004,** *43*, 1591–1598.
4. Alkan, M.; Doğan, M. Dissolution Kinetics of Colemanite in Oxalic Acid Solutions. *Chem. Eng. Process.* **2004,** *43* (7), 867–872.

5. Ata, O. N.; Çolak, S.; Çopur, M.; Çelik, C. Determination of the Optimum Conditions for Boric Acid Extraction with Carbon Dioxide Gas in Aqueous Media from Colemanite Containing Arsenic. *Ind. Eng. Chem. Res.* **2000**, *39*, 488–493.

6. Bay, K. *Boric Acid Production from Colemanite Ores by the Use of Weak Acids*, Master of Science Thesis, İstanbul Technical University, İstanbul, 2002.

7. Bixler, G. H.; Sawyer, D. L. Boron Chemicals from Searles Lake Brines. *Ind. Eng. Chem. Res.* **1957**, *49*, 322–333.

8. Boncukcuoğlu, R.; Yılmaz, M. T.; Kocakerim, M. M.; Tosunoğlu, V. Utilization of Borogypsum as Set Retarder in Portland Cement Production. *Cem. Concr. Res.* **2002**, *32* (3), 471–475.

9. Budak, A.; Gönen, M. Extraction of Boric Acid from Colemanite Mineral by Super-critical Carbon Dioxide. *J. Supercrit. Fluids* **2014**, *92*, 183–189.

10. Bulutcu, A. N.; Ertekin, C. O.; Celikoyan, M. B. K. Impurity Control in the Production of Boric Acid from Colemanite in the Presence of Propionic Acid. *Chem. Eng. Process.* **2008**, *47* (12), 2270–2274.

11. Çavuş, F.; Kuşlu, S. Dissolution Kinetics of Colemanite in Citric Acid Solutions Assisted by Mechanical Agitation and Microwaves. *Ind. Eng. Chem. Res.* **2005**, *44*, 8164–8170.

12. Çolak, M.; Gemici, Ü.; Tarcan, G. The Effects of Colemanite Deposits on the Arsenic Concentrations of Soil and Ground Water in Igdeköy-Emet, Kütahya, Turkey. *Water, Air, Soil Pollut.* **2003**, *149* (1), 127–143.

13. Deepak, F. L.; Vinod, C. P. K.; Govindaraj, M. A.; Rao, C. N. R.; Boron Nitride Nano-tubes and Nanowires. *Chem. Phys. Lett.* **2002**, *353* (5–6), 345–352.

14. Delfini, M.; Ferrini, M.; Manni, A.; Massacci, P.; Piga, L. Arsenic Leaching by Na₂S to Decontaminate Tailings Coming from Colemanite Processing. *Miner. Eng.* **2003**, *16* (1), 45–50.

15. Doğan, H. T.; Yartaşı, A. Kinetic Investigation of Reaction between Ulexite Ore and Phosphoric Acid. *Hydrometallurgy* **2009**, *96* (4), 294–299.

16. Cavdar, A. D. Effect of Various Wood Preservatives on Limiting Oxygen Index Levels of Fir Wood. *Measurement* **2014**, *50*, 279–284.

17. Edwards, J. O. Detection of Anionic Complexes by pH Measurements. II. Some Evidence for Peroxyborates. *J. Am. Chem. Soc.* **1953**, *75* (24), 6154–6155.

18. Ekinci, Z.; Şayan, E.; Beşe, A. V.; Ata, O. N. Optimization and Modeling of Boric Acid Extraction from Colemanite in Water Saturated with Carbon Dioxide and Sulphur Dioxide Gases. *Int. J. Miner. Process* **2007**, *82*, 187–194.

19. Ekmekyapar, A.; Demirkıran, N.; Künkül, A. Dissolution Kinetics of Ulexite in Acetic Acid Solutions. *Chem. Eng. Res. Des.* **2008**, *86* (9), 1011–1016.

20. Elbeyli, İ. Y. Production of Crystalline Boric Acid and Sodium Citrate from Borax Decahydrate. *Hydrometallurgy* **2015**, *158*, 19–26.

21. Emil, F. *Process for the Manufacture of Boric Acid from Sodium Tetraborate*, U.S. Patent 195,010,6A, March 06, 1934.

22. Gönen, M.; Nyankson, E.; Gupta, R. B. Boric Acid Production from Colemanite Together with Ex-Situ CO₂ Sequestration. *Ind. Eng. Chem. Res.* **2016**, *55*, 5116–5124.

23. Gülensoy, H.; Kocakerim, M. M. Solubility of Colemanite Mineral in CO₂-Containing Water and Geological Formation of This Mineral. *Bull. Miner. Res. Explor. Inst. Turk.* **1978**, *90*, 1–19.

24. Gür, A. Dissolution Mechanism of Colemanite in Sulfuric Acid Solutions. *Korean J. Chem. Eng.* **2007**, *24*, 588–591.

25. Hagenson, L. C. *Sonochemical Reactions: Mass Transfer and Kinetic Studies of a Solid–Liquid System*. Ph.D. Dissertation, Iowa State University: Iowa, 1997.

26. Hagenson, L. C.; Doraiswamy, L. K. Comparison of the Effects of Ultrasound and Mechanical Agitation on a Reacting Solid–Liquid System. *Chem. Eng. Sci.* **1998,** *53* (1), 131–148.

27. Hartman, G. J. Fort Cady In-Situ Borate Mining Project. *Ceram. Eng. Sci. Proc.* **1997,** *18* (21), 167–172.

28. Hilal, N.; Kim, G. J.; Somerfield, C. Boron Removal from Saline Water: A Comprehensive Review. *Desalination* **2011,** *273* (1), 23–35.

29. Hollander, M.; Rieman, W. Titration of Boric Acid in Presence of Mannitol. *Ind. Eng. Chem. Anal. Ed.* **1945,** *17,* 602–603.

30. Houston, T. A.; Wilkinson, B. L.; Blanchfield, J. T. Boric Acid Catalyzed Chemoselective Esterification of α-Hydroxycarboxylic Acids. *Org. Lett.* **2004,** *6* (5), 679–681.

31. Jun, L.; Shuping, X.; Shiyang, G. FT-IR and Raman Spectroscopic Study of Hydrated Borates. *Spectrochim. Acta A* **1995,** *51* (4), 519–532.

32. Kocakuşak, S.; Köroglu, H. J.; Tolun, R. Drying of Wet Boric Acid by Microwave Heating. *Chem. Eng. Process.* **1998,** *37* (2), 197–201.

33. Koç, C. Effects of Boron Pollution in the Lower Buyuk Menderes Basin (Turkey) on Agricultural Areas and Crops. *Environ. Prog. Sustain Energy* **2011,** *30* (3), 347–357.

34. Kuskay, B.; Bulutcu, A. N. Design Parameters of Boric Acid Production Process from Colemanite Ore in the Presence of Propionic Acid. *Chem. Eng. Process.* **2011,** *50* (4), 377–383.

35. Lide, D. R. *CRC Handbook of Chemistry and Physics*, 84th ed., CRC Press: Boca Raton, FL, 2003.

36. Lyczko, N.; Espitalier, F; Louisnard, O.; Schwartzentruber, J. Effect of Ultrasound on the Induction Time and the Metastable Zone Widths of Potassium Sulphate. *Chem. Eng. J.* **2002,** *86* (3), 233–241.

37. Mergen, A.; Demirhan, M. H.; Bilen, M. Processing of Boric Acid from Borax by a Wet Chemical Method. *Adv. Powder Technol.* **2003,** *14* (3), 279–293.

38. Nelson, N. P. *Production of Boric Acid*, U.S. Patent 303,126,4, 1962.

39. Okay, O.; Güçlü, H.; Soner, E.; Balkaş, T. Boron Pollution in the Simav River, Turkey and Various Methods of Boron Removal. *Water Res.* **1985,** *19* (7), 857–862.

40. Okur, H.; Tekin, T.; Ozer, A. K.; Bayramoglu, M. Effect of Ultrasound on the Dissolution of Colemanite in H_2SO_4. *Hydrometallurgy* **2002,** *67* (1–3), 79–86.

41. Orhan, Y.; Fehmi, U.; Goknur, K.; Derya, M.; Soner, G.; Ebru, K.; Hamdi, D. M.; Emrah, C. C.; Omer, I. *Production of Boric Carbonate and Sodium Bicarbonate from Sodium Borate Solvents*. W.I.P. Organization: Turkey. WO2010/140989A1, 2010.

42. Özmetin, C.; Kocakerim, M. M.; Yapıcı, S.; Yartaşı, A. A Semiempirical Kinetic Model for Dissolution of Colemanite in Aqueous CH_3COOH Solutions. *Ind. Eng. Chem. Res.* **1996,** *35* (7), 2355–2359.

43. Pankajavalli, R.; Anthonysamy, S.; Ananthasivan, K.; Vasudeva Rao, P. R. Vapour Pressure and Standard Enthalpy of Sublimation of H_3BO_3. *J. Nucl. Mater.* **2007,** *362* (1), 128–131.

44. Parks, J. L.; Edwards, M. Boron in the Environment. *Crit. Rev. Environ. Sci. Technol.* **2005,** *35* (2), 81–114.

45. Patnaik, P. *Handbook of Inorganic Chemicals*; McGraw-Hill Handbooks: New York, 2002.

46. Power, I. M.; Harrison, A. L.; Dipple, G. M.; Southam, G. Carbon Sequestration via Carbonic Anhydrase Facilitated Magnesium Carbonate Precipitation. *Int. J. Greenhouse Gas Control* **2013**, *16*, 145–155.

47. Roskill, I. S. *Boron: Global Industry Markets and Outlook*, 12th ed., Roskill Information Services Limited: London, 2010.

48. Santos, H. M.; Lodeiro, C.; Capelo-Martínez, J. L. *The Power of Ultrasound. Ultrasound in Chemistry*, Wiley-VCH Verlag GmbH & Co. KGaA: Weinheim, Germany, 2009.

49. Schubert, D. Boron Oxides, Boric Acid, and Borates. In *Kirk–Othmer Encyclopedia of Chemical Technology*; John Wiley & Sons, Inc.: Weinheim, 2011; pp 1–68.

50. Schubert, D. M.; Alam, F.; Visi, M. Z.; Knobler, C. B. Structural Characterization and Chemistry of the Industrially Important Zinc Borate, $Zn[B_3O_4(OH)_3]$. *Chem. Mater.* **2003**, *15* (4), 866–871.

51. Seifritz, W. CO_2 Disposal by Means of Silicates. *Nature* **1990**, *345*, 486.

52. Shen, K. K.; Kochesfahani, S.; Jouffret, F. Zinc Borates as Multifunctional Polymer Additives. *Polym. Adv. Technol.* **2008**, *19* (6), 469–474.

53. Shiloff, J. C. *Boric Acid Production*, U.S. Patent 365,069,0A, March 21, 1972.

54. Sinirkaya, M.; Kocakerim, M. M.; Boncukçuoğlu, R.; Küçük, Ö.; Öncel, S. Recovery of Boron from Tincal Wastes. *Ind. Eng. Chem. Res.* **2005**, *44*, 427–433.

55. Smith, R. A. Boric Oxide, Boric Acid, and Borates. *Ullmann's Encyclopedia of Industrial Chemistry*; Wiley-VCH Verlag GmbH & Co. KGaA: Weinheim, 2000.

56. Tanaka, M.; Fujiwara, T. Physiological Roles and Transport Mechanisms of Boron: Perspectives from Plants. *Eur. J. Physiol.* **2008**, *456*, 671–677.

57. Tans, P.; Keeling, R. *Trends in Atmospheric Carbon Dioxide*; NOAA Earth System Research Laboratory: Boulder, CO, USA. Available online: http://www.esrl.noaa.gov/gmd/ccgg/trends/ (accessed September 25, 2015).

58. Taylan, N.; Gürbüz, H.; Bulutcu, A. N. Effects of Ultrasound on the Reaction Step of Boric Acid Production Process from Colemanite. *Ultrason. Sonochem.* **2007**, *14* (5), 633–638.

59. Teipel, U.; Mikonsaari, I. Size Reduction of Particulate Energetic Material. *Propellants, Explos., Pyrotechnol.* **2002**, *27* (3), 168–174.

60. Temur, H.; Yartaşı, A.; Çopur, M.; Kocakerim, M. M. The Kinetics of Dissolution of Colemanite in H_3PO_4 Solutions. *Ind. Eng. Chem. Res.* **2000**, *39*, 4114–4119.

61. Thompson, L. H.; Doraiswamy, L. K. Sonochemistry: Science and Engineering. *Ind. Eng. Chem. Res.* **1999**, *38*, 1215–1249.

62. Tolun, R.; B. D. Emir; I. E. Kalafatoglu; S. Kocakusak, N. Yalaz. Production of Sodium Hydroxide and Boric Acid by the Electrolysis of Sodium Borate Solutions. U.S. Patent 444,463,3A, April 24, 1984.

63. Tunc, M.; Kocakerim, M. M.; Yapici, S.; Bayrakçeken, S. Dissolution Mechanism of Ulexite in H_2SO_4 Solution. *Hydrometallurgy* **1999**, *51* (3), 359–370.

64. Vengosh, A.; Kloppmann, W.; Marei, A.; Livshitz, Y.; Gutierrez, A.; Banna, M.; Guerrot, C.; Pankratov, I.; Raanan, H. Sources of Salinity and Boron in the Gaza Strip: Natural Contaminant Flow in the Southern Mediterranean Coastal Aquifer. *Water Resour. Res.* **2005**, *41*, 1–19.

65. Weinthal, E.; Parag, Y.; Vengosh, A.; Muti, A.; Kloppmann, W. The EU Drinking Water Directive: The Boron Standard and Scientific Uncertainty. *Eur. Environ.* **2005**, *15*, 1–12.

66. Wyness, A. J.; Parkman, R. H.; Neal, C. A Summary of Boron Surface Water Quality Data throughout the European Union. *Sci. Total Environ.* **2003**, *314–316*, 255–269.

67. Elbeyli, İ. Y.; Turan, A. Z.; Kalafatoğlu, İ. E. The Electrochemical Production of Boric Acid. *J. Chem. Technol. Biotechnol.* **2015,** *90* (10), 1855–1860.

68. Yeşilyurt, M. Determination of the Optimum Conditions for the Boric Acid Extraction from Colemanite Ore in HNO_3 Solutions. *Chem. Eng. Process.* **2004,** *43* (10), 1189–1194.

69. Yeşilyurt, M.; Çolak, S.; Çalban, T.; Genel, Y. Determination of the Optimum Conditions for the Dissolution of Colemanite in H_3PO_4 Solutions. *Ind. Eng. Chem. Res.* **2005,** *44*, 3761–3765.

70. ZareNezhad, B. Production of Crystalline Boric Acid through the Reaction of Colemanite Particles with Propionic Acid. *Dev. Chem. Eng. Mineral Process.* **2003,** *11* (3–4), 363–380.

CHAPTER 3

DUAL ROLE PLAYED BY IONIC LIQUIDS TO MODULATE THE INTERFACIAL AND MICELLAR PROPERTIES OF THE SINGLE-CHAIN CATIONIC SURFACTANTS IN AQUEOUS SOLUTION

UTKARSH U. MORE[1,3], OMAR A. EL SEOUD[2*], and NAVED I. MALEK[1,2*]

[1]Applied Chemistry Department, Sardar Vallabhbhai National Institute of Technology, Ichchhanath, Surat, Gujarat 395007, India

[2]Institute of Chemistry, The University of São Paulo, Box 26077, São Paulo, SP 05513-970, Brazil

[3]Department of Chemistry, Marwadi University, Rajkot, Gujarat 360003, India

*Corresponding author. E-mail: navedmalek@chem.svnit.ac.in, navedmalek@yahoo.co.in

CONTENTS

ABSTRACT

Present chapter demonstrates the dual role played by imidazolium based ionic liquids (ILs) having short to long alkyl chain in the head group on modulating the interfacial and micellar properties of the single chain cationic surfactants in aqueous solution. The ILs used are 1-alkyl-3-methylimidazolium bromide (alkyl = 1-butyl, $[C_4mim][Br]$; 1-hexyl, $[C_6mim]$ [Br], and 1-octyl, $[C_8mim][Br]$) and the single chain cationic surfactants are 1-C_n-trimethylammonium bromide ($C_n = C_{12}$, C_{14}, and C_{16}). Various state of the art analytical techniques such as surface tension, conductance, spectral change of a dye by UV–vis, fluorescence, and dynamic light scattering (DLS) were used to characterize the interfacial, spectral, and micellar size change. All the measurements were performed at controlled temperature at 298.15 K. Results indicate that surface properties are influenced positively by the type and concentration of the added ILs. For shorter chain ILs, that is $[C_nmim][Br]$ ($n = 4,6$), cmc of aqueous C_nTAB ($n = 12,16$) decreases with $c(IL)$, whereas for the higher alkyl chain length IL, that is for $[C_8mim][Br]$, cmc decreases up to 0.8 wt% and then slightly increases, while for $C_{14}TAB$, *cmc* decreases with $c(ILs)$. The results obtained with the ILs are compared by the addition of common electrolyte NaBr at the identical composition to that of ILs. DLS data revealed an increase in micellar size of surfactants aggregates with decrease in the aggregation number (N_{agg}), determined by the fluorescence quenching method. We observed that N_{agg} increases with increasing the alkyl chain length of surfactants. We have compared our data with the literature data with various ILs as the additives.

3.1   INTRODUCTION

Physical universe is best understood by understanding the "Science" through application of various scientific methods including keen observations, proposing several hypotheses to explain the observations, and finally approving those hypotheses in the best reliable and valid ways of experimentation. Contribution of the branch of natural science that deals with the constituting components of the substances (atoms and molecules) and named "Chemistry" is incredible. Chemistry plays an important role in our daily life in various forms such as in form of medicines, insecticides, textiles, cement, polymers, surfactants, and food materials to name a few. Among these, surfactants are the earliest known and the most used chemicals, dates back to around 2800 BC in ancient Babylon.[1] Newer discoveries and inventions originating from progress in scientific arena have been proved advantageous in the versatility of applications and thus popularity of these substances. Further, environmentally cooperative nature and wide range of applications makes the surfactant-based systems a topic of major research interest in both academia and industry. Surface-active molecules self-assemble in aqueous media to form nanosized aggregates called micelles, which are used as media for various synthetic and catalytic reactions,[2,3] drug delivery,[4,5] in separations,[6] and in numerous pharmaceutical formulations showing how essential and diverse these systems are.

To tune the shape and size of these nano-aggregates for better properties and applicability, several strategies were experimented. One of these is the addition of additives, such as inorganic[7-19] and organic additives,[20-22] addition of solvents,[23-30] and recently rediscovered ionic liquids (ILs).[31-44] The role played by the additive includes manipulating the (1) repulsion between the charged head groups and (2) hydrophobic–hydrophobic interaction in controlled manner to get various structural aggregates. Among the additives, ILs have emerged as potentially the best possible additives due to their unusual physicochemical properties and immense technological applications.[45-47] These properties includes very low vapor pressure, high thermal stability, wide liquidious range, tunable polarity, ability to dissolve various compounds, and recently reported higher interfacial and bulk properties.[45-47] As ILs have the properties of both molten salts and organic liquids, their behavior will be more complex than aqueous solution. Therefore, one can expect different micellization behavior of aqueous surfactant solution in ILs. Hence, it is of great interest to employ the ILs in concert with other environmentally friendly systems thus forming new types of "hybrid" environmentally benign systems. These "hybrid" systems will have massive potential in various fields of science and technology. Various research groups including our own group have reported the formation of micelles,[48-50] microemulsions,[51-53] liquid crystals,[54] gels,[55] and vesicles[56] by using ILs as either additives, cosurfactants, or cosolvents in aqueous surfactant systems.

Among the possible 10^{18} available combinations of the cations and anions to form ILs (Scheme 3.1), imidazolium-based ILs are studied extensively in the field of colloid and interface science due to their characteristic properties such as ability to form weaker hydrogen bonding and π-stacking, which are absent in other ILs as well as in organic solvents. Several reports including reviews are published recently reporting the use of ILs as the modifiers for the aqueous surfactant solution.[4,32,33,35,42,51,52,56-79]

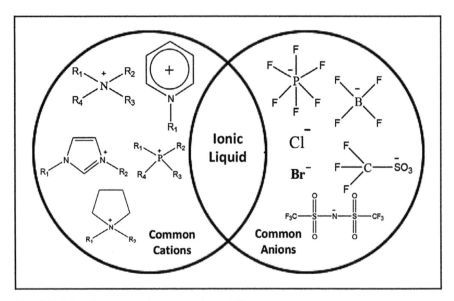

SCHEME 3.1   Common cations and anions of ILs.

Modaressi and his coworkers have reported the aggregation behavior of C_{16}TAB in aqueous solution of ammonium-based ILs (propyl-(2-hydroxyethyl)-dimethyl-ammonium bromide [C_3Br], butyl-(2-hydroxyethyl)-dimethylammonium bromide [C_4Br]) through conductometric studies. The higher critical micelle concentration (cmc) is observed with more hydrophobic C_4Br. cmc of C_{16}TAB increases with increasing the temperature.[31] cmc of the C_{16}TAB, C_{16}TAC surfactants increases with increasing the concentration of ILs ([C_5mim][PF$_6$] (3-methyl-1-pentyl-imidazolium hexafluorophosphate). cmc of the C_{12}TAB increases with increasing the concentration of 1-butyl-2,3-dimethylimidazolium chloride ([bdmim][Cl]).[33,34] Influence of 1-tetradecyl-3-methylimidazolium bromide [C_{14}mim][Br], a surface-active IL on the micellar morphological changes of cationic surfactants such as C_{14}TAB, dimethylditetradecylammonium bromide (DTDAB), alkane-bis(tetradecyldimethyl ammonium bromide) (14-2-14 and 14-4-14 gemini) was investigated through conductivity, surface tension, fluorescence, and[1]H-NMR techniques. The cmc of C_{14}TAB decreases with increasing the concentration of IL and increases with the increase in concentration of IL for DTDAB and gemini surfactants. On altering the hydrocarbon chain of the surfactants, a momentous change in the behavior of IL + cationic surfactant mixtures occurs, that is, unfavorable interactions for IL + C_{14}TAB systems becomes favorable (synergistic), while C_{14}TAB is replaced by twin tail surfactants.[4] Physicochemical properties (cmc, aggregation number, aggregate size, and polydispersivity) of aqueous C_{16}TAB changes more significantly in case of IL [C_6mim][Br] than cosurfactant n-hexyltrimethylammonium bromide (HeTAB) addition. Both [C_6mim][Br] and HeTAB show electrolytic as well as cosurfactant-type behavior within aqueous C_{16}TAB when present at low concentrations. At higher concentrations, only [C_6mim][Br] behaves as a cosolvent toward altering the physicochemical properties of aqueous C_{16}TAB.[38] cmc of aqueous dodecyltrimethylammonium bromide (C_{12}TAB) decreases with increasing concentrations of ILs (1,2-dimethyl-3-octylimidazolium chloride [Odmim][Cl]). The properties of C_{12}TAB solution changes in different way at higher concentration of IL.[40]

Here, we present the role of ILs on changing the micellar and interfacial properties of the single-chain cationic surfactants. The variables studied here are (1) c(IL), (2) alkyl chain length of IL and surfactant. The ILs and surfactants studied in the present investigation are [$C_n$mim][Br] (n = 4, 6, and 8), and (C_nTAB, n = 12, 14, and 16). The reasons for the selection of these ILs are the relatively higher interfacial tension of the imidazolium bromide ILs and the possible cation–π interactions in the surfactant–ILs system. Further, bromide anion containing surfactant in aqueous solution adopt the vertical orientation at the interface, which increases the chance of the surfactant molecules to get absorbed in higher number as compared to the chloride and iodide anion containing surfactants. This increases the population of the surfactant molecules on the interface and ultimately increases the chance to change the morphology of the surfactant molecules. We have selected the series of cationic surfactants to study the effect of ILs on the hydrophobicity of the surfactant molecules. We have investigated the tendency of ILs to modulate the properties of the aqueous surfactant solution through measuring conductance and surface tension. Dye–surfactant interaction was also studied in the presence of ILs. Fluorescence quenching and dynamic light-scattering measurements were performed to investigate the aggregation number and size of the aggregates in the absence and presence of ILs. This will help in determining packing of the micellar aggregates of the surfactants in aqueous solution in the presence of ILs. The data obtained from the present investigation will be helpful in finding the surfactant mixed system for the possible commercial applications.

3.2 EXPERIMENTAL

3.2.1 MATERIALS

ILs used in the present investigation were synthesized in the laboratory employing the procedures reported previously and were dried under reduced pressure to remove the traces amount of the water.[54,75–77] Cationic single-chain surfactants (C_{12}TAB, C_{14}TAB, and C_{16}TAB) were purchased from Aldrich and used without further purification. Other solvents and reagents used in the synthesis of ILs were from Merck and used with proper purification and drying. Water used throughout the study was doubly distilled and having conductivity of 6.1–6.4 μS cm^{-1}. 8-Anilino-1-napthalenesulfonic acid (ANS) was purchased from Aldrich with purity better than 97% and used as received.

3.2.2 SOLUTION PREPARATION

Aqueous IL solutions were prepared by weighing the ILs on Metler Toledo Analytical balance (B 204-S), which is operated in dry box and measuring the weight with precision of ±0.0001 g. Calculated amount of surfactant solution from the stock solution was then added and ensured the homogeneous solution. Care was taken to minimize the exposure of the solution with air to minimize the moisture absorption. All the measurements were performed in triplicate and the mean value was taken into consideration.

3.2.3 METHODS

3.2.3.1 CONDUCTANCE MEASUREMENTS

EUTECH PC 6000 digital microprocessor-based conductivity meter with a sensitivity of 0.1 μS cm^{-1} was used to measure the specific conductivities of the surfactant solution. The uncertainty in the measurement was of 0.5%. All the measurements were performed at 298.15 K and the circulating water temperature bath was used to maintain the temperature within 0.1 K. EC-CONSEN 21B conductivity cell with an inbuilt temperature probe was used for the measurements. Aqueous KCl solutions (0.01–1.0 mol kg^{-1}) were used to calibrate the conductivity cell.

3.2.3.2   SURFACE TENSION MEASUREMENTS

Kruss K9 tensiometer with platinum ring detachment method was used to determine the surface tension of the surfactant solution at controlled temperature of 298.15 K with a precision of ±0.1 K. Double distilled water with surface tension value of 72.2 mN m^{-1} was used throughout the experiment. Surface tension values were measured with an accuracy of 0.1 mN m^{-1}.

3.2.3.3   UV–VIS ABSORBANCE MEASUREMENTS

Varian Carry 50 absorption spectrophotometer was used to determine the interaction between the dye–surfactant interaction in the absence and presence of ILs in the range of 200–500 nm. The change in the λ_{max} of 8-ANS at two different wavelengths (382 and 272 nm)[33,34] was considered for the dye–surfactant interaction in absence and presence of ILs.

3.2.3.4   FLUORESCENCE SPECTROSCOPY

Aggregation number of the surfactant aggregates was determined through Cary Eclipse fluorescence spectrophotometer using cetylpyridinium chloride (CPC) as static quencher at 298.15 ±0.1 K using constant temperature bath.

3.2.3.5   DYNAMIC LIGHT SCATTERING

Size of the micellar aggregates were determined by Spectro Size 300, dynamic light scattering (DLS) instrument from Malvern Instruments, United Kingdom using He–Ne lesser (633 nm, 4 M$_w$) at 298.15 ±0.1 K.

3.3   RESULTS AND DISCUSSION

Notes:

1) The variables in the present study are c(IL), alkyl chain length in the head group of IL and surfactants.
2) We present our data in the order of sequence of events, that is, adsorption of C$_n$TAB (n = 12, 14, 16) at air/water interface and its micellization.

3) It is noteworthy that at higher concentration (>1.0 wt%) of the additives, the cmc measurements become unreliable, rendering it impossible to extract any meaningful conclusion.

3.3.1 INTERFACIAL PARAMETERS OF SURFACTANTS AT AIR/WATER INTERFACE

Surface tension measurement at 298.15 K was performed to characterize the change in the interfacial parameters of the studied surfactants in the absence and presence of the various additives (ILs and NaBr). The results of various adsorption parameters such as surface pressure at cmc (Π_{cmc}), maximum surface excess (Γ_{max}), minimum area per molecule at air/water interface (A_{min}), and packing parameter (p) for the studied surfactant systems are recorded in Table 3.1.

For neat and mixed surfactant systems (C_nTAB + ILs), Π_{cmc} increases with the alkyl chain length of the C_nTAB ($n = 12$, 14, 16), which confirms the highest efficiency of C_{16}TAB.[80] Π_{cmc} increases with increasing the c(ILs) and c(NaBr), which indicate the higher efficiency of the system. In the presence of additives (ILs and NaBr), the repulsion between the head group decreases; hence, more surfactant molecules will be adsorbed at the air/water interface as compared to pure solutions, which increases the Π_{cmc}.[2] The Π_{cmc} is higher in [C_8mim][Br] as compared to [C_6mim][Br] and [C_4mim][Br], which confirms the better efficiency of [C_8mim][Br].

The observed values of Γ_{max} for pure C_nTAB (1.15_{C12TAB}, 1.21_{C14TAB}, and $1.88_{C16TAB} \times 10^{-6}$ mol m^{-2}) and A_{min} (144_{C12TAB}, 137_{C14TAB}, and 89_{C16TAB} Å^2) are in good agreement with the literature.[3-5] For [C_4mim][Br] and [C_6mim][Br] as an additive, Γ_{max} and packing parameter (p) increases with c(ILs) and for [C_8mim][Br], Γ_{max} and p increases up to 0.8 wt% and then small decrease at 1.0 wt% of C_nTAB ($n = 12$, 14, 16) surfactant systems. Reverse is the trend for A_{min}. The increasing A_{min} at higher concentration of the ILs suggest its cosurfactant like behavior. With the addition of NaBr, monotonic increase in Γ_{max} and p, whereas A_{min} decreases were observed, which is expected for an inorganic electrolyte.[2,6,81,82] The presence of counter ions near the polar heads of surfactant molecules decreases the electrostatic repulsion between the head groups. This reduction in the repulsion makes it possible for the surfactant molecules to approach each other more closely. Thus, the maximum adsorption increases and the area occupied per surfactant molecule decreases. Low values of A_{min} suggest the close packing at the air/water interface and the orientation of the surfactant molecules at

TABLE 3.1 Interfacial Parameters: Surface Pressure at cmc (Π_{cmc}), Surface Excess Adsorption (Γ_{max}), Minimum Area Per Molecule (A_{min}), and Packing Parameter (p) for Aqueous C_nTAB ($n = 12, 14, 16$) in Absence and Presence of Different Additives at 298.15 K.

Additives % (w/w)	$C_{12}TAB$				$C_{14}TAB$				$C_{16}TAB$			
	Π_{cmc} (mN m⁻¹)	Γ_{max} 10⁶ (mol m⁻²)	A_{min} (Å²)	p	Π_{cmc} (mN m⁻¹)	Γ_{max} 10⁶ (mol m⁻²)	A_{min} (Å²)	p	Π_{cmc} (mN m⁻¹)	Γ_{max} 10⁶ (mol m⁻²)	A_{min} (Å²)	p
[C₄mim][Br]												
0.0	32.6	1.15	144	0.15	34.7	1.21	137	0.15	39.9	1.88	89	0.24
0.2	32.9	1.21	137	0.15	35.2	1.25	133	0.16	38.4	1.96	85	0.25
0.4	33.8	1.25	133	0.16	36.1	1.30	127	0.17	39.9	2.03	82	0.26
0.6	34.1	1.28	130	0.16	36.6	1.34	124	0.17	40.6	2.10	79	0.27
0.8	35.1	1.31	127	0.17	37.5	1.40	119	0.18	41.4	2.14	77	0.27
1.0	35.5	1.33	124	0.17	38.0	1.43	116	0.18	41.4	2.23	75	0.28
[C₆mim][Br]												
0.2	33.5	1.25	132	0.16	36.9	1.30	127	0.17	40.0	2.00	83	0.25
0.4	34.0	1.31	127	0.16	37.1	1.33	125	0.17	40.4	2.05	81	0.26
0.6	34.5	1.33	124	0.17	38.6	1.40	119	0.18	40.8	2.11	79	0.27
0.8	35.1	1.39	119	0.18	40.7	1.44	115	0.18	41.2	2.19	76	0.28
1.0	35.5	1.43	116	0.18	41.1	1.50	111	0.19	41.4	2.25	74	0.29
[C₈mim][Br]												
0.2	34.9	1.28	129	0.16	37.3	1.34	124	0.17	40.4	2.06	80	0.26
0.4	35.5	1.32	125	0.17	37.4	1.40	119	0.18	41.1	2.15	77	0.27
0.6	37.8	1.36	122	0.17	39.1	1.48	112	0.19	41.5	2.21	75	0.28
0.8	38.4	1.40	119	0.18	41.4	1.52	109	0.19	42.0	2.33	71	0.30
1.0	39.8	1.37	121	0.17	42.9	1.45	114	0.18	43.5	2.19	76	0.28

TABLE 3.1 *(Continued)*

Additives % (w/w)	$C_{12}TAB$				$C_{14}TAB$				$C_{16}TAB$			
	Π_{cmc} (mN m^{-1})	Γ_{max} 10^6 (mol m^{-2})	A_{min} (Å^2)	p	Π_{cmc} (mN m^{-1})	Γ_{max} 10^6 (mol m^{-2})	A_{min} (Å^2)	p	Π_{cmc} (mN m^{-1})	Γ_{max} 10^6 (mol m^{-2})	A_{min} (Å^2)	p
NaBr												
0.2	33.6	1.22	136	0.15	37.1	1.28	130	0.16	37.3	1.97	84	0.25
0.4	35.3	1.25	132	0.16	37.2	1.31	126	0.17	37.6	2.07	80	0.26
0.6	36.4	1.29	129	0.16	38.2	1.37	121	0.17	38.5	2.12	78	0.27
0.8	37.5	1.33	124	0.17	38.5	1.42	117	0.18	38.7	2.17	77	0.27
1.0	38.5	1.35	123	0.17	39.2	1.44	115	0.18	39.6	2.24	74	0.29

Standard uncertainties u are $u(\Pi_{cmc}) = \pm0.1$ mN m^{-1}, $u(\Gamma_{max}) = \pm0.01 \times 10^{-6}$ (mol m^{-2}), $u(A_{min}) = \pm2.00$ Å^2, $u(p) = \pm0.02$, and $u(T) = 0.01$ K.

the interface is almost perpendicular to the interface.[83] For the higher alkyl chain length ILs, at higher concentration, population of the IL molecule at the interface increases which prevents the surfactant molecule to enter the surface monolayer and hence Γ_{max} decreases and loose packing density at the air/water interface.[4,5] As presented in Table 3.1, the values of Γ_{max} are lower than those of the electrolyte NaBr at the same concentration. The results suggest that surfactant adsorption at the air/water interface decreases more in the presence of ILs than NaBr. This shows that ILs are more effective than NaBr. Javadian et al. reported similar observation for the $C_{16}TAB$ and SDS surfactants in the presence of ILs.[5,84]

Table 3.1 shows that for pure C_nTAB (n = 12, 14, 16) and IL + C_nTAB systems, $p < 0.33$ which shows the formation of spherical micelles for the pure and mixed systems.[85] The p values for pure C_nTAB amphiphiles are in agreement with the literature.[3,5,85,48]

Γ_{max} and p increase and those of A_{min} decrease, when the alkyl chain length of C_nTAB increases[49] due to more hydrophobic surfactant molecules. The surface parameters obtained by the surface tension measurements suggest possible interaction between surfactant and the ILs, which can modify the interfacial and micellization behavior of the aqueous C_nTAB surfactant for better applications.

3.3.2 MICELLAR AGGREGATION BEHAVIOR OF C_nTAB SURFACTANT WITH ADDITIVES IN AQUEOUS MEDIUM

The concentration above which the surfactant changes its behavior in physical and spectral properties (conductance, surface tension, dye solubilization, etc.) is known as cmc. These methods provide reliable information regarding the micellization behavior as well as the surface properties of the surfactants before and after formation of micelles.[50–56] In the current investigation, we have used several physical (conductance, surface tension), spectral (dye solubilization through UV–vis absorption spectroscopy), and scattering (DLS) techniques to characterize the aggregation behavior of aqueous surfactant solution in the absence and presence of the additives (ILs and NaBr). In all the methods, there is a clear change of slope in the cmc region of the property plotted as a function of increasing $c(C_nTAB)$.

Briefly, as per the Onsager theory of electrolyte conductivity, we observed a change in the slope between the pre- and postmicellar region in conductance measurements and the intersection point of the two straight lines represents the cmc of surfactants. In case of surface tension plots, surface tension

(γ) of the solution decreases with increasing the surfactant concentration in the solution due to the monomer adsorption at the air/water interface and then becomes constant after micelle formation. Dye solubilization method works on the principle of solution polarity. With increasing concentration of surfactant monomers, dye molecules starts associating with surfactant monomer which decreases the microscopic polarity near the dye molecules, at the micellar pseudo phase, dye molecules are solubilized and the absorbance remain practically constant or small increase in absorbance was observed. The CMC is defined as that value of concentration at which the rate of absorbance alters its sign, that is, in pre-cmc, the rate of absorbance decreases and in post-cmc it tends to increase. The cmc of aqueous pure C_nTAB at 298.15 K determined from various analytical techniques (conductance, surface tension, and dye solubilization) are presented in Figure 3.1.

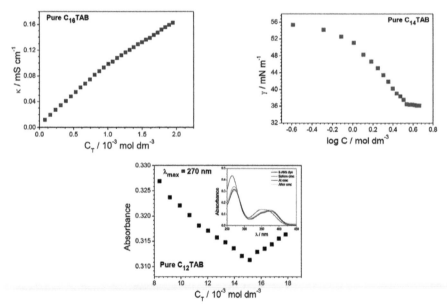

FIGURE 3.1   Specific conductance, surface tension, and absorbance plots of pure aqueous $C_{12}TAB$ (■), $C_{14}TAB$ (■), and $C_{16}TAB$ (■) solution at 298.15 K.

Similarly, we have investigated the effect of ILs on the aggregation behavior of cationic surfactants by employing the same techniques used for the pure surfactant systems. The study was performed at various concentration of the ILs. We compare all the results with NaBr as the additives at similar concentration of the ILs. The results of cmc are summarized in Table 3.2.

TABLE 3.2 Comparison of cmc Obtained from Different Methods of Aqueous C_nTAB (n = 12, 14, 16) in Absence and Presence of Additives at 298.15 K.

Additives % (w/w)	cmc 10^3 (mol dm^{-3})								
	C_{12}TAB			C_{14}TAB			C_{16}TAB		
	Conductance	Surface tension	UV–vis	Conductance	Surface tension	UV–vis	Conductance	Surface tension	UV–vis
[C$_4$mim][Br]									
0.0	19.6	16.4	15.1	3.40	3.40	3.55	0.98	0.94	0.97
0.2	15.3	14.1	14.0	3.33	3.36	3.46	0.92	0.88	0.92
0.4	13.0	12.9	12.7	3.22	3.28	3.24	0.85	0.81	0.88
0.6	10.8	9.60	11.3	3.11	3.20	2.90	0.78	0.74	0.77
0.8	8.50	8.10	9.90	2.76	2.80	2.65	0.65	0.66	0.67
1.0	7.30	6.20	8.40	2.52	2.40	2.24	0.51	0.52	0.55
[C$_6$mim][Br]									
0.2	14.2	12.9	13.3	3.32	3.32	3.40	0.91	0.79	0.87
0.4	11.9	10.7	11.5	3.24	3.20	3.20	0.80	0.74	0.76
0.6	9.60	8.50	9.20	3.00	3.00	2.80	0.67	0.66	0.67
0.8	7.25	6.18	7.60	2.60	2.50	2.60	0.58	0.57	0.61
1.0	6.12	5.00	5.90	2.40	2.20	2.20	0.45	0.38	0.61
[C$_8$mim][Br]									
0.2	11.8	10.9	12.1	3.24	3.24	3.10	0.83	0.72	0.55
0.4	9.84	8.70	9.20	3.20	3.00	3.00	0.71	0.64	0.43
0.6	8.45	6.22	5.90	2.80	2.80	2.60	0.56	0.51	a
0.8	6.11	5.30	3.10	2.40	2.40	2.40	0.42	0.35	a
1.0	8.40	6.40	a	2.20	2.13	2.10	0.53	0.51	a

TABLE 3.2 (Continued)

Additives % (w/w)	$cmc\ 10^3$ (mol dm^{-3})								
	C$_{12}$TAB			C$_{14}$TAB			C$_{16}$TAB		
	Conductance	Surface tension	UV–vis	Conductance	Surface tension	UV–vis	Conductance	Surface tension	UV–vis
NaBr									
0.2	14.1	14.0	12.7	3.24	3.30	3.50	0.93	0.86	0.91
0.4	11.9	11.9	10.6	3.20	2.90	3.30	0.87	0.79	0.83
0.6	7.31	8.90	7.40	3.00	2.80	3.00	0.64	0.72	0.72
0.8	4.90	7.30	5.70	2.40	2.70	2.60	0.51	0.44	0.61
1.0	3.69	5.00	4.10	2.30	2.30	2.20	0.37	0.36	0.43

Standard uncertainties u are $u(cmc) = 0.03 \times 10^{-3}$ mol dm^{-3} and $u(T) = 0.01$ K.w

[a]At these high concentrations of the additives, UV–vis measurements fail to determine the cmc.

Pertaining to above, following information was obtained:

1) Table 3.2 represents the comparison of cmc observed from various techniques (conductance, surface tension, and UV–vis) of aqueous C_nTAB (n = 12, 14, 16) with and without additives (ILs and NaBr) at 298.15 K. cmc of the pure C_nTAB (n = 12, 14, 16) are in excellent agreement with the literature.[3,21,48,86–88]

2) The agreement between the results of distinct techniques is satisfactory, taking into account that these are sensitive to different aspects of the micellization process. The reason for the observed dependence of *cmc* on the techniques employed have been discussed by the work of Mukerjee and Mysels, where they have compiled 54 cmcs for $C_{16}TAB$ (at 298.15 K), differing, for the same technique, by 22%.[89,90]

3) Representative specific conductance (κ) plots for aqueous $C_{12}TAB$ and $C_{14}TAB$ in various compositions of ILs and NaBr mixed media at 298.15 K are depicted in Figures 3.2 and 3.3, respectively. For a fixed concentration of surfactant, as the concentration of the ILs

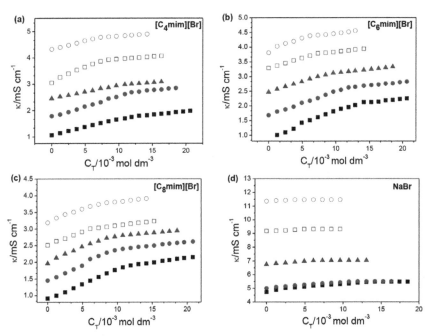

FIGURE 3.2 Representative specific conductance plots of aqueous $C_{12}TAB$ in presence of different wt% [(■) 0.2, (●) 0.4, (▲) 0.6, (□) 0.8, and (○) 1.0 wt%] of additives (ILs/NaBr) mixed media at 298.15 K.

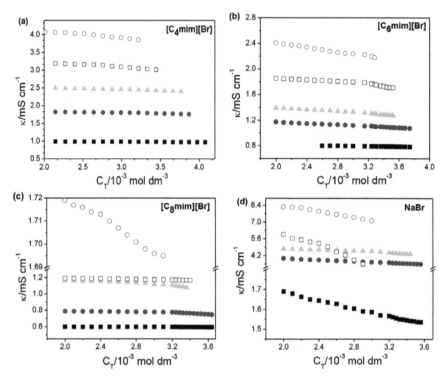

FIGURE 3.3   Representative specific conductance plots of aqueous $C_{14}TAB$ in presence of different wt% [(■) 0.2, (●) 0.4, (▲) 0.6, (□) 0.8, and (○) 1.0 wt%] of additives (ILs/NaBr) mixed media at 298.15 K.

increases (from 0.2 to 1.0 wt%), the specific conductance increases. The reason is the availability of the free cations ([$C_n$mim$^+$] with $n = 4$, 6, and 8) and anions [Br$^-$] in the solution after dissociation. Various research groups have reported similar behavior for cationic,[35,38,65] anionic,[32,33,36,44,64] and nonionic surfactants.[43,91]

4)   At fixed concentration, [C_4mim][Br] containing surfactant solutions have higher conductance as compared to [C_6mim][Br] and [C_8mim][Br]. This is due to the higher limiting ionic conductivity of [C_4mim$^+$] (15.0 S cm^2 mol^{-1}) than [C_6mim$^+$] (0.56 S cm^2 mol^{-1}) and [C_8mim$^+$] (0.05 S cm^2 mol^{-1}) at ambient conditions. It was observed that at fixed concentration of IL and NaBr, the conductivity of NaBr is significantly higher due to the smaller size of cation and higher transport number (t[Na+] = 0.74, t[C_4mim+] = 0.58, t[C_6mim+] = 0.52, and t[C_8mim+] = 0.50).[92]

5) Interestingly, in the presence of ILs, with increasing concentration of C_nTAB, n = 14 and 16, conductance decreases, whereas for C_{12}TAB, in the presence of additives (ILs and NaBr), we observe increase in conductance (Fig. 3.2). Limiting ionic conductivity for the surfactant cations $[C_{12}TA^+]$, $[C_{14}TA^+]$, and $[C_{16}TA^+]$ (64.0,[92] 0.94, and 0.11[93,94] S cm^2 mol^{-1}, respectively) and IL cation ($[C_4mim^+]$, $[C_6mim^+]$ and $[C_8mim^+]$ are (15.0, 0.56, and 0.05 S cm^2 mol^{-1}, respectively) deciding factors in overall conductance of the surfactants solution.

6) cmc of aqueous C_nTAB (n = 12, 16) decreases with c(IL), except for $[C_8mim][Br]$, cmc decreases up to 0.8 wt% and then small change in cmc was observed, while for C_{14}TAB, cmc decreases with c(ILs).[4] Similar results were reported for aqueous cationic,[4,5,88,33,38,40,95,96] anionic,[41,84,97] and zwitterionic surfactants.[39]

7) The behavior of ILs was compared with NaBr as the additive. The purpose was to study (1) the impact of the size of the additives and (2) to check whether the ILs behave as the electrolytes. For NaBr as the additives, monotonic decrease in cmc was observed, which is the known behavior of the inorganic electrolyte in case of the ionic surfactants. Pankaj et al. investigated the effect of various inorganic/organic salts on the micellization behavior of C_{16}TAB. Results suggest that cmc of the C_{16}TAB decreases with the concentration of additives (NaBu, NaBenz, and NaSal).[88] Smirnova et al. also studied the effect of ILs with varying cations and anions as well as NaCl on the cmc of the SDS, and the results suggest that the cmc decreases with increasing concentration of ILs/NaCl and the effect produced by the ILs is stronger than NaCl; such behavior was explained by the amphiphilic nature of ILs and electrostatic interaction, as well as mixed micelle formation.[41]

8) cmc decreases monotonically for the inorganic electrolyte NaBr due to the fact that with increasing the concentration of the electrolyte, the head group repulsion decreases, which resulted in lower cmc.[2,4] For $[C_8mim][Br]$, decrease in cmc at lower concentration (i.e., <0.8 wt%) also attributed to the role of ILs as an electrolyte. At higher concentration (i.e., at 1.0 wt%), an increase in the cmc may be due to the increased availability of the hydrophobic chain, which enhances the electrostatic repulsion between the intermolecular head groups and increases the transfer of the surfactant tail from the bulk phase into the micellar core. This phenomenon decreases the hydrophobic–hydrophobic interaction and eventually increases the cmc.

Such favorable interactions of the amphiphic molecules suggest the formation of mixed micelle.[98]

9) The *cmc* decreases with increasing the alkyl chain length of the ILs for the identical surfactants. The order is [C$_8$mim][Br] > [C$_6$mim][Br] > [C$_4$mim][Br] due to increasing hydrophobicity of the later. Smirnova et al. have reported decrease in cmc of anionic surfactant on addition of small amounts of [C$_n$mim][X] (n = 4, 6, and 10, X = Br, PF$_6$, and BF$_4$); the higher the n value, the steeper is the decrease.[41]

10) Regarding the role of the anions to modulate the micellar properties of the aqueous surfactant systems, Pal et al. have reported that for the same alkyl chain length, the cmc values for the anionic surfactants (SDS) decrease in order of the anion's basicity and the hydration radius of the anion, that is, the *cmc* decreases in the order Br$^-$ > BF$_4^-$ ≥ PF$_6^-$.[63]

11) cmc of C$_n$TAB (n = 12, 14, 16) decreases with increasing the alkyl chain length of C$_n$TAB surfactants and the order is C$_{12}$TAB < C$_{14}$TAB < C$_{16}$TAB, cmc decreases more in C$_{16}$TAB.[3,88] This is due to the hydrophobic–hydrophobic interaction between chain lengths of surfactants. To seek better understanding of the effect of the added ILs on the micellization behavior of the cationic surfactants, we compared our results for identical concentration of C$_{14}$TAB + [C$_6$mim][Br] (0.2 wt% 80 mmol dm^{-3}) with the C$_{16}$TAB + [C$_6$mim][Br] (80 mmol dm^{-3}).[38] The analysis showed that for C$_{14}$TAB and C$_{16}$TAB, the decrease in *cmc* is about 2.35% and 56%, respectively, relative to the neat cmc values of both cationic surfactants. The role of the cationic chain length of the ILs on the aggregation behavior of the surfactant was analyzed by comparing the results obtained in the present investigation with the literature,[4] that is, for 0.2 wt% addition, the *cmc* decreases around 2.35, 4.70, and 9.25% for [C$_6$mim][Br], [C$_8$mim][Br], and [C$_{14}$mim][Br],[4] respectively. These results are in line with the results reported by several other researchers.[5,84,39] The overall effect of the investigated ILs ([C$_n$mim][Br] with n = 4, 6, and 8) on the cmc of the aqueous C$_n$TAB is the combination of mainly two major factors, the electrolyte effect and the hydrophobic effect. The dominance of the above factors depends on the various physical parameters such as concentration and the alkyl chain length of the added ILs.

12) From the conductance data, we calculate the micellar parameters, that is, degree of counterion binding (β) and Gibbs free energy

of micellization (ΔG_m^o). For C_{14}TAB and C_{16}TAB, conductance decreases with increasing concentration of surfactants in presence of ILs. The slope becomes negative in pre and postmicellar region and the calculation of the β is not feasible. Figure 3.2 indicates that the conductance of aqueous C_{12}TAB increases with $c(C_{12}$TAB) in presence of ILs. The β and ΔG_m^o values for neat C_{12}TAB (0.77 and −34.9) are presented in Table 3.3 and are in good agreement with literature.[3,19,95] The counterion binding increases with increasing the c(ILs) due to the binding of the surfactant molecules more tightly with IL and the micelles are close packed. The ΔG_m^o values of C_{12}TAB increases negatively with increasing the concentration of ILs, which is due to the strong hydrophobic–hydrophobic interaction and spontaneous micellization.

TABLE 3.3 Counterion Binding (β) and Gibbs Free Energy of Micellization (ΔG_m^o) of Aqueous C_{12}TAB in Presence and Absence of ILs.

Additives % (w/w)	β	ΔG_m^o (kJ mol⁻¹)
	C$_{12}$TAB	
[C$_4$mim][Br]		
0.0	0.77	−34.9
0.2	0.79	−36.3
0.4	0.83	−37.9
0.6	0.87	−39.6
0.8	0.88	−41.0
1.0	0.90	−42.2
[C$_6$mim][Br]		
0.2	0.84	−37.8
0.4	0.86	−39.0
0.6	0.89	−40.5
0.8	0.90	−42.1
1.0	0.93	−43.5
[C$_8$mim][Br]		
0.2	0.87	−39.1
0.4	0.89	−40.5
0.6	0.91	−41.6
0.8	0.93	−43.7
1.0	0.96	−42.6

13) Surface tension isotherms (Fig. 3.4) reveal that with increasing the concentration and alkyl chain length of the ILs, surface tension decreases. The surface tension decreases more for the [C$_8$mim] [Br], due to the higher hydrophobicity. Addition of ILs causes the compression of the electric double layer, which reduces the repulsion between the surfactant head groups and reduces the cmc. Similar synergistic effects for mixtures of cationic gemini surfactants with different aromatic hydrotropes (benzoate, salicylate) and salt anions (Br$^-$, NO$_3^-$, SCN$^-$) also have been reported.[2,99] For present surfactant systems, cmc of C$_n$TAB (n = 12, 16) decreases with increasing c([C$_4$mim][Br] and [C$_6$mim][Br]), whereas for [C$_8$mim] [Br], cmc decreases up to 0.8 wt% and then increases, same as per the conductance results. For C$_{14}$TAB, *cmc* values decreases linearly with c(IL).[5,84] *cmc* of C$_n$TAB decreases linearly with increasing c(NaBr).[6,31,100–104]

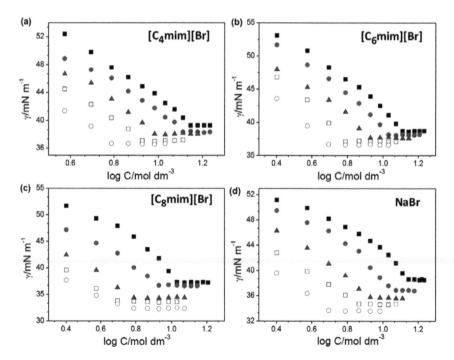

FIGURE 3.4   Representative surface tension plots of aqueous C$_{12}$TAB in presence of different concentrations [(■) 0.2, (●) 0.4, (▲) 0.6, (□) 0.8, and (○) 1.0 wt%] of (a) [C$_4$mim] [Br], (b) [C$_6$mim][Br], (c) [C$_8$mim][Br], and (d) NaBr at ambient conditions.

14) The *cmc* decreases more in the presence of [C_8mim][Br] than [C_6mim][Br] and [C_4mim][Br] due to the higher hydrophobicity of [C_8mim][Br]. Beyaz et al. have reported that the cmc decreases from 170 to 1.9 mmol dm^{-3} with increase of the chain length from methyl to octyl.[97] The cmc of C_nTAB (n = 12, 14, 16) decreases with increasing the alkyl chain length of C_nTAB surfactants and the order is C_{12}TAB < C_{14}TAB < C_{16}TAB. This is due to the hydrophobic–hydrophobic interaction between chain lengths of surfactants.[3]

3.3.3 SOLUBILIZATION OF 8-ANS IN MICELLAR SOLUTION OF AQUEOUS C_nTAB IN THE PRESENCE OF ILS

UV–vis absorption measurement using 8-ANS dye as the absorption probe was used to study the effect of added ILs on the micellization behavior of C_nTAB. Addition of a small amount of oppositely charged dye should change the aggregation behavior of the ionic surfactants. The dyes show different spectroscopic characteristics in the pre and postmicellar regions.[105,106] This property has been utilized in the spectrophotometric determination of metal ions and to improve spectral characteristics of colored systems.

Initially as the concentration of neat aqueous C_nTAB (n = 12, 14, 16) increases, λ_{max} shifted toward red shift with decrease in the absorption intensity up to the *cmc* due to dye–surfactant interactions. The association of the 8-ANS with aqueous C_nTAB shifts the λ_{max} toward higher wavelength as compared to the λ_{max} of the dye in aqueous medium. Once the solution becomes dye saturated, that is, beyond the cmc, further addition of surfactant causes increase in the absorption intensity due to the free dye molecules. Such observation leads to the conclusion that at concentration >*cmc*, the size of the micelles increases which resulted in more spaces available for the dye molecule to be accommodated in the core of micelle. In the presence of ILs, as the concentration of surfactant increase, the absorption maxima observed hyperchromic as well as bathochromic shift, similar to crystal violet dye with the anionic surfactant.[84]

Absorption spectra of 8-ANS (an anionic dye) was observed in the pre and postmicellar region of C_nTAB (n = 12, 14, 16) (with and without additives; ILs and NaBr). The change in spectrum before and after micellization gives clear indication of cmc. The distinction in the absorption spectra of the dye observed in the pre and postmicellar region was designated as the cmc of the surfactant as was observed for the anionic single-chain surfactant in the presence of IL.[107,108] The representative plots of absorbance versus [C_{12}TAB] with [C$_8$mim][Br] are shown in Figure 3.5 and the compiled data for the effect of added ILs/NaBr are reported in Table 3.2.

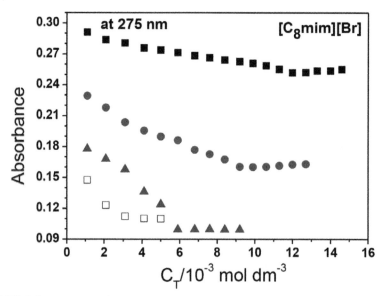

FIGURE 3.5 Representative absorbance plots of 8-ANS dye in aqueous C$_{12}$TAB in the presence of different concentrations [(■) 0.2, (●) 0.4, (▲) 0.6, and (□) 0.8 wt%] of [C$_8$mim][Br] at 298.15 K.

From the above observation, it is concluded that the *cmc* of C$_n$TAB (n = 12, 14, 16) decreases with increasing the concentration of ([C$_n$mim] [Br] n = 4, 6, 8) ILs. The trend of decrease in cmc with added ILs was found in the following order: [C$_8$mim][Br] > [C$_6$mim][Br] > [C$_4$mim][Br]. cmc values of C$_n$TAB decreases monotonically with c(NaBr). As the hydrophobicity of the medium increases (water < [C$_4$mim][Br] < [C$_6$mim][Br] < [C$_8$mim][Br]), λ_{max} shifts to the higher wavelength due to the electrostatic attractive interaction between anionic dye (8-ANS) and the cationic micelle of C$_n$TAB and the hydrophobic–hydrophobic interaction between surfactant and IL molecule. Several researchers have studied a range of dyes to study the aggregation behavior of the ionic surfactants.[109–111] The aggregation behavior of SDS, in a range of RTILs with varied alkyl chain lengths ranging from methyl to octyl, was studied using 8-ANS as the dye.[97] Using crystal violet dye, Javadian et al.[84] reported a similar shift for single-chain anionic surfactants.

3.3.4   EFFECT OF ILs ON THE MICELLAR AGGREGATION NUMBER OF C_nTAB

Pyrene is used as a fluorescence probe to measure various important micellar parameters such as the cmc, aggregation numbers (N_{agg}), dipolarity, and microfluidity to name a few.[52,112–114] The aggregation number, N_{agg} of C_nTAB ($n = 12, 14, 16$) micelle in the presence of different concentrations of ILs ([C_nmim][Br], $n = 4, 6, 8$) were obtained by fluorescence quenching of pyrene by a cosurfactant CPC as a quencher. The representative fluorescence emission spectra (Fig. 3.6a) of pyrene were obtained from fixed [$C_{12}TAB$] (80 mM micellar solution, five times more than *cmc*) and varying [CPC] (quencher) solution in the presence of 0.6 wt% [C_8mim][Br]. The N_{agg} were obtained from the slope of linear correlation ($0.9947 \leq R^2 \leq 0.9961$) plots of $\ln(I_o/I_Q)$ versus [CPC]. Representative plots obtained for $C_{12}TAB$ in [C_8mim][Br] at various concentration of [C_8mim][Br] are shown in Figure 3.6b and the values calculated, from such plots for the other surfactant systems, are listed in Table 3.4.

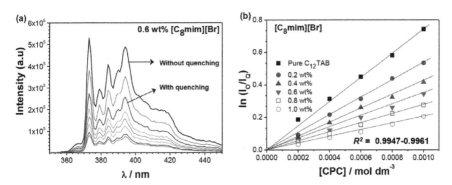

FIGURE 3.6   (a) Representative fluorescence (emission) spectra of 1 µM pyrene in 80 mM aqueous micellar solution of $C_{12}TAB$ in 0.6 wt% [C_8mim][Br] at different quencher concentrations. (b) Representative plots of $\ln(I_o/I_Q)$ versus [CPC] in 80 mM micellar solution of $C_{12}TAB$ in absence and presence of different composition of [C_8mim][Br] at 298.15 K.

The N_{agg} values of neat C_nTAB (38_{C12TAB}, 41_{C14TAB}, and 77_{C14TAB}) are in excellent agreement with literature.[115–119] It is noteworthy to mention here that N_{agg} of C_nTAB decreases with c(additives), similar to single-chain cationic surfactants in ILs[38] and dimeric cationic surfactants in alcohol.[83] The observed decrease in N_{agg} is more significant for [C_8mim][Br] addition, which may be due to the higher hydrophobic–hydrophobic interactions.[116]

The smaller means looser micelle than in pure water, which easies the incorporation of IL molecules into the interfacial region. In binary mixture of $C_nTAB + IL$, the decreasing trend of N_{agg} indicates that the electrostatic repulsion dominates over the hydrophobic interactions, which leads to the formation of loosely packed micellar structure and due to the ion–dipole interaction between the head groups of the surfactant and ILs but still the less efficient packing in mixed micelles is observed than in pure micelles leading to a decrease in N_{agg}. The interactions involving IL ions and the surfactant are very different for a nonionic or a zwitterionic compared to an ionic surfactant.[39] The major electrostatic attractive interaction occurs between Br⁻ of IL and $[-N^+(CH_3)_2]_2$ of C_nTAB. The alkyl chain length of IL at position 1 of C_nmim^+ renders its involvement as cosurfactants.

TABLE 3.4 Values of N_{agg} for Aqueous C_nTAB (n = 12, 14, 16) in Absence and Presence of Different Additives Obtained from Fluorescence Quenching Measurements at 298.15 K.

Additives % (w/w)	N_{agg}			
	[C₄mim][Br]	[C₆mim][Br]	[C₈mim][Br]	NaBr
		$C_{12}TAB$		
0.0	38	38	38	38
0.2	36	29	26	32
0.4	33	25	23	30
0.6	31	22	18	25
0.8	28	20	16	23
1.0	24	19	12	20
		$C_{14}TAB$		
0.0	41	41	41	41
0.2	25	18	20	28
0.4	–	–	–	–
0.6	21	14	14	22
0.8	–	–	–	–
1.0	16	11	12	19
		$C_{16}TAB$		
0.0	77	77	77	77
0.2	73	69	64	72
0.4	70	65	59	68
0.6	67	59	55	63
0.8	64	56	52	60
1.0	62	53	49	56

Standard uncertainties u are $u(N_{agg}) = \pm 1.0$.

3.3.5   SIZE OF MICELLAR AGGREGATES BY DLS MEASUREMENTS

The aggregate size of the surfactants was determined with the help of DLS measurements. The representative CONTIN plots of pure $C_{14}TAB$ (35 mM) and $C_{14}TAB$ + $[C_8mim][Br]$ at various concentrations are depicted in Figure 3.7. The hydrodynamic radii (R_h) of C_nTAB (n = 12, 14, 16) with $[C_nmim][Br]$ (n = 4, 6, 8) ILs and NaBr are presented in Table 3.5. The size of $C_{16}TAB$ micelles in the absence of additives (R_h close to 2.23 nm) is very close to the reported value.[21,120]

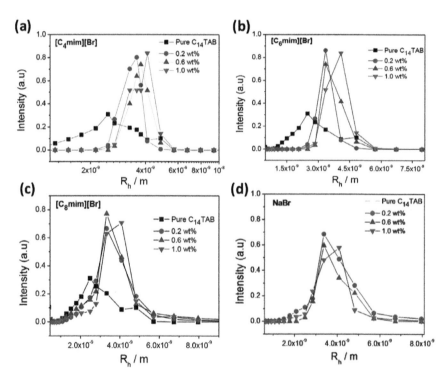

FIGURE 3.7   Representative CONTIN plots of size distribution for aqueous $C_{14}TAB$ at a concentration 35 mM in various compositions of ILs and NaBr at 298.15 K.

For the neat surfactant systems, size of micellar aggregates (R_h) decreases with increasing the alkyl chain length of C_nTAB (n = 12, 14, 16) and the order is $C_{12}TAB < C_{14}TAB < C_{16}TAB$, which may be due to the steric hindrance created by the large increase in the number hydrophobic group of surfactant, which may destabilize the large micelle. The R_h of surfactant aggregates increases with increasing the concentration of ILs due to the incorporation of IL molecules into the micellar structure and the packing intensity of the aggregates decreases, thus forming the loosely packed structures.

TABLE 3.5 Hydrodynamic Radii, R_h of Aqueous C_nTAB (n = 12, 14, 16) Aggregates in Absence and Presence of Different Additives at 298.15 K.

Additives % (w/w)	R_h (nm)			
	[C$_4$mim][Br]	[C$_6$mim][Br]	[C$_8$mim][Br]	NaBr
C$_{12}$TAB				
0.0	2.87	2.87	2.87	2.87
0.2	3.54	3.51	3.49	3.89
0.6	3.89	3.92	3.61	4.54
1.0	4.36	4.29	4.17	4.60
C$_{14}$TAB				
0.0	2.76	2.76	2.76	2.76
0.2	3.49	3.46	3.46	3.80
0.6	3.81	3.76	3.47	4.08
1.0	4.13	4.09	4.09	4.47
C$_{16}$TAB				
0.0	2.23	2.23	2.23	2.23
0.2	2.43	2.06	1.48	2.09
0.6	2.88	2.48	1.74	2.43
1.0	3.37	2.84	2.45	3.39

Standard uncertainties u are $u(R_h) = \pm 0.03$ nm.

Behera et al. reported similar results in which the average size of C$_{16}$TAB aggregates increases in aqueous solutions with increasing concentration of [C$_6$mim][Br].[38] The R_h of C$_n$TAB increases with increasing the concentration of NaBr. The addition of the inorganic electrolyte (KBr) has a similar effect on cationic gemini surfactants.[121] The hydrodynamic radii (R_h) decreases with increasing the alkyl chain length of ILs. The trend observed for decreasing R_h as a function of added ILs is [C$_8$mim][Br] > [C$_6$mim][Br] > [C$_4$mim][Br]. This is due to an increase in hydrophobic–hydrophobic interactions between surfactant tail and ILs molecule.

3.4 CONCLUSIONS

We conclude this chapter based on the following points:

1) The effect of 1-alkyl-3-methylimidazolium bromide [$C_n$mim][Br] (n = 4, 6, and 8) ILs on the micellization properties of single-chain cationic surfactants (C_{12}TAB, C_{14}TAB, and C_{16}TAB) of different chain length was studied using different analytical techniques.

2) Surface adsorption parameters were calculated from adsorption isotherm plots. Results suggest that Π_{cmc} values increases with increasing concentration of additives (ILs and NaBr) and alkyl chain length of the C_nTAB (n = 12, 14, 16), which confirms the better efficiency of the ILs and surfactants.

3) λ_{max} and p values increase with increasing c(ILs). The reverse trend was observed for A_{min} due to the closer packing of the surfactant ions on the surface and the orientation of the surfactant molecules is almost perpendicular to the surface.

4) The results from the conductance, surface tension, and UV–vis spectroscopy measurements observed that cmc of aqueous C_nTAB (n = 12, 16) decreases with increasing the concentration of [C_4mim][Br] and [C_6mim][Br]. For [C_8mim][Br], cmc of C_nTAB (n = 12, 16) decreases up to 0.8 wt% and increase at the 1.0 wt%, while for C_{14}TAB, cmc decreases linearly with c(ILs and NaBr).

5) The cmc decreases with increasing the alkyl chain length of ILs and C_nTAB surfactants. This may be due to the higher hydrophobic–hydrophobic interactions between alkyl chain of ILs and surfactants.

6) N_{agg} decreases with increasing the concentration and alkyl chain length of ILs. We observed that N_{agg} decreases with increasing the concentration of NaBr. The N_{agg} increases with increasing the alkyl chain length of C_nTAB.

7) R_h increases with increasing the concentration of ILs, which is most likely due to the incorporation of IL molecules into the micellar structure and the packing intensity of the micellar aggregates decreases thus forming the loosely packed structures. R_h decreases with increasing the alkyl chain length of ILs and surfactants, which may be due to the steric hindrance created by the large increase in the number hydrophobic group of surfactant, which may destabilize the large micelle.

Finally, we conclude that ILs shows dual behavior, electrolyte as well as cosurfactant. From the above results, we observed that at lower concentration, ILs acts as electrolyte and cosurfactant like behavior at higher concentration and the formation of mixed micelle with surfactant.

ACKNOWLEDGMENTS

N. I. M. acknowledges financial assistances through Department of Science and Technology, New Delhi (SR/FT/CS-014/2010), Institute Research Grants to the assistant professors by SVNIT and Council of Scientific and Industrial Research (CSIR), New Delhi (Grant No. 01 (2545)/11/EMR-II). TEQIP fellowship to U. More is kindly acknowledged here. O. A. El Seoud thanks FAPESP and CNPq for financial support (2014/22136-4) and Research Productivity fellowship (307022/2014-5), respectively.

KEYWORDS

- microemulsions
- surfactant
- hydrophobic interaction
- aqueous solution
- liquid crystals

REFERENCES

1. Butler, H. *Poucher's Perfumes, Cosmetics and Soaps*, 10th ed.; Kluwer Academic Publishers: Dordrecht, 2000.
2. Khan, F.; Siddiqui, U. S.; Khan, I. A.; Kabir-ud-Din. Physicochemical Study of Cationic Gemini Surfactant Butanediyl-1,4-bis(dimethyldodecylammonium bromide) with Various Counterions in Aqueous Solution. *Colloids Surf. A: Physicochem. Eng. Aspects* **2012**, *394*, 46–56.
3. Basu Ray, G.; Chakraborty, I.; Ghosh, S.; Moulik, S. P.; Palepu, R. Self-Aggregation of Alkyl Trimethyl Ammonium Bromides (C_{10}-, C_{12}-, C_{14}-, and C_{16}TAB) and their Binary Mixtures in Aqueous Medium: A Critical and Comprehensive Assessment of Interfacial Behavior and Bulk Properties with Reference to Two Types of Micelle Formation. *Langmuir* **2005**, *21*, 10958–10967.

4. Sharma, R.; Mahajan, S.; Mahajan, R. K. Surface Adsorption and Mixed Micelle Formation of Surface Active Ionic Liquid in Cationic Surfactants: Conductivity, Surface Tension, Fluorescence and NMR Studies. *Colloids Surf. A: Physicochem. Eng. Aspects* **2013**, *427*, 62–75.

5. Javadian, S.; Ruhi, V.; Shahir, A. A.; Heydari, A.; Akbari, J. Imidazolium-Based Ionic Liquids as Modulators of Physicochemical Properties and Nanostructures of CTAB in Aqueous Solution: The Effect of Alkyl Chain Length, Hydrogen Bonding Capacity, and Anion Type. *Ind. Eng. Chem. Res.* **2013**, *52*, 15838–15846.

6. Ghasemian, E.; Najafi, M.; Rafati, A. A.; Felegari, Z. Effect of Electrolytes on Surface Tension and Surface Adsorption of 1-Hexyl-3-methylimidazolium Chloride Ionic Liquid in Aqueous Solution. *J. Chem. Thermodyn.* **2010**, *42*, 962–966.

7. Iyota, H.; Tomimitsu, T.; Shimada, K.; Ikeda, N.; Motomura, K.; Aratono, M. Charge Number Effect on the Miscibility of Inorganic Salt and Surfactant in Adsorbed Film and Micelle: Inorganic Salt–Octyl Methyl Sulfoxide Mixtures. *J. Colloid Interface Sci.* **2006**, *299*, 428–434.

8. Sarac, B.; Bester-Rogac, M. Temperature and Salt-Induced Micellization of Dodecyltri-methylammonium Chloride in Aqueous Solution: A Thermodynamic Study. *J. Colloid Interface Sci.* **2009**, *338*, 216–221.

9. Espito, C.; Colicchio, P.; Facchiano, A.; Ragone, R. Effect of a Weak Electrolyte on the Critical Micellar Concentration of Sodium Dodecyl Sulfate, *J. Colloid Interface Sci.* **1998**, *200*, 310–312.

10. Hao, L.-S.; Hu, P.; Nan, Y.-Q. Salt Effect on the Rheological Properties of the Aqueous Mixed Cationic and Anionic Surfactant Systems. *Colloids Surf. A: Physicochem. Eng. Aspects* **2010**, *361*, 187–195.

11. Kumar, B.; Tikariha, D.; Ghosh, K. K. Effects of Electrolytes on Micellar and Surface Properties of Some Monomeric Surfactants. *J. Dispers. Sci. Technol.* **2012**, *33*, 265–271.

12. Cookey, G. A.; Obunwo, C. C. Effects of Sodium Bromide Salt and Temperature on the Behaviour of Aqueous Solution of Cetyltrimethylammonium Bromide. *IOSR J. Appl. Chem.* **2014**, *7*, 34–38.

13. Hooshyar, H.; Sadeghi, R. Influence of Sodium Salts on the Micellization and Interfa-cial Behavior of Cationic Surfactant Dodecyltrimethylammonium Bromide in Aqueous Solution. *J. Chem. Eng. Data* **2015**, *60*, 983–992.

14. Kumar, S.; Mandal, A. Studies on Interfacial Behavior and Wettability Change Phenomena by Ionic and Nonionic Surfactants in Presence of Alkalis and Salt for Enhanced Oil Recovery. *Appl. Surf. Sci.* **2016**, *372*, 42–51.

15. Rub, M. A.; Azum, N.; Asiri, A. M.; Alfaifi, S. Y. M.; Alharthi, S. S. Interaction between Antidepressant Drug and Anionic Surfactant in Low Concentration Range in Aqueous/Salt/Urea Solution: A Conductometric and Fluorometric Study. *J. Mol. Liq.* **2017**, *227*, 1–14.

16. Dar, A. A.; Chatterjee, B.; Rather, G. M.; Das, A. R. Mixed Micellization and Interfacial Properties of Dodecyltrimethyl Ammonium Bromide and Tetraethyleneglycol Mono-*n*-dodecyl Ether in Absence and Presence of Sodium Propionate. *J. Colloid Interface Sci.* **2006**, *298*, 395–405.

17. Bhat, M. A.; Dar, A. A.; Amin, A.; Rather, G. M. Co- and Counterion Effect on the Micellization Characteristics of Dodecylpyridinium Chloride. *J. Dispers. Sci. Technol.* **2008**, *29*, 514–520.

18. Naskar, B.; Dan, A.; Ghosh, S.; Aswal, V. K.; Moulik, S. P. Revisiting the Self-Aggregation Behavior of Cetyltrimethylammonium Bromide in Aqueous Sodium Salt Solution with Varied Anions. *J. Mol. Liq.* **2012,** *170,* 1–10.
19. Chauhan, S.; Kaur, M.; Kumar, K.; Chauhan, M. S. Study of the Effect of Electrolyte and Temperature on the Critical Micelle Concentration of Dodecyltrimethylammonium Bromide in Aqueous Medium. *J. Chem. Thermodyn.* **2014,** *78,* 175–181.
20. Fang, L.; Tan, J.; Zheng, Y.; Li, H.; Li, C.; Feng, S. Effect of Organic Salts on the Aggregation Behavior of Tri-(trimethylsiloxy)silylpropylpyridinium Chloride in Aqueous Solution. *Colloids Surf. A: Physicochem. Eng. Aspects* **2016,** *509,* 48–55.
21. Bharmoria, P.; Vaneet; Banipal, P. K.; Kumar, A.; Kang, T. S. Modulation of Micellization Behavior of Cetyltrimethylammonium Bromide (CTAB) by Organic Anions in Low Concentration Regime. *J. Sol. Chem.* **2015,** *44,* 16–33.
22. Hassan, P. A.; Yakhmi, J. V. Growth of Cationic Micelles in the Presence of Organic Additives. *Langmuir* **2000,** *16,* 7187–7191.
23. Adane, D. Fenta, Surface and Thermodynamic Studies of Micellization of Surfactants in Binary Mixtures of 1,2-Ethanediol and 1,2,3-Propanetriol with Water. *J. Phys. Sci.* **2015,** *10,* 276–288.
24. Naorem, H.; Devi, S. D. Conductometric and Surface Tension Studies on the Micellization of Some Cationic Surfactants in Water-Organic Solvent Mixed Media. *J. Surf. Sci. Technol.* **2006,** *22,* 89–100.
25. Dubey, N. CTAB Aggregation in Solutions of Higher Alcohols: Thermodynamic and Spectroscopic Studies. *J. Mol. Liq.* **2013,** *184,* 60–67.
26. Bielawska, M.; Chodzinska, A.; Janczuk, B.; Zdziennicka, A. Determination of CTAB CMC in Mixed Water + Short-Chain Alcohol Solvent by Surface Tension, Conductivity, Density and Viscosity Measurements. *Colloids Surf. A: Physicochem. Eng. Aspects* **2013,** *424,* 81–88.
27. Li, W.; Han, Y. C.; Zhang, J. L.; Wang, L. X.; Song, J. Thermodynamic Modeling of CTAB Aggregation in Water–Ethanol Mixed Solvents. *Colloid J.* **2006,** *68,* 304–310; Li, W.; Han, Y. C.; Zhang, J. L.; Wang, B. G. Effect of Ethanol on the Aggregation Properties of Cetyltrimethylammonium Bromide Surfactant. *Colloid J.* **2005,** *67,* 159–163.
28. Anderson, M. T.; Martin, J. E.; Odinek, J. G.; Newcomer, P. P. Effect of Methanol Concentration on CTAB Micellization and on the Formation of Surfactant-Templated Silica (STS). *Chem. Mater.* **1998,** *10,* 1490–1500.
29. Benito, I.; Garcia, M. A.; Monge, C.; Saz, J. M.; Marina, M. L. Spectrophotometric and Conductimetric Determination of the Critical Micellar Concentration of Sodium Dodecyl Sulfate and Cetyltrimethylammonium Bromide Micellar Systems Modified by Alcohols and Salts. *Colloids Surf. A: Physicochem. Eng. Aspects* **1997,** *125,* 221–224.
30. Nazir, M.; Ahanger, S.; Akbar, A. Micellization of Cationic Surfactant Cetyltrimethylammonium Bromide in Mixed Water–Alcohol Media. *J. Dispers. Sci. Technol.* **2009,** *30,* 51–55.
31. Modaressi, A.; Sifaoui, H.; Grzesiak, B.; Solimando, R.; Domanska, U.; Rogalski, M. CTAB Aggregation in Aqueous Solutions of Ammonium Based Ionic Liquids: Conductimetric Studies. *Colloids Surf. A: Physicochem. Eng. Aspects* **2007,** *296,* 104–108.
32. Pal, A.; Chaudhary, S. Effect of Hydrophilic Ionic Liquid on Aggregation Behavior of Aqueous Solutions of Sodium Dodecylsulfate (SDS). *Fluid Phase Equilibr.* **2013,** *352,* 42–46.

33. Chaudhary, S.; Pal, A. Conductometric and Spectroscopic Study of Interaction of Cationic Surfactants with 3-Methyl-1-pentylimidazolium Hexafluorophosphate. *J. Mol. Liq.* **2014**, *190*, 10–15.

34. Pal, A.; Pillania, A. Modulating Effect of Ionic Liquid 1-butyl-2,3-Dimethylimidazolium Chloride on Micellization Behaviour of Cationic Surfactant Dodecyltrimethyl Ammonium Bromide in Aqueous Media. *Fluid Phase Equilibr.* **2015**, *389*, 67–73.

35. Pal, A. Chaudhary, S. Ionic Liquid Induced Alterations in the Physicochemical Properties of Aqueous Solutions of Sodium Dodecylsulfate (SDS). *Colloids Surf. A: Physicochem. Eng. Aspects* **2013**, *430*, 58–64.

36. Behera, K.; Pandey, S. Modulating Properties of Aqueous Sodium Dodecyl Sulfate by Adding Hydrophobic Ionic Liquid. *J. Colloid Interface Sci.* **2007**, *316*, 803–814.

37. Behera, K.; Pandey, S. Ionic Liquid Induced Changes in the Properties of Aqueous Zwitterionic Surfactant Solution. *Langmuir* **2008**, *24*, 6462–6469.

38. Behera, K.; Om, H.; Pandey, S. Modifying Properties of Aqueous Cetyltrimethyl Ammonium Bromide with External Additives: Ionic Liquid 1-Hexyl-3-Methylimidazolium Bromide versus Cosurfactant *n*-Hexyltrimethyl Ammonium Bromide. *J. Phys. Chem. B* **2009**, *113*, 786–793.

39. Behera, K.; Pandey, S. Interaction between Ionic Liquid and Zwitterionic Surfactant: A Comparative Study of Two Ionic Liquids with Different Anions. *J. Colloid Interface Sci.* **2009**, *331*, 196–205.

40. Pal, A.; Pillania, A. Thermodynamic and Aggregation Properties of Aqueous Dodecyltrimethyl Ammonium Bromide in the Presence of Hydrophilic Ionic Liquid 1,2-Dimethyl-3-octylimidazolium Chloride. *J. Mol. Liq.* **2015**, *212*, 818–824.

41. Smirnova, N. A.; Vanin, A. A.; Safonova, E. A.; Pukinsky, I. B.; Anufrikov, Y. A.; Makarov, A. L. Self-assembly in Aqueous Solutions of Imidazolium Ionic Liquids and their Mixtures with an Anionic Surfactant. *J. Colloid Interface Sci.* **2009**, *336*, 793–802.

42. Zhang, S.; Gao, Y.; Dong, B.; Zheng, L. Interaction between the Added Long-Chain Ionic Liquid 1-Dodecyl-3-Methylimidazolium Tetrafluoroborate and Triton X-100 in Aqueous Solutions. *Colloids Surf. A: Physicochem. Eng. Aspects* **2010**, *372*, 182–189.

43. Behera, K.; Pandey, M. D.; Porel, M.; Pandey, S. Unique Role of Hydrophilic Ionic Liquid in Modifying Properties of Aqueous Triton X-100. *J. Chem. Phys.* **2007**, *127*, 184501–184510.

44. Behera, K.; Pandey, S. Concentration-Dependent Dual Behavior of Hydrophilic Ionic Liquid in Changing Properties of Aqueous Sodium Dodecyl Sulfate. *J. Phys. Chem. B* **2007**, *111*, 13307–13315.

45. Welton, T. Room-Temperature Ionic Liquids. Solvents for Synthesis and Catalysis. *Chem. Rev.* **1999**, *99*, 2071–2084.

46. Wasserscheid, P.; Keim, W. Ionic Liquids—New "Solutions" for Transition Metal Catalysis. *Angew. Chem. Int. Ed.* **2000**, *39*, 3772–3789.

47. Wasserscheid, P.; Welton, T. *Ionic Liquids in Syntheses*; Wiley: Weinhein, Germany, 2003.

48. Rajput, S. M.; More, U. U.; Vaid, Z. S.; Prajapati, K. D.; Malek, N. I. Impact of Organic Solvents on the Micellization and Interfacial Behavior of Ionic Liquid Based Surfactants. *Colloids Surf. A: Physicochem. Eng. Aspects* **2016**, *507*, 182–189.

49. Das, C.; Das, B. Thermodynamic and Interfacial Adsorption Studies on the Micellar Solutions of Alkyltrimethylammonium Bromides in Ethylene Glycol (1) + Water (2) Mixed Solvent Media. *J. Chem. Eng. Data* **2009**, *54*, 559–565.

50. Tanford, C. *The Hydrophobic Effect: Formation of Micelles and Biological Membranes*; Wiley–Interscience: New York, 1973.

51. Holmberg, K.; Johnsson, B.; Kronberg, B.; Lindman, B. *Surfactants and Polymers in Aqueous Solution*, 2nd ed.; Wiley: Chichester, 2003.

52. Baker, G. A.; Pandey, S.; Pandey, S.; Baker, S. N. A New Class of Cationic Surfactants Inspired by *N*-Alkyl-*N*-methyl Pyrrolidinium Ionic Liquids. *Analyst* **2004**, *129*, 890–892.

53. Buzzeo, M. C.; Evans, R. G.; Compton, R. G. Non-Haloaluminate Room-Temperature Ionic Liquids in Electrochemistry—A Review. *Chem. Phys. Chem.* **2004**, *5*, 1106–1120.

54. More, U.; Kumari, P.; Vaid, Z.; Behera, K.; Malek, N. I. Interaction between Ionic Liquids and Gemini Surfactant: A Detailed Investigation into the Role of Ionic Liquids in Modifying Properties of Aqueous Gemini Surfactant. *J. Surfact. Deterg.* **2016**, *19*, 75–89.

55. Chen, H.; Han, L.; Luo, P.; Ye, Z. The Interfacial Tension between Oil and Gemini Surfactant Solution. *Surf. Sci.* **2004**, *552*, L53–L57.

56. Jones, M. J.; Chapman, D. *Micelles, Monolayers, and Biomembranes*; Wiley: New York, 1995.

57. Bloom, H.; Reinsborough, V. C. Cryoscopy in Molten Pyridinium Chloride. *Aust. J. Chem.* **1967**, *20*, 2583–2587.

58. Evans, D. F.; Yamauchi, A.; Jason, W. G.; Bloomfield, V. A. Micelle Formation in Ethylammonium Nitrate, a Low-Melting Fused Salt. *J. Phys. Chem.* **1983**, *87*, 3537–3541.

59. Patrascu, C.; Gauffre, F.; Nallet, F.; Bordes, R.; Oberdisse, J.; de Lauth-Viguerie, N.; Mingotaud, C. Micelles in Ionic Liquids: Aggregation Behavior of Alkyl Poly (ethyleneglycol)-ethers in 1-Butyl-3-methyl-imidazolium Type Ionic Liquids. *ChemPhysChem* **2006**, *7*, 99–101.

60. Chen, L. G.; Bermudez, H. Solubility and Aggregation of Charged Surfactants in Ionic Liquids. *Langmuir* **2012**, *28*, 1157–1162.

61. Beyaz, A.; Oh, S. W.; Reddy, V. P. Synthesis CMC Studies of 1-Methyl-3-(pentafluorophenyl)imidazolium Quaternary Salts. *Colloids Surf. B: Biointerfaces* **2004**, *36*, 71–74.

62. Fernandez-Castro, B.; Mendez-Morales, T.; Carrete, J.; Fazer, E.; Cabeza, O.; Rodriguez, J. R.; Turmine, M.; Varela, L. M. Surfactant Self-Assembly Nanostructures in Protic Ionic Liquids. *J. Phys. Chem. B* **2011**, *115*, 8145–8154.

63. Pal, A.; Chaudhary, S. Ionic Liquids Effect on Critical Micelle Concentration of SDS: Conductivity, Fluorescence and NMR Studies. *Fluid Phase Equilibr.* **2014**, *372*, 100–104.

64. Pal, A.; Pillania, A. Effect of Trisubstitutedimidazolium Based Ionic Liquid 1-Butyl-2,3-dimethylimidazolium Chloride on the Aggregation Behaviour of Sodium Dodecylsulphate in Aqueous Media. *Colloids Surf. A: Physicochem. Eng. Aspects* **2014**, *452*, 18–24.

65. Pal, A.; Kumar, B.; Kang, T. S. Effect of Structural Alteration of Ionic Liquid on their Bulk and Molecular Level Interactions with Ethylene Glycol. *Fluid Phase Equilibr.* **2013**, *358*, 241–249.

66. Pal, A.; Pillania, A. Modulating the Aggregation Behavior of Aqueous Sodium Dodecylsulphate (SDS) with Addition of Trisubstitutedimidazolium Based Ionic Liquid 1-Butyl-2,3-dimethylimidazolium Tetrafluoroborate [bdmim][BF$_4$]. *Fluid Phase Equilibr.* **2014**, *375*, 23–29.

67. Misono, T.; Sakai, H.; Sakai, K.; Abe, M.; Inoue, T. Surface Adsorption and Aggregate Formation of Nonionic Surfactants in a Room Temperature Ionic Liquid, 1-Butyl-3-methylimidazolium Hexafluorophosphate (bmimPF$_6$). *J. Colloid Interface Sci.* **2011,** *358*, 527–533.

68. Inoue, T.; Iwasaki, Y. Cloud Point Phenomena of Polyoxyethylene-type Surfactants in Ionic Liquid Mixtures of emimBF$_4$ and hmimBF$_4$. *J. Colloid Interface Sci.* **2010,** *348*, 522–528.

69. Javadian, S.; Ruhi, V.; Heydari, A.; Shahir, A. A.; Yousefi, A.; Akbari, J. Self-Assembled CTAB Nanostructures in Aqueous/Ionic Liquid Systems: Effects of Hydrogen Bonding. *Ind. Eng. Chem. Res.* **2013,** *52*, 4517–4526.

70. Shi, L.; Jing, X.; Gao, H.; Gu, Y.; Zheng, L. Ionic Liquid-Induced Changes in the Properties of Aqueous Sodium Dodecyl Sulfate Solution: Effect of Acidic/Basic Functional Groups. *Colloid Polym. Sci.* **2013,** *291*, 1601–1612.

71. Firestone, M. A.; Dzielawa, J. A.; Zapol, P.; Curtiss, L. A.; Seifert, S.; Dietz, M. L. Lyotropic Liquid-Crystalline Gel Formation in a Room-Temperature Ionic Liquid. *Langmuir* **2002,** *18*, 7258–7260.

72. Bowers, J.; Butts, C. P.; Martin, P. J.; Vergara-Gutierrez, M. C. Aggregation Behavior of Aqueous Solutions of Ionic Liquids. *Langmuir* **2004,** *20*, 2191–2198.

73. Dong, B.; Li, N.; Zheng, L. Q.; Yu, L.; Inoue, T. Surface Adsorption and Micelle Formation of Surface Active Ionic Liquids in Aqueous Solution. *Langmuir* **2007,** *23*, 4178–4182.

74. Mahajan, K.; Sharma, R. Analysis of Interfacial and Micellar Behavior of Sodium Dioctylsulphosuccinate Salt (AOT) with Zwitterionic Surfactants in Aqueous Media. *J. Colloid Interface Sci.* **2011,** *363*, 275–283.

75. Malek, N. I.; Ijardar, S. P.; Oswal, S. L. Excess Molar Properties for Binary Systems of C$_n$MIM-BF$_4$ Ionic Liquids with Alkylamines in the Temperature Range (298.15 to 318.15) K. Experimental Results and Theoretical Model Calculations. *J. Chem. Eng. Data* **2014,** *59*, 540–553.

76. Ijardar, S. P.; Malek, N. I. Experimental and Theoretical Excess Molar Properties of Imidazolium Based Ionic Liquids with Molecular Organic Solvents—I. 1-Hexyl-3-Methylimidazlouim Tetrafluoroborate and 1-Octyl-3-methylimidazlouim Tetraflouroborate with Cyclic Ethers. *J. Chem. Thermodyn.* **2014,** *71*, 236–248.

77. Malek, N. I.; Singh, A.; Surati, R.; Ijardar, S. P. Study on Thermophysical and Excess Molar Properties of Binary Systems of Ionic Liquids. I: [C$_n$mim][PF$_6$] (*n* = 6, 8) and Alkyl Acetates. *J. Chem. Thermodyn.* **2014,** *74*, 103–118.

78. Pal, A.; Kumar, B. Densities, speeds of sound and ^{1}H NMR Spectroscopic Studies for Binary Mixtures of 1-Hexyl-3-methylimidazolium Based Ionic Liquids with Ethylene Glycol Monomethyl Ether at Temperature from *T* = (288.15–318.15) K. *Fluid Phase Equilibr.* **2012,** *334*, 157–165.

79. Wang, H.; Wang, J.; Zhang, S. Apparent Molar Volumes and Expansivities of Ionic Liquids [C$_n$mim]Br (*n* = 4, 8, 10, 12) in Dimethyl Sulfoxide. *J. Chem. Eng. Data* **2012,** *57*, 1939–1944.

80. Mata, J.; Varade, D.; Bahadur, P. Aggregation Behavior of Quaternary Salt Based Cationic Surfactants. *Thermochim. Acta* **2005,** *428*, 147–155.

81. Shang, Y.; Wang, T.; Han, X.; Peng, C.; Liu, H. Effect of Ionic Liquids C$_n$mimBr on Properties of Gemini Surfactant 12-3-12 Aqueous Solution. *Ind. Eng. Chem. Res.* **2010,** *49*, 8852–8857.

82. Yousuf, S.; Mohd. Akram; Kabir-ud-Din. Effect of Salt Additives on the Aggregation Behavior and Morphology of 14-E2-14. *Colloids Surf. A: Physicochem. Eng. Aspects* **2014,** *463,* 8–17.

83. Mohammad, R.; Khan, I. A.; Kabir-ud-Din; Schulz, P. C. Mixed Micellization and Interfacial Properties of Alkanediyl-α,ω-bis (Dimethylcetylammonium Bromide) in the Presence of Alcohols. *J. Mol. Liq.* **2011,** *162,* 113–121.

84. Javadian, S.; Nasiri, F.; Heydari, A.; Yousefi, A.; Asadzadeh Shahir, A. Modifying Effect of Imidazolium-Based Ionic Liquids on Surface Activity and Self-assembled Nanostructures of Sodium Dodecyl Sulfate. *J. Phys. Chem. B* **2014,** *118,* 4140–4150.

85. Israelachvili, J. N.; Mitchell, D. J.; Ninham, B. W. Theory of Self-assembly of Hydrocarbon Amphiphiles into Micelles and Bilayers. *J. Chem. Soc. Faraday Trans.* **1976,** *2,* 1525–1568.

86. Berr, S. S. Solvent Isotope Effects on Alkytrimethylammonium Bromide Micelles as a Function of Alkyl Chain Length. *J. Phys. Chem.* **1987,** *91,* 4760–4765.

87. Bakshi, M. S.; Sachar, S. Influence of Hydrophobicity on the Mixed Micelles of Pluronic F127 and P103 Plus Cationic Surfactant Mixtures. *Colloids Surf. A: Physicochem. Eng. Aspects* **2006,** *276,* 146–154.

88. Barry, B. W.; Morrison, J. C.; Russell, G. F. J. Prediction of the Critical Micelle Concentration of Mixtures of Alkyltrirnethylammonium Salts. *J. Colloid Interface Sci.* **1970,** *33,* 554–561.

89. Mukerjee, P. Nature of the Association Equilibriums and Hydrophobic Bonding in Aqueous Solutions of Association Colloids. *Adv. Colloid Interf. Sci.* **1967,** *1,* 242–275.

90. Mukerjee, P.; Mysels, K. J. *Critical Micelle Concentrations of Aqueous Surfactant Systems*; National Bureau of Standards of NSRDS-NBS 36: Washington DC, 1972; p 20234.

91. Behera, K.; Dahiya, P.; Pandey, S. Effect of Added Ionic Liquid on Aqueous Triton X-100 Micelles. *J. Colloid Interface Sci.* **2007,** *307,* 235–245.

92. Sadasivuni, K. K.; Ponnamma, D.; Kim, J.; Cabibihan, J.-J.; AlMaadeed, M. A. Thermoelectric Properties of Biopolymer Composites, Chapter 5. In *Biopolymer Composites in Electronics;* Elsevier, Matthew Deans, 2016; pp 155–184. ISBN: 978-0-12-809261-3.

93. Ruiz, C. C. Thermodynamics of Micellization of Tetradecyltrimethyl Ammonium Bromide in Ethylene Glycol–Water Binary Mixtures. *Colloid Polym. Sci.* **1999,** *277,* 701–707.

94. Manna, K.; Panda, A. K. Physicochemical Studies on the Interfacial and Micellization Behavior of CTAB in Aqueous Polyethylene Glycol Media. *J. Surfact. Deterg.* **2011,** *14,* 563–576.

95. Paul, B. K.; Moulik, S. P. Chapter 10- Ionic liquids in colloidal regime. In *Ionic Liquid Based Surfactant Science: Formation, Characterization and applications;* Mukherjee, I., Moulik, S. P., Eds.; John Wiley and Sons, Inc.: Hoboken, NJ, 2015; pp 207–238.

96. Cunha, E.; Passos, M. L. C.; Pinto, P. C. A. G.; Lúcia, M.; Saraiva, M. F. S. Improved Activity of α-Chymotrypsin in Mixed Micelles of Cetyltrimethylammonium Bromide (CTAB) and Ionic Liquids: A Kinetic Study Resorting to Sequential Injection Analysis. *Colloids Surf. B: Biointerfaces* **2014,** *118,* 172–178.

97. Beyaz, A.; Oh, W. S.; Reddy, V. P. Ionic Liquids as Modulators of the Critical Micelle Concentration of Sodium Dodecyl Sulfate. *Colloids Surf. B: Biointerfaces* **2004,** *35,* 119–124.

98. Koya, P. A.; Kabir-ud-Din. Studies on the Mixed Micelles of Alkyltrimethylammonium Bromides and Butanediyl-1,4-Bis(alkyldimethylammonium Bromide) Dimeric

Surfactants in the Presence and Absence of Ethylene Glycol at Different Temperatures. *J. Colloid Interface Sci.* **2011**, *360*, 175–181.

99. Wattebled, L.; Laschewsky, A. Effects of Organic Salt Additives on the Behavior of Dimeric ("Gemini") Surfactants in Aqueous Solution. *Langmuir* **2007**, *23*, 10044–10052.

100. Jakubowska, A. Interactions of Different Counterions with Cationic and Anionic Surfactants. *J. Colloid Interface Sci.* **2010**, *346*, 398–404.

101. Yu, D.; Huang, X.; Deng, M.; Lin, Y.; Jiang, L.; Huang, J.; Wang, Y. Effects of Inorganic and Organic Salts on Aggregation Behavior of Cationic Gemini Surfactants. *J. Phys. Chem. B* **2010**, *114*, 14955–14964.

102. Chakraborty, I.; Moulik, S. P. Self-aggregation of Ionic C10 Surfactants Having Different Headgroups with Special Reference to the Behavior of Decyltrimethylammonium Bromide in Different Salt Environments: A Calorimetric Study with Energetic Analysis. *J. Phys. Chem. B* **2007**, *111*, 3658–3664.

103. Rosen, M. J. *Surfactants and Interfacial Phenomena*, 3rd ed.; Wiley: New York, 2004.

104. Rodrıguez, A.; Graciani, M.; Munoz, M.; Robina, I.; Moya, M. L. Effects of Ethylene Glycol Addition on the Aggregation and Micellar Growth of Gemini Surfactants. *Langmuir* **2006**, *22*, 9519–9525.

105. Colichman, E. Spectral Study of Long Chain Quaternary Ammonium Salts in Brom Phenol Blue Solutions. *J. Am. Chem. Soc.* **1951**, *73*, 3385–3388.

106. Rosendorfova, J.; Cermakova, L. Spectrophotometric Study of the Interaction of Some Triphenylmethane Dyes and 1-Carbethoxypentadecyltrimethylammonium Bromide. *Talanta* **1980**, *27*, 705–708.

107. El-Daly, S. A.; Abdel-Kader, M. H.; Issa, R. M.; el-Sherbini el-Sayed, A. Influence of Solvent Polarity and Medium Acidity on the UV–Vis Spectral Behavior of 1-Methyl-4-[4-amino-styryl] Pyridinium Iodide. *Spectrochim. Acta A: Mol. Biomol. Spectrosc.* **2003**, *59*, 405–411.

108. Mandal, A. B.; Dhathathreyan, A.; Jayakumar, R.; Ramasami, T. Characterisation of Boc-Lys(Z)-Tyr-NHNH2 Dipeptide. Part 1—Physico-chemical Studies on the Micelle Formation of a Dipeptide in the Absence and Presence of Ionic Surfactants. *J. Chem. Soc. Faraday Trans.* **1993**, *89*, 3075–3079.

109. Micheau, J. C.; Zakharova, G. V.; Chibisov, A. K. Reversible Aggregation, Precipitation and Re-dissolution of Rhodamine 6G in Aqueous Sodium Dodecyl Sulfate. *Phys. Chem. Chem. Phys.* **2004**, *6*, 2420–2425.

110. Gohain, B.; Sarma, S.; Dutta, R. K. Protonated Dye–Surfactant Ion Pair Formation between Neutral Red and Anionic Surfactants in Aqueous Submicellar Solutions. *J. Mol. Liq.* **2008**, *142*, 130–135.

111. Bielska, M.; Sobczynska, A.; Prochaska, K. Dye–Surfactant Interaction in Aqueous Solutions. *Dyes Pigment* **2009**, *80*, 201–205.

112. Acree, W. E.; Meyer, R. A. *Encyclopedia of Analytical Chemistry*; John Wiley & Sons Ltd.: Chichester, 2000.

113. Lakowicz, J. R. *Principles of Fluorescence Spectroscopy*, 3rd ed.; Springer: Berlin, 2006.

114. Pandey, S.; Redden, R. A.; Fletcher, K. A.; Palmer, C. P. Characterization of Solvation Environment Provided by Dilute Poly(Sulfonyl Maleic Anhydride-*co*-Dodecyl Vinyl Ether) Solutions at Various pH Using Pyrene and 1,3-Bis(1-pyrenyl)propane as Fluorescence Probes. *Macromol. Chem. Phys.* **2003**, *204*, 425–435.

115. Rodríguez, A.; Junquera, E.; del Burgo, P.; Aicart, E. Conductometric and Spectrofluorimetric Characterization of the Mixed Micelles Constituted by

Dodecyltrimethylammonium Bromide and a Tricyclic Antidepressant Drug in Aqueous Solution. *J. Colloid Interface Sci.* **2004**, *269*, 476–483.

116. Shivaji Sharma, K.; Patil, S. R.; Rakshit, A. K.; Glenn, K.; Doiron, M.; Palepu, R. M.; Hassan, P. A. Self-Aggregation of a Cationic–Nonionic Surfactant Mixture in Aqueous Media: Tensiometric, Conductometric, Density, Light Scattering, Potentiometric, and Fluorometric Studies. *J. Phys. Chem. B* **2004**, *108*, 12804–12812.

117. Pisárčik, M.; Devínsky, F.; Pupák, M. Determination of Micelle Aggregation Numbers of Alkyltrimethylammonium Bromide and Sodium Dodecyl Sulfate Surfactants Using Time-Resolved Fluorescence Quenching. *Open Chem.* **2015**, *13*, 922–931.

118. Baker, G.; Bright, A. F. V.; Pandey, S. Quantifying Critical Micelle Concentration and Nonidealities within Binary Mixed Micellar Systems: An Upper-Level Undergraduate Laboratory. *Chem. Educ.* **2001**, *6*, 223–226.

119. Evans, D. F.; Allen, M.; Ninham, B. W.; Fouda, A. Critical Micelle Concentrations for Alkyltrimethylammonium Bromides in Water from 25 to 160°C. *J. Solut. Chem.* **1984**, *13*, 87–101.

120. Feitosa, E.; Brazolin, M. R. S.; Georgetto Naal, R. M. Z.; Freire de Morais Del Lama, M. P.; Lopes, J. R.; Loh, W.; Vasilescu, M. Structural Organization of Cetyltrimethyl-ammonium Sulfate in Aqueous Solution: The effect of Na_2SO_4. *J. Colloid Interface Sci.* **2006**, *299*, 883–889.

121. Kabir-ud-Din; Siddiqui, U. S.; Ghosh, G. Growth of Gemini Surfactant Micelles under the Influence of Additives: DLS Studies. *J. Dispers. Sci. Technol.* **2009**, *30*, 1310–1319.

CHAPTER 4

THE MODELING POSSIBILITIES FOR ORGANIC REACTIVITY: POLARIC AND STERIC EFFECTS

V. I. KODOLOV[*]

Chemistry and Chemical Technology Department, M. T. Kalashnikov Izhevsk State Technical University, Izhevsk, Russia

[*]*E-mail: kodol@istu.ru*

CONTENTS

ABSTRACT

This chapter includes the information about semi-empiric notions in physical organic chemistry (organic reactivity) and their applications for prognosis of reactions flowing. The physical nature of following notions as electronegativity, polarity, and steric effects are discussed. The parameters of polarity as well as electronegativity are used in the correlation equations for the estimation of reactivity and properties of phosphorus organic compounds as also for polymerization and copolymerization processes. The influence of electronegativity or polarity constants on the bond (P=O) vibration wave number in IR spectra is considered in detail. The semi-empiric constants are given according to mesoscopic physics definitions.

4.1 INTRODUCTION

The estimation of chemical particles reactivity and the determination of chemical reactions direction are the actual trend in physical chemistry. Usually, the polarity constants and values of electronegativity of atoms or atoms groups are used for the estimation of chemical particles (reagents) reactivity or their activity in chemical reactions.

In physical organic chemistry, the following notions as inductive, inductomeric, mesomeric, and electromeric effects are considered. According to the theory of free-energy linear relation,[1-5] all reactions are divided on different types because of the conditions (including reagents nature) reactions conducting. In this chapter, the Taft constants are proposed for the chemical particles reactivity estimation in corresponding reactions. The Taft constants correspond to the comparative values of electrostatic and steric (spaced) effects during the interaction of chemical reagents. The development of such new trends, as synergetic, fractal theory, nanochemistry, and mesoscopic physics, leads to the new directions on analysis of chemical particles reactivity in different types of chemical reactions.

4.2 PHYSICAL MEANING CORRELATIONS OF POLARITY CONSTANTS AND ELECTRONEGATIVITIES

In physical chemistry, the correlation equations are widely applied for the prognosis of chemical reagents characteristics with the using of electronegativity and Taft polarity constants. It also applies the correlations between polarity constants as electronegativity values and the ionization potentials, dipole moments, electron density values as well as wave numbers in IR spectra and other parameters in EPR, NMR spectra, etc.[6,7] There are many empiric equations about the correlation of electronegativity with force constants which characterize the defined chemical bonds. One from these empiric equations is Gordy formula[8]:

$$\chi = r/(an)^{2/3}\sqrt{(K^{4/3} - b^{4/3})}, \tag{4.1}$$

where a and b are constants.

When the ionic molecule is considered, the formula for force constant is transformed in simple appearance:

$$K = Z^2/r^3, \tag{4.2}$$

where Z is the charge on element surface and r is the distance from symmetry place to element surface.

The values of force constants are calculated on formula (4.2) with using values of charges, determined by Slayter rule, and r values, and also are compared with force constants from Kross table (Table 4.1).[8]

TABLE 4.1 The Comparison of Force Constants, Calculated on Formula (4.2), and Force Constants.

Chemical bonds	Charge (el. st. units)	Distance, r (Å)	Force constant, K from eq. (4.2) (dyn/cm)	Force constant from Kross table
C–C	$Z_{C-} = 3.25$	1.54	6.5×10^5	5×10^5
C=C	$Z_{C=} = 3.6$	1.35	12.5×10^5	12×10^5
C–H	$Z_{C-} = 3.25$	1.07	4.4×10^5	5×10^5
	$Z_{H-} = 0.7$			
C–O	$Z_{C-} = 3.25$	1.8	6.0×10^5	5×10^5
	$Z_{O-} = 4.55$			
C=O	$Z_{C=} = 3.6$	1.5	12.3×10^5	12×10^5
	$Z_{O=} = 4.9$			
P=O	$Z_{P=} = 5.15$	1.9	9×10^5	9×10^5
	$Z_{O=} = 4.9$			

According to Ref. [8], the force constants of chemical bonds are parameters for the correspondent bonds wave numbers calculations. Therefore, there are many correlation equations that reflect the dependence of chemical bonds wave numbers with electronegativity values or polar constants. The high range of wave numbers changes are especially determined by the great inductive effect of additive atoms groups and the bond big polarization. One of that bond with high polarization is P=O bond. The following equations establish the relations between wave numbers or bond wave length and electronegativity values or Taft polar constants:

$$\hat{v}_{P=O} = 1198 + 16.8\Sigma\sigma, \tag{4.3}$$

$$\hat{v}_{P=O} = 930 + 40\Sigma\pi, \tag{4.4}$$

$$\lambda_{P=O} = (39.96 - \Sigma\chi)/3.995, \tag{4.5}$$

where σ is the Taft polar constant,[1] π is the Thomas electronegativity value,[7] χ is the Bell electronegativity value.[6] The first equation is proposed by Griffin.[8]

Let us compare the Griffin equation with the Herzberg–Kondrat'ev equation, which is obtained on the basis of quantum mechanics[10,11]:

$$\hat{v} = \hat{v}_v + B \,[J'(J' + 1) - J(J + 1)], \tag{4.6}$$

where $\hat{v}_v$ is the wave number correspondent to the bond vibration; B is the parameter, which is connected to the group (atom) rotation; and J' and J are the rotation quantum numbers for upper and inner levels, respectively.

In this case, $\hat{v}_v$ can be approximately calculated on formula:

$$\hat{v}_v = 1/2\pi c\sqrt{K/\mu}, \tag{4.7}$$

where c is the light velocity (cm/s); K is the force constant, calculated on formula (4.2); μ is the approved mass equaled to $(m_P \times m_O)/(m_P + m_O)$.

If the values K and μ stand to eq. (4.7), then the meaning of $\hat{v}_v$ will be equaled ~1200 cm^{-1}.

The parameter B is calculated on formula:

$$B = h/(8\pi^2 c\mu r^2), \tag{4.8}$$

where h is the Plank constant; μ is the approved mass; and r is the distance from the symmetry place to the surface of correspondent pair of atoms.

Value B can be transformed in the case, if the meanings h, c and the dimension units of mass and distance would stand in formula as

$$B = 16.8(5)/\mu r^2. \tag{4.9}$$

When eqs. (4.7) and (4.8) will be put in eq. (4.6) as well as the meaning of correspondent parameters, the next equation for the P=O bond wave number will be received:

$$\ddot{\upsilon}_v = (1190\text{–}1200) + 16.8\ [J'(J' + 1) - J(J + 1)]/\mu r^2. \tag{4.10}$$

The meanings of J' and J correspond to $h/2\pi\hat{J}'$ and $h/2\pi\hat{J}$, where $\hat{J}'$ and $\hat{J}$ reflect the moments of motion quantities—$(mVr)'$ and (mVr). From the comparison between eq. (4.3) and eq. (4.10), the following formula (4.11) is received:

$$\sigma = [J'(J' + 1) - J(J + 1)]/\mu r^2 = (h^2/4\pi^2\mu r^2)[(\hat{J}' - \hat{J})(\hat{J}' + \hat{J} + 1)] \tag{4.11}$$

On the basis of the parameter dimension analysis in formula (4.11), the Taft polarity constants may be written as

$$\sigma = h^2/4\pi^2\ (mVr\cdot\Delta mVr)/\mu r^2 = h^2/4\pi^2\omega\cdot\Delta(I\omega) \tag{4.12}$$

Thus, Taft polarity constant is explained as part of kinetic energy, and since $\Delta\chi = \gamma\Delta\sigma$,[12] the electronegativity is the change of kinetic energy during the electron density displacement with the atom. This part of energy can be identical to the surface energy of atom or atoms group and has link with Taft constants and the reactivity of compounds, containing the definite atom groups.

4.3   CORRELATIONS OF TAFT CONSTANTS AND ORGANIC REACTIVITY

The organic reactivity in reactions is determined by sum of inductive and steric factors with correspondent corrections on resonance effect and on activity of hydrogen atoms. This is illustrated by the Taft equation[1]:

$$\Delta\Delta F = -2.3RT(\sigma^*\rho + E_s\delta + \Delta nh + \psi), \tag{4.13}$$

In eq. (4.13), the value $\Delta\Delta F$ is the change of Helmholtz free energy in reactions series with participation of compounds containing changing atoms groups: $\Delta\Delta F = \Delta F_i - \Delta F_0$, where ΔF_i is the free energy change in reaction with reagent containing i atoms group, and ΔF_0 is the change of this parameter when there is standard group of atoms, for example, CH_3 group.

Parameters in eq. (4.13) correspond to

1. σ^* is the Taft constant, characterized polarity or inductive effect of corresponded atoms group (for standard group CH_3, $\sigma^* = 0$).
2. ρ is the constant, characterized reaction series.
3. E_s is the Taft constant, characterized steric effect of atoms group.
4. δ is the constant, characterized the influence of steric factors in reaction series.
5. Δnh is the correction on the influence of hydrogen atoms on delocalization of π electrons.
6. ψ is the correction on resonance effect of atoms group.

In case, if the inductive effect has advantageous influence in process, eq. (4.13) is written as

$$\Delta\Delta F = -2.3RT(\sigma^* \cdot \rho) \tag{4.14}$$

According to Ref. [12], ρ is defined as

$$\rho = (46.1 \cdot \gamma^2/2.3RT) \cdot \Delta\sigma_r, \tag{4.15}$$

where $\Delta\sigma_r$ is the change of reaction center polarity value, which can be expressed by σ^* values.

As obtained from eq. (4.14):

$$\sigma^* \cdot \rho = (\Delta F_0 - \Delta F_i)/2.3RT. \tag{4.16}$$

Since the values σ and ρ have no dimensions, these values can be presented as ratio of surface energies changes:

$$\sigma^* = (\varepsilon_0 - \varepsilon_i)/\varepsilon_0, \tag{4.17}$$

where ε_0 is the surface energy of i fragment, ε_i is the surface energy of standard fragment. On analogy with this expression, let us write the ratio of volume energies changes for fragments or the ratio of fragments volumes. These relations are identical to E_s values

$$E_s = (V_0 - V_i)/V_0, \tag{4.18}$$

where V_0 is the volume value for standard fragment, V_i is the volume of i-fragment.

The calculation of vibration volumes for several atoms groups is carried out on following scheme:

- the center of charge gravity is determined;
- the rotation (vibration) of atoms group around axis, which connect the center of charge gravity with "main atom" of reaction center, is proposed; spherical segment or spherical sector is formed during that rotation (vibration); and
- the volume of body formed at rotation is calculated on base of known linear and angle values.

For instance, let us assume that groups CH_3 and CH_3O are presented as volumes of spherical segments (Fig. 4.1). In this case, the center of gravity for CH_3 group is found on axis $OC°$ (Fig. 4.1a). The volume of spherical segment obtained at the CH_3-group rotation (vibration) equals 37.4 Å³.[13] The center of gravity for CH_3O group is found at point L on line OC (Fig. 4.1b). The distance OL is calculated from following equation:

$$Z_O \cdot OL = Z(CH_3)(OC - OL), \qquad (4.19)$$

where Z_O is the Slater charge for Oxygen (4.55) and $Z(CH_3)$ is the charge of CH_3 group (2.72).

The spherical segment obtained at the rotation of CH_3O group around axis $LC°$ has volume that equals 50 Å³ at $a \sim 2.45$ Å and $h \sim 3.29$ Å.

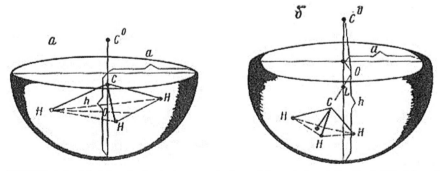

FIGURE 4.1 Models of spherical segments formed by rotation (vibration) CH_3 (a) and CH_3O (b) groups.

The comparison of the relative volumes for several atoms groups and Taft constants E_s is given in Table 4.2.

TABLE 4.2 The Comparison of Calculated Values E_s with Correspondent Taft Constants.

Atom or atoms group	Vibration volume (Å^3)	Relative change of volume	E_s	Difference between E_s calc. and tabl.
H	7.2	0.21	0.25	0.04
CH_3	37.4	0.00	0.00	0
C_3H_7	58.0	−0.56	−0.56	0
$(CH_3)_2CH$	70.0	−0.88	−0.85	0.03
$(CH_3)_3C$	117.0	−2.14	−2.14	0
$(C_2H_5)_3C$	198.0	−4.4	−4.4	0
CF_3	78	−1.1	−1.16	0.06
CCl_3	114 (yellow)	−2.06	2.06	0
CBr_3	130 (yellow)	−2.5	2.43	0.07

Thus, the main factors including energetic and steric parameters expressed as relative changes of surface energies and volumes for correspondent atoms groups are identical to Taft equation. Therefore, these notions may be used at consideration of reaction series, for example, vinyl monomer polymerization processes.

4.4 TAFT CONSTANTS IN ESTIMATION OF VINYL MONOMERS ACTIVITIES IN POLYMERIZATION

For the polymerization process analysis, let us consider the changes of chain growth constant k_g at the polymerization of styrene and methylacrylate. The following correlation equation for k_g is known[14]:

$$\lg k_p = 0.7\sigma^* + 0.09\sigma_r' + 0.03E_s^0 + 2.15 \qquad (4.20)$$

In this case, coefficient of correlation corresponds to 0.994. Value σ_r' has meaning as in chapter,[14] and value E_s^0 is given with correction on hyperconjugation. Parameters σ_r' and E_s^0 can increase the chain growth constant only in case when these parameters are increased during process. Therefore, it's interesting to observe the changes of polarity for the growing polymeric chain using the equation for σ^* of electro-acceptor atoms groups:

$$\sigma^*(XCH_2CH(R)CH_2) = 0.36[0.36(\sigma_0^* + \sigma_R^*) - 0.02\sigma_0^*\sigma_R^*], \qquad (4.21)$$

In atoms groups $XCH_2CH(R)CH_2$, fragment X corresponds to initiator atoms group which connect with monomer, and fragment R is the atoms group connected with vinyl bond of monomer. In future, $\sigma^*(XCH_2)$ is denoted as σ^*_0. Then, the polarity parameters values for growing polymeric chain will be changed according to following formulas:

$$\sigma^*_1 = \sigma^*(XCH_2CH(R)CH_2)$$

$$\sigma^*_2 = 0.36[0.36(\sigma^*_1 + \sigma^*_R) - 0.02\sigma^*_1\sigma^*_R] \tag{4.22}$$

$$\sigma^*_3 = 0.36[0.36(\sigma^*_2 + \sigma^*_R) - 0.02\sigma^*_2\sigma^*_R]$$

etc.

Using these formulas, it is established that the polarity of growing polymeric chain for polystyrene, if $X \equiv H$, changes are as follows: $\sigma^*_0 = 0.0000$; $\sigma^*_1 = 0.0777$; $\sigma^*_2 = 0.0874$; $\sigma^*_3 = 0.0887$; $\sigma^*_4 = 0.0888$; $\sigma^*_5 = 0.0888$ at $\sigma^*_R = 0.6$. The growing polymeric chain for polymethylacrylate, the values of polarity constant will be the following: $\sigma^*_0 = 0.0000$; $\sigma^*_1 = 0.2592$; $\sigma^*_2 = 0.2891$; $\sigma^*_3 = 0.2925$; $\sigma^*_4 = 0.2928$; $\sigma^*_5 = 0.2929$; $\sigma^*_6 = 0.2929$ at $\sigma^*_R = 2.0$.

In other words, the changes of polarity for growing chain are decreased to zero after fourth–fifth step of polymerization. This is correspondent to data of Alfrey–Borer,[15] according which polarity of growing polymeric chain is constant. If $\Delta\sigma^* = 0$, eq. (4.20) can be simplified as

$$\Delta \lg k_p = \Delta\sigma'_r r + \Delta E^0_s \delta \tag{4.23}$$

The values σ'_r are correlated with σ^* and E^0_s. The linear dependence between ($E_s - E^0_s$) and σ'_r is found for several aliphatic atoms groups (Fig. 4.2). Therefore, it can be proposed that these parameters are defined by conformation effects.

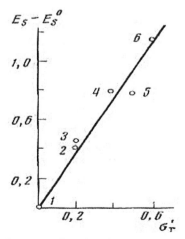

FIGURE 4.2   The correlation ($E_s - E^0_s$) and σ'_r for several aliphatic atoms groups. 1—CH_3, 2—C_2H_5, 3—C_3H_7, 4—iso-C_3H_7, and 5—tret-C_4H_9.

The increase of influence of resonance and steric factors during the polymeric chain growth leads to development of self-organization and self-similarity in process of submolecular form creation.

4.5 MESOSCOPICS NOTIONS AND KOLMOGOROV–AVRAMI EQUATION IN ORGANIC REACTIVITY

The development of such new trends, as synergetic, fractal theory, nanochemistry, and mesoscopic physics, lead to new directions on analysis of chemical particles reactivity in different types of chemical reactions. In the trends indicated above, the size and energetic characteristics of chemical particles take the main place. The size of particles is denoted as approximately 10 nm, and the motion freedom of them is limited by the vibration with high frequency and electron transport across them. In above example, the growing polymeric chain after fourth or fifth step of process has size near to 10 nm that is reason of start for self-organization process with reservation conformation order, which determine the finished product structure. For the process explanation, the equation of Kolmogorov–Avrami[16] can be used:

$$W = 1 - \exp(-k\tau^n), \tag{4.24}$$

where W is the part of obtained product (for instance, polymer), k is the process rate constant, τ is the duration of process, n is the fractal dimension (for one measured process $n = 1$).

For the comparative estimation of reagents (or nanostructures) in one reaction series, the application of the theory of free energy linear dependence is possible. In this case, the reactions are considered by using one of the reagents as the standard compound for which W is fixed W_0. The estimation of reactivity can be proposed on the difference, $W - W_0$, where W is calculated based on formula (4.24), and W_0 is defined based on analogous formula with changes $k_0\tau_0^n$. It is noted that the fractal dimension n do not change because the comparison is carried out for one type of reactions. The following equation for difference $W - W_0$ can be written:

$$W - W_0 = \exp(-k_0\tau_0) - \exp(-k\tau^n), \tag{4.25}$$

and after eq. (4.24) transformation:

$$\lg(W - W_0) = k/k_0 \, (\tau/\tau_0)^n - 1 \tag{4.26}$$

If lg k/k_0 is defined as

$$\lg k/k_0 = -2.3RT\{[(\varepsilon_0 - \varepsilon)/\varepsilon_0]a + [(V_0 - V)/V_0]b\}, \qquad (4.27)$$

and then if this expression after transformation stands in eq. (4.26), then eq. (4.28) is received:

$$\lg(W - W_0) = (\tau/\tau_0)^n \exp\{[(\varepsilon_0 - \varepsilon)/\varepsilon_0]a + [(V_0 - V)/V_0]b\}, \qquad (4.28)$$

where values a and b are the parameters, which correct the influence polar and steric effects on reactivity.

This is new trend as variant of organic reactivity theory and the possibility of chemical particles reactivity comparative estimation, including nanostructures reactivity.

4.6 CONCLUSION

The variant of organic reactivity theory development is proposed. This variant is based on the consideration of physical notions as electronegativity and Taft constants (polarity constant σ^* and steric constant E_s). The relative changes of surface energies $[(\varepsilon_0 - \varepsilon)/\varepsilon_0]$ and analogous changes of volumes $[(V_0 - V)/V_0]$ for activity characteristics of atoms groups are defined as parameters for the estimation of reactivity in the reactions. In the variant of theory, there is the application of new scientific trends: synergetics, chemistry in nanoreactors, fractal theory, and mesoscopics. For the estimation of nanostructures reactivity, the using of Kolmogorov–Avrami equation is discussed.

KEYWORDS

- electronegativity
- Taft constant of polarity
- Taft steric constant
- organic reactivity
- Griffin and Herzberg–Kondrat'ev equations
- Kolmogorov–Avrami equation
- polymerization and copolymerization processes

REFERENCES

1. Taft, R. U. The Division of Polar, Steric and Resonance Factors influences on Organic Reactivity. In *The Steric Effects in Organic Chemistry*; 1960; p 719, pp 562–693.
2. Pal'm, V. A. The Fundamentals of Organic Reactions Quantities' Theory. In *Chemistry*; 1967; 356 p.
3. Pal'm, V. A. The Introduction in Theoretical Organic Chemistry. *High School*, 1974; p 446.
4. Kodolov, V. I.; Spasskiy, S. S. The Structure and Properties of Phosphorus Organic Compounds. *Usp. Chimii* **1964,** *33* (2), 1501–1525.
5. Kodolov, V. I. Possibilities of Modeling in Organic Chemistry. *Org. React.* **1965,** *2* (4, 6), 11–17.
6. Batsanov, S. S. *The Electro Negativity of Elements and Chemical Bond*; SD AS USSR: Novosibirsk, 1962; p 320.
7. Kodolov, V. I.; Semerneva, G. A.; Spasskiy, S. S. About Change of P=O Bond Wave Number for Phosphorus Organic Compounds. In *Proceeding of Chemistry Institute UB AS USSR: Element Organic Compounds*; UB AS USSR: Sverdlovsk, 1966, Issue 13; pp 99–101.
8. Vol'kenshteyn, L. V.; El'yashevitch, M. A.; Stepanov, V. I. *Molecules Vibration*; AS USSR: Sverdlovsk, 1943; p 510.
9. Herzberg, L.; Herzberg, G. Molecules Spectra. In *Fundamental Formulas of Physics*; Menzel, D., Ed.; 1957.
10. Kondrat'ev, V. N. *Structures of Atoms and Molecules*; Physmatgiz, 1959.
11. Herzberg, G. *Vibration and Rotation Spectra of Many Atomic Molecules*. 1949.
12. Ritchie, C. D.; Sager, W. F. *Progress in Physical Organic Chemistry;* J. Wiley & Sons: New York–London–Sydney, 1964; Vol. 2, pp 323–400.
13. Kodolov, V. I. Calculation Possibility of E_s Values from Taft Equation. *J. Phys. Chem.* **1966,** *40* (1), 56–57.
14. Kodolov, V. I.; Spasskiy, S. S. Parameters in Equations of Alfrey–Price and Taft. *Vysokomol. Soed.* **1976,** *18A* (9), 1986–1992.
15. Alfrey, T.; Borer, J.; Mark, G. *Copolymerization*. 1963.
16. Kodolov, V. I.; Trineeva, V. V. Theoretical Fundamentals of Nanomaterials Science. *Chem. Phys. Mesosc.* **2016,** *18* (3), 390–404.

CHAPTER 5

AN INSIGHT INTO NEARLY IDEAL BEHAVIOR OF IONIC LIQUID MIXTURES

SUSHMA P. IJARDAR[1,2*]

[1]*Department of Chemistry, Veer Narmad South Gujarat University, Surat 395007, Gujarat, India*

[2]*Salt and Marine Chemicals Division, CSIR—Central Salt and Marine Chemicals Research Institute (CSIR-CSMCRI), Council of Scientific & Industrial Research (CSIR), Bhavnagar 364002, Gujarat, India*

*E-mail: sushmaijardar@yahoo.co.in

CONTENTS

ABSTRACT

The ionic liquids (ILs) are well known as "designer solvent" for their ability to fine-tune physicochemical properties by simply exchanging cation or anion without altering its chemical nature. Mixing of two simple ILs together results into IL mixtures are named "double-salt ionic liquids" (DSILs) or "mixed ionic liquids" in literature defined as a homogeneous liquid at room temperature having two or more than two different ions. DSILs provide new opportunity to fine-tune physicochemical properties using proper combination of cation–anion. Depending upon the nature of ions present in DSILs, their physicochemical properties are slightly different than simple ILs, which exhibits ideal to nonideal mixing behavior. Specially, nonideal mixing behavior of DSILs seems to be advantageous for further advancement in the field of ILs which is reported and analyzed here for new DSILs.

5.1 INTRODUCTION

The ionic liquids (ILs) are distinct class of salts composed entirely of ions with numerous combinations of bulky organic cation and tiny organic or inorganic anion; most of them are liquid at room temperature. The huge interest and growth that have been observed in the research of ILs during the last two decades are attributed to their bunch of properties; undetectable vapor pressure at room temperature, thermal stability, chemical stability, high conductivity, broad liquid range, wide electrochemical windows, ability to dissolve solutes, and recycling nature.[1-4] They have been successfully employed in widespread industrial applications in catalysis, polymer synthesis, organic synthesis, analytical chemistry, membrane technology, electrodeposition, dye-sensitized solar cell, lubricant, extraction, and biomass processing to name a few.[5-12]

ILs are tagged as "designer solvents" for ability to fine-tune physicochemical properties by choosing different combination of cation and anion suitable for given application. If binary, ternary of quaternary systems of ILs are considered, approximately 10^{18} types of ILs are possible.[13] The majority of research has been focused on mixtures obtained by mixing simple IL with one cation and one anion. One approach is addition of molecular organic solvent (MOS) in simple ILs forming binary or ternary liquid mixtures. The past studies confirmed that addition of MOS greatly alters physicochemical properties of simple ILs resulting nonideal behavior of mixtures. The nonideal behavior have been explained by comparing behavior of mixing component in pure state and in mixtures resulting in interstitial accommodation and variation in molecular interactions exists within mixtures.[14–19]

The second approach that has introduced new strategy of mixing two simple ILs resulting IL mixtures receives considerable interest in recent years.[20–24] The reviews provide fruitful information about basic physical (density, viscosity), chemical (polarity), thermal properties, structural behavior, and their successful applications. The resulting IL mixtures that expected fine-tune physicochemical properties provide guidelines in formulation of task-specific ILs for particular applications. Roger and his group have suggested term "double-salt ionic liquids" (DSILs) for the mixtures obtained by mixing two or more simple ILs which exhibits nonideal properties.[20] He also suggested that each binary or ternary composition of ILs mixture can be considered as unique DSILs with properties different than simple ILs. When DSILs are formed, the association of ions in individual ILs are disturbed due to different extent of molecular interactions exists in mixtures.[20] DSILs have obvious advantageous over IL mixtures with MOS as synthesis and properties of individual simple ILs are well established. The ideal mixing of DSIL provides further opportunity to fine-tune the physicochemical properties between properties of individual ILs.[23]

In 2012, Welton has proposed the nomenclature of DSILs considering (1) no. of constitute and (2) no. of component present in ILs mixture.[21] [A][B] represents cation and [X][Y] represents anion of the simple ILs, when two simple ILs [A][X] and [B][Y] are mixed, the resulting IL mixture can be abbreviated as [A][B][X][Y]. This represents binary system with two compounds or ternary system with three constituents. In this category, there can be possibility of formation of IL mixtures abbreviated as [A][X][Y] or [A][B][X] with common cation or anion, respectively. If three ILs [A][X], [A][Y], and [A][Z] or [A][X][B][X] and [C][X] are mixed, it can be abbreviated as [A][X][Y][Z] or [A][B][C][X]. It is a ternary system with three compounds or quaternary system. For simplicity, in present study, we have adopted resulting DSILs as binary systems that always contain two different ILs with common cation.

Detailed literature reveals that as compared to IL mixtures with MOS, very limited publications are amiable related to ideal/nonideal behavior of DSILs. DSILs have been treated in similar manner as mixtures of ILs + MOS.[25–30] Based on variation in physicochemical properties, DSILs exhibit nearly ideal even nonideal behavior that has also been reported for DSILs. In many IL mixtures, unexpected physicochemical properties are obtained upon mixing,[20] and even relevant features at the nanoscopic level are obtained, although apparently ideal behavior is obtained for certain macroscopic properties.[23] The structure investigation of DSILs has been recently reported.[24] Therefore, systematical study is required for new DSILs exhibit nonideal behavior, which could provide different and tunable properties to those of pure ILs for new developments.

This chapter focuses on insights into volumetric behavior of three new binary IL mixtures (DSILs). DSILs were obtained by mixing two pure ILs namely 1-ethyl-3-methylimidazolium bis(trifluoromethanesulfonylimide [C$_2$mim][NTf$_2$] + 1-ethyl-3-methylimidazolium trifluoromethanesulfonate [C$_2$mim][OTf], +1-ethyl-3-methylimidazolium methanesulfonate [C$_2$mim] [MeSO$_3$], +1-ethyl-3-methylimidazolium thiocyanate [C$_2$mim][SCN]. The volumetric behavior was examined using measured density of pure simple ILs and corresponding DSILs at 0.1 MPa, over entire range of compositions and temperature ranges from 293.15 to 323.15 K. Excess molar volumes (deviation from ideality) were calculated from measured values of density to check mixing behavior of DSILs. The mixing behavior will be discussed in terms of size difference of anion and molecular interaction present within the mixtures.

5.2 EXPERIMENTAL

5.2.1 CHEMICALS

The common name, abbreviation, molecular formula, and molecular weight of simple ILs used in present study is given in Table 5.1. [C$_2$mim][NTf$_2$], [C$_2$mim][OTf], [C$_2$mim][MeSO$_3$], and [C$_2$mim][SCN] were purchased from Sigma Aldrich with certified purity of ≥98.0%, ≥98.0%, ≥95.0%, and ≥99.0%, respectively. The CAS No., source, initial, and final purity of ILs are listed in Table 5.2. The mentioned impurity of water was ≤0.5%. To reduce water content, all ILs was kept under reduced pressure at 313.15 K for 3–4 days. The final purity of ILs were 98.5, 98.5, 97.2, and 99.3 mass%, respectively. The purity was confirmed through ^{1}H NMR, recorded by advanced Brucker DPX 200. The water content for the dried ILs was determined by Karl Fisher titrator (Metrohm, 890 Titrando) and found to be 200, 200, 400, and 150 ppm, respectively. Table 5.4 represents the comparison between experimental and literature values of densities ρ for pure ILs used in present work.[31–49] The literature values have been collected from ThemoIL (IL database updated Nov 2016). The experimental values of density for all ILs are in good agreement with literature for [C$_2$mim][NTf$_2$],[31,33,34] [C$_2$mim] [OTf],[37,41] [C$_2$mim][MeSO$_3$],[43,46] and [C$_2$mim][SCN].[49]

TABLE 5.1 Details of Simple ILs Used in DSIL Formation.

Name: 1-Ethyl-3-methylimidazolium bis(trifluoromethylsulfonyl)imide Abbreviation: $[C_2mim][NTf_2]$ Molecular formula: $C_8H_{11}F_6N_3O_4S_2$ Molecular weight: 391.31	
Name: 1-Ethyl-3-methylimidazolium trifluoromethanesulfonate Abbreviation: $[C_2mim][OTf]$ Molecular formula: $C_7H_{11}F_3N_2O_3S$ Molecular weight: 260.23	
Name: 1-Ethyl-3-methylimidazolium methanesulfonate Abbreviation: $[C_2mim][MeSO_3]$ Molecular formula: $C_7H_{14}N_2O_3S$ Molecular weight: 206.26	
Name: 1-Ethyl-3-methylimidazolium thiocynate Abbreviation: $[C_2mim][SCN]$ Molecular formula: $C_7H_{11}N_3S$ Molecular weight: 169.25	

5.2.2 EXPERIMENTAL SECTION

5.2.2.1 IL MIXTURES PREPARATION

The formulations for DSILs and pure ILs used in preparation of DSILs are given in Table 5.3. The DSILs (binary mixtures) were prepared by mass on Mettler Toledo analytical balance B 204-S with an uncertainty of $\pm 1.10^{-7}$ kg. The mixtures were prepared in tightly capped glass vials just prior to measurements. An average uncertainty in the mole fraction was found around $\pm 1.10^{-4}$. Due to considerable viscosity, the mixing was enhanced using homogenizer. The homogeneous mixture was transferred to syringe and introduce in densimeter for density measurement.

TABLE 5.3 DSILs Composed of Three Ions Used in Present Study.

Simple IL$_1$	Simple IL$_2$	DSILs
[C$_2$mim][NTf$_2$]	[C$_2$mim][OTf]	[C$_2$mim][NTf$_2$][OTf]
[C$_2$mim][NTf$_2$]	[C$_2$mim][MeSO$_3$]	[C$_2$mim][NTf$_2$][MeSO$_3$]
[C$_2$mim][NTf$_2$]	[C$_2$mim][SCN]	[C$_2$mim][NTf$_2$][SCN]

5.2.2.2 DENSITY

The densities for pure ILs and corresponding DSILs were recorded using high precision vibrating tube densimeter DSA 5000 from Anton-Paar. The DSA automatically performed viscosity corrections for density during measurements up to 700 mPa s.[50] The temperature in DSA 5000 was controlled by built-in solid-state thermostat with precision of ±0.005 K. The DSA was calibrated by measuring densities of double-distilled water and dry air at 298.15 K. The performance of DSA was checked by comparing density and speed of sound measurements for binary mixture of cyclohexane + benzene.[51] The uncertainty in density and speed of sound was ±0.05 kg m^{-3}.

5.3 RESULTS AND DISCUSSION

5.3.1 DENSITY OF PURE ILS

Table 5.4 represent comparison between experimental values of densities for pure [C$_2$mim][NTf$_2$], [C$_2$mim][OTf], [C$_2$mim][MeSO$_3$], and [C$_2$mim][SCN].[31-49] As per general observation, density of ILs increases with increase in molecular weight of anions.[20] The density of simple ILs decreases linearly with increasing temperature. The linear decrease in densities of simple ILs can be explained in terms of volume expansively or thermal expansion coefficient. The thermal expansion coefficient α_p is an indicator for expansion or contraction of fluid in all directions with temperature.

The measured density values in the temperature range from 298.15 to 323.15 K were fitted to the following equation:

$$\rho = \sum_{i=1}^{2} C_i (T - 273.15)^{i-1} \qquad (5.1)$$

TABLE 5.4 Comparison between Experimental and Literature Values of Densities ρ (g cm^{-3}) for Simple ILs at 0.1 MPa.

		[C$_2$mim][NTf$_2$]	
Temp.	Experimental value	Literature value	
298.15	1.518230	1.5185[a], 1.517[b], 1.51874[c], 1.5164[d], 1.518[e], 1.5193[f]	0.651
303.15	1.513011	1.5134[a], 1.512[b], 1.51368[c], 1.5113[d], 1.512[e], 1.5142[f]	0.653
308.15	1.508415	1.5084[a], 1.50866[c], 1.5063[d], 1.507[e]	0.655
313.15	1.503383	1.5033[a], 1.502[b], 1.50364[c], 1.5012[d], 1.502[e], 1.5043[f]	0.657
318.15	1.498853	1.4983[a], 1.49864[c], 1.4962[d]	0.659
323.15	1.493141	1.4933[a], 1.491[b], 1.49366[c], 1.4912[d], 1.4943[f]	0.6619
		[C$_2$mim][OTf]	
298.15	1.385710	1.3836[g], 1.384[h], 1.3859[i], 1.3859[j], 1.3853[k]	0.653
303.15	1.380910	1.3702[g], 1.3816[i], 1.3818[j], 1.3811[k], 1.3799[l], 1.3818[m]	0.645
308.15	1.376131	1.3752[g], 1.375[h], 1.3773[i], 1.3773[j], 1.3769[k]	0.647
313.15	1.372101	1.3611[g], 1.3731[i], 1.3732[j], 1.3727[k], 1.3716[l], 1.3717[m]	0.649
318.15	1.367902	1.3669[g], 1.367[h], 1.3689[i], 1.3689[j], 1.3686[k]	0.651
323.15	1.363151	1.352[2], 1.3648[i], 1.3647[j], 1.3632[l]	0.653
		[C$_2$mim][MeSO$_3$]	
298.15	1.240020	1.2424[i], 1.2415[n], 1.241[o], 1.23996[p], 1.234[q]	0.498
303.15	1.237742	1.239[i], 1.2382[n], 1.237[o], 1.231[q]	0.499
308.15	1.234951	1.2356[i], 1.2348[n], 1.234[o], 1.23322[p], 1.228[q]	0.500
313.15	1.231574	1.2322[i], 1.2314[n], 1.230[o], 1.225[q]	0.502
318.15	1.228587	1.2288[i], 1.2282[n], 1.227[o], 1.22651[p], 1.221[q]	0.503
323.15	1.225561	1.2255[i], 1.224[o], 1.218[q]	0.504
		[C$_2$mim][SCN]	
298.15	1.117231	1.1168[r], 1.1167[s], 1.117[i], 1.1155[a], 1.116[o]	0.579
303.15	1.113962	1.1139[i], 1.1124[a], 1.113[o]	0.581
308.15	1.110520	1.1107[r], 1.1104[s], 1.1108[i], 1.1094[a], 1.110[o]	0.583
313.15	1.107351	1.1078[i], 1.1063[a], 1.107[o]	0.584
318.15	1.104082	1.1047[r], 1.1048[s,i], 1.1033[a], 1.104[o]	0.586
323.15	1.101152	1.1018[i], 1.1003[4], 1.101[o]	0.588

Standard uncertainties $u(x) = 1 \times 10^{-4}$, $u() = 0.05$ kg m^{-3}, $u(T) = 0.05$ K.

[a]Ref. [31], [b]Ref. [32], [c]Ref. [33], [d]Ref. [34], [e]Ref. [35], [f]Ref. [36], [g]Ref. [37], [h]Ref. [38], [i]Ref. [39], [j]Ref. [40], [k]Ref. [41], [l]Ref. [42], [m]Ref. [43], [n]Ref. [44], [o]Ref. [45], [p]Ref. [46], [q]Ref. [47], [r]Ref. [48], and [s]Ref. [49].

The thermal expansion coefficient α_p defined as

$$\alpha_p = V^{-1}\ (\partial\ V/\partial\ T)_p = -\rho^{-1}\ (\partial\rho/\partial T)_p \tag{5.2}$$

At each temperature, α_p was obtained by using coefficient of eq. (5.1). The values of thermal expansion coefficient α_p for simple ILs at studied temperatures are included in Table 5.1. A very small deviation in α_p was observed due to linear drop in density of simple ILs with temperature. The values of α_p for simple ILs decrease in the order of $[C_2mim][MeSO_3] > [C_2mim][SCN] > [C_2mim][OTf] > [C_2mim][NTf_2]$ indicates that $[C_2mim][NTf_2]$ contracts less as compared to other studied ILs at fixed temperature.

5.3.2 DENSITY AND EXCESS MOLAR VOLUME OF IL MIXTURES (DSILS)

Tables 5.5–5.7 represent measured values of densities, ρ for investigated IL mixtures: $[C_2mim][NTf_2][OTf]$, $[C_2mim][NTf_2][MeSO_3]$, and $[C_2mim][NTf_2]$ [SCN] over entire composition range and at 5 K interval between $T = 298.15$ and 323.15 K. All three DSILs form homogeneous mixtures at all compositions. All IL mixtures are binary systems that contain common cation C_2mim^+ and two different anions NTf_2–OTf, NTf_2–$MeSO_3$, or NTf_2–SCN. Any change in single molecular entity will lead to change in volumetric behavior.[25] It can be seen from Table 5.6–5.8 that the density of DSILs increase with increasing mole fraction of $[C_2mim][NTf2]$ in IL mixtures. As per general trend, the density of IL mixtures decreases with increase in temperature and decreases with decrease in molecular weight of added anions.[20]

Ideal to nearly ideal behavior has been observed in DSILs composed of more than two ions. Ideal solutions obey Raoult's, that is, chemical potential of component of mixture (molar volume) show linear dependence on mole fraction of pure component.[21] The calculated values of molar volume V_m for studied mixtures are given in Table 5.5–5.7. The linear dependence of molar volume as a function of $[C_2mim][NTf_2]$ mole fraction is graphically presented in Figure 5.1a–c. It follows the same trend reported in literature. However, Roger[20] suggested to consider excess molar volume to check ideal or nonideal behavior of IL mixtures.

The excess molar volume, V_m^E was derived using eq. (5.2)

$$V_m^E = \sum_{i=1}^{2} x_i M_i (\rho^{-1} - \rho_i^{-1}) \tag{5.3}$$

TABLE 5.5 Experimental Values of Densities ρ, Molar Volumes V_m, and Excess Molar Volumes V_m^E for [C$_2$mim][NTf$_2$]$_X$[OTf]$_{(1-X)}$ at (293.15–313.15) K at 0.1 MPa.

$X_{[C_2mim][NTf_2]}$	298.15 K	303.15 K	308.15 K	313.15 K	318.15 K	323.15 K
			ρ (g cm^{-3})			
0	1.385710	1.380910	1.376130	1.372100	1.367900	1.363150
0.1054	1.402351	1.397652	1.392971	1.388843	1.384670	1.379886
0.2036	1.417286	1.412494	1.407838	1.403641	1.399439	1.394509
0.2995	1.431370	1.426540	1.421907	1.417682	1.413456	1.408430
0.3992	1.445592	1.440673	1.436081	1.431727	1.427436	1.422313
0.4985	1.459121	1.454166	1.449561	1.445080	1.440770	1.435580
0.5964	1.471838	1.466895	1.462303	1.457702	1.453359	1.448086
0.7024	1.485021	1.480021	1.475442	1.470750	1.466320	1.460921
0.7792	1.494209	1.489152	1.484564	1.479791	1.475373	1.469891
0.9021	1.507972	1.502844	1.498321	1.493455	1.488960	1.483320
1	1.518230	1.513011	1.508415	1.503383	1.498853	1.493141
			V_m (cm^3 mol^{-1})			
0	188	188	189	190	190	191
0.1054	195	196	197	197	198	199
0.2036	202	203	204	204	205	206
0.2995	209	210	211	211	212	213
0.3992	216	217	218	218	219	220
0.4985	223	224	225	225	226	227
0.5964	230	231	231	232	233	234
0.7024	237	238	239	240	240	241
0.7792	243	243	244	245	246	247
0.9021	251	252	253	253	254	255
1	258	259	259	260	261	262
			V_m^E (m^3 mol^{-1})			
0	0	0	0	0	0	0
0.105	0.251	0.231	0.221	0.217	0.208	0.196
0.204	0.406	0.392	0.382	0.373	0.362	0.356
0.300	0.488	0.473	0.462	0.441	0.429	0.422
0.399	0.496	0.487	0.473	0.454	0.447	0.438
0.499	0.467	0.456	0.446	0.432	0.421	0.408
0.596	0.410	0.391	0.381	0.369	0.359	0.344
0.702	0.311	0.294	0.284	0.271	0.269	0.258
0.779	0.218	0.205	0.198	0.186	0.178	0.170
0.902	0.091	0.082	0.067	0.052	0.051	0.051
1	0	0	0	0	0	0

TABLE 5.6 Experimental Values of Densities ρ, Molar Volumes V_m, and Excess Molar Volumes V_m^E for $[C_2mim][NTf_2]_X[MeSO_3]_{(1-X)}$ at (293.15–313.15) K at 0.1 MPa.

$X_{[C_2mim][NTf_2]}$	298.15 K	303.15 K	308.15 K	313.15 K	318.15 K	323.15 K
			ρ (g cm^{-3})			
0	1.24102	1.23774	1.23495	1.231574	1.228587	1.225561
0.1154	1.285754	1.282281	1.279242	1.275695	1.272568	1.269161
0.2043	1.316977	1.313274	1.310132	1.306413	1.303092	1.299449
0.3124	1.351987	1.348081	1.344698	1.340754	1.33725	1.333297
0.3967	1.377247	1.373102	1.369563	1.365513	1.361923	1.357686
0.5047	1.407271	1.402926	1.399165	1.394918	1.391111	1.386614
0.5904	1.429356	1.424826	1.420928	1.416536	1.412655	1.407922
0.6998	1.45568	1.451009	1.446896	1.442321	1.438262	1.433244
0.7751	1.472643	1.467824	1.463623	1.458942	1.454732	1.449571
0.8865	1.496152	1.49118	1.48677	1.481896	1.477535	1.472111
1	1.51823	1.513011	1.508415	1.503383	1.498853	1.493141
			V_m (cm^3 mol^{-1})			
0	166	167	167	167	168	168
0.1154	177	177	178	178	179	179
0.2043	185	186	186	187	187	188
0.3124	195	196	196	197	197	198
0.3967	203	204	204	205	205	206
0.5047	213	214	214	215	215	216
0.5904	221	221	222	223	223	224
0.6998	231	231	232	233	233	234
0.7751	237	238	239	240	240	241
0.8865	248	248	249	250	251	252
1	258	259	259	260	261	262
			V_m^E (m^3 mol^{-1})			
0	0	0	0	0	0	0
0.1154	0.263	0.251	0.248	0.238	0.225	0.224
0.2043	0.422	0.412	0.397	0.386	0.377	0.368
0.3124	0.522	0.508	0.495	0.487	0.477	0.466
0.3967	0.551	0.546	0.532	0.517	0.499	0.494
0.5047	0.533	0.526	0.516	0.503	0.492	0.481
0.5904	0.491	0.488	0.476	0.464	0.445	0.436
0.6998	0.394	0.382	0.375	0.366	0.350	0.344
0.7751	0.305	0.297	0.285	0.276	0.268	0.256
0.8865	0.155	0.142	0.138	0.135	0.129	0.120
1	0	0	0	0	0	0

TABLE 5.7 Experimental Values of Densities ρ, Molar Volumes V_m, and Excess Molar Volumes V_m^E for $[C_2mim][NTf_2]_X[SCN]_{(1-X)}$ at (293.15–313.15) K at 0.1 MPa.

$X_{[C_2mim][NTf_2]}$	298.15 K	303.15 K	308.15 K	313.15 K	318.15 K	323.15 K
			ρ (g cm^{-3})			
0	1.117231	1.113962	1.110520	1.107351	1.104082	1.101152
0.1148	1.187610	1.184150	1.180552	1.177252	1.173804	1.170552
0.2055	1.236471	1.232791	1.229125	1.225576	1.222140	1.218523
0.3011	1.282910	1.279110	1.275280	1.271540	1.267940	1.264015
0.3980	1.325445	1.321455	1.317463	1.313488	1.309722	1.305571
0.5014	1.366370	1.362162	1.358059	1.353984	1.350082	1.345602
0.5883	1.397852	1.393466	1.389297	1.384943	1.380905	1.376281
0.6895	1.431470	1.426825	1.422586	1.418096	1.413920	1.409053
0.7753	1.457682	1.452981	1.448582	1.443921	1.439745	1.434604
0.8921	1.490530	1.485626	1.481126	1.476302	1.471929	1.466435
1	1.518230	1.513011	1.508415	1.503383	1.498853	1.493141
			V_m (cm^3 mol^{-1})			
0	151	152	152	153	153	154
0.1148	164	164	165	165	166	166
0.2055	174	174	175	175	176	176
0.3011	184	185	185	186	186	187
0.3980	194	195	196	196	197	197
0.5014	205	206	207	207	208	209
0.5883	215	215	216	217	217	218
0.6895	225	226	227	227	228	229
0.7753	234	235	236	236	237	238
0.8921	246	247	248	249	250	251
1	258	259	259	260	261	262
			V_m^E (m^3 mol^{-1})			
0	0	0	0	0	0	0
0.1148	0.290	0.274	0.268	0.244	0.239	0.225
0.2055	0.459	0.442	0.426	0.407	0.379	0.372
0.3011	0.562	0.530	0.518	0.496	0.470	0.463
0.3980	0.593	0.559	0.552	0.536	0.515	0.497
0.5014	0.590	0.557	0.550	0.519	0.498	0.486
0.5883	0.539	0.509	0.497	0.485	0.469	0.444
0.6895	0.445	0.427	0.411	0.394	0.383	0.355
0.7753	0.348	0.317	0.314	0.303	0.277	0.261
0.8921	0.179	0.152	0.149	0.138	0.127	0.127
1	0	0	0	0	0	0

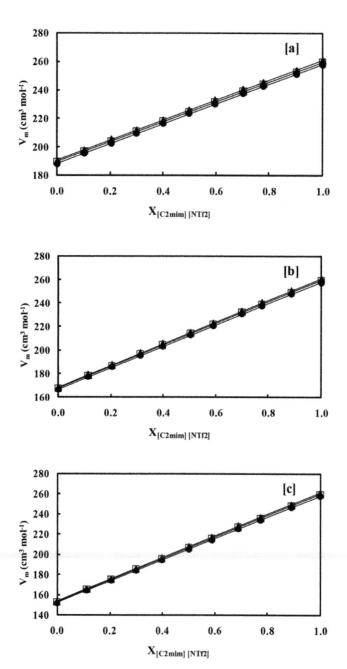

FIGURE 5.1   Molar volume as a function composition of [C$_2$mim][NTf$_2$] at 298.15 K at 298.15 K (●), 308.15 (□), 318.15 K (▲). (a) [C$_2$mim][NTf$_2$][OTf], and (b) [C$_2$mim][NTf$_2$] [MeSO$_3$], (c) [C$_2$mim][NTf$_2$][SCN].

where ρ is the density of mixtures and M_i, x_i, and ρ_i are the molar mass, mole fraction, and density of pure component i, respectively.

The values of V_m^E were correlated by Redlich–Kister equation:

$$V_m^E = x_1(1-x_1)\sum_{i=1}^{n} A_i(1-2x_1)^{i-1} \tag{5.4}$$

Table 5.8 represents coefficients A_i and standard deviations σ of eq. (5.4). The standard deviation was obtained by the following equation:

$$\sigma = \left[\sum_{i=1}^{M}\left(V_{m(\text{calc})}^E - V_{m(\text{exp})}^E\right)^2 / (N-m)\right]^{1/2} \tag{5.5}$$

where N is the number of experimental points and m is the number of parameters.

TABLE 5.8 Temperature Dependence of A_i of eq. (5.4) for Excess Molar Volume, V_m^E (cm^3 mol^{-1}), and their Standard Deviation (σ).

Temp.	A_1	A_2	A_3	A_4	σ
		$[C_2\text{mim}][NTf_2]_x[OTf]_{(1-x)}$			
298.15	2.17112	0.75397	−0.14419	−0.12307	0.006165
303.15	2.15034	0.73455	−0.30896	−0.08341	0.008108
308.15	2.09682	0.68881	−0.30064	0.01329	0.004786
313.15	2.04794	0.70336	−0.32517	−0.09266	0.005606
318.15	1.99183	0.74527	−0.36574	−0.22964	0.007606
323.15	1.95523	0.71541	−0.40355	−0.10421	0.006141
		$[C_2\text{mim}][NTf_2]_x[MeSO_3]_{(1-x)}$			
298.15	2.17112	0.75397	−0.14419	−0.12307	0.006165
303.15	2.15034	0.73455	−0.30896	−0.08341	0.008108
308.15	2.09682	0.68881	−0.30064	0.01329	0.004786
313.15	2.04794	0.70336	−0.32517	−0.09266	0.005606
318.15	1.99183	0.74527	−0.36574	−0.22964	0.007606
323.15	1.95523	0.71541	−0.40355	−0.10421	0.006141
		$[C_2\text{mim}][NTf_2]_x[SCN]_{(1-x)}$			
298.15	2.35918	0.76764	0.02845	−0.20104	0.00534
303.15	2.24403	0.71082	−0.12394	0.04067	0.009548
308.15	2.1983	0.70156	−0.13388	0.00236	0.005686
313.15	2.13037	0.68074	−0.28233	−0.0817	0.009935
318.15	2.03715	0.58191	−0.30979	0.14839	0.005963
323.15	1.9602	0.75641	−0.30402	−0.28931	0.004673

The variation in excess molar volumes, V_m^E as a function of [C$_2$mim] [NTf$_2$] concentration and temperature are graphically presented in Figure 5.2a–c. The variation in V_m^E with temperature is very small. The V_m^E of presently studied IL mixtures of [C$_2$mim][NTf$_2$][OTf], [C$_2$mim][NTf$_2$][MeSO$_3$], and [C$_2$mim][NTf$_2$][SCN] shows positive deviation. The positive deviation indicates that larger volume was observed during mixing of two different ILs than calculated by Raoult's law; hence, no strong chemical interaction between ions.[20] Maximum positive V_m^E = 0.59 × 10^{-6} m^3 mol^{-1} have been observed for [C$_2$mim][NTf$_2$][SCN] and least for [C$_2$mim][NTf$_2$][OTf] at 298.15 K. Figure 5.3 depicted that V_m^E increases in order of [C$_2$mim] [NTf$_2$][SCN] (0.59 × 10^{-6} cm^3 mol^{-1}) > [C$_2$mim][NTf$_2$][MeSO$_3$] (0.55 × 10^{-6} cm^3 mol^{-1}) > [C$_2$mim][NTf$_2$][OTf] (0.5 cm^3 mol^{-1}). The negative V_m^E was reported for [C$_2$mim][N(CN)$_2$][BF$_4$] and [C$_2$mim][Cl][SCN].[26,52] In the present study, V_m^E values are relatively higher than reported in literature.[25–30] However, large positive V_m^E around 0.76 cm^3 mol^{-1} have been reported for [C$_2$mim][EtSO$_4$][NTf$_2$].[53]

The possible reasons for positive deviation for studied IL mixtures may be attributed combined effects[23]: (1) *Difference in size of anions*: [NTf$_2$] > [OTf] > [MeSO$_3$] > SCN]. Lopes et al.[25] reported the very small positive V_m^E (0.1% of mixtures molar volume) for [C$_4$mim]([NTf$_2$] + [PF$_6$]), [C$_4$mim] ([NTf$_2$] + [BF$_4$]) and [C$_4$mim]([BF$_4$] + [PF$_6$]) binary IL mixtures. The very small positive V_m^E around 0.12 cm^3 mol^{-1} have been reported for [NTf$_2$] + [PF$_6$] and [BF$_4$] + [PF$_6$] as compared to [NTf$_2$] + [BF$_4$] (0.28 cm^3 mol^{-1}); they concluded linear correlation between variations in V_m^E with size difference of anions. The small positive V_m^E (0.1%) was reported for [C$_2$mim][N(CN)$_2$] [BF$_4$], [C$_4$pyr][BF$_4$][NTf$_2$], and [C$_4$mim][BF$_4$][PF$_6$] by Stoppa et al.[26] and Larriba et al.[30] also found same correlation with size difference of anions.

(2) *Interaction between anions*: The positive deviation for liquid mixtures with common cation was supported by measurement of enthalpy of mixing by Navia et al.[28] They suggested that stronger interaction between two unlike anions as compared to like anions gave negative H^E for [C$_4$mim]([BF$_4$] + [MeSO$_4$]) and [C$_4$mim]([BF$_4$] + [PF$_6$]). The small negative H^E was calculated for [C$_4$mim][NO$_3$][ClO$_4$] using molecular dynamics simulations by Castejon and Lashock.[53]

(3) *Change of fluorinated interactions in mixtures*: [NTf$_2$]$^-$ is highly fluorinated anions which further extend fluorinated interaction when mixed

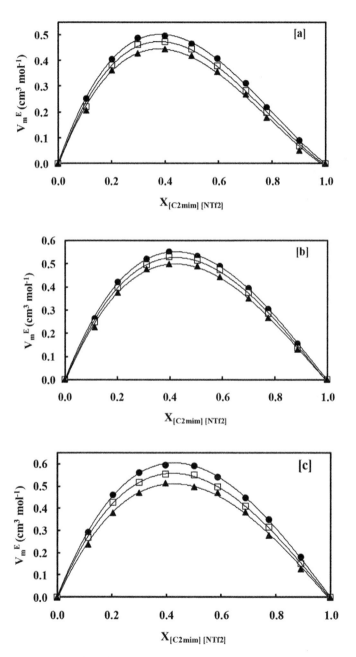

FIGURE 5.2 Excess molar volumes as a function composition of $[C_2mim][NTf_2]$ at 298.15 K (○), 308.15 (□), 318.15 K (▲). (a) $[C_2mim][NTf_2][OTf]$, (b) $[C_2mim][NTf_2][MeSO_3]$, and (c) $[C_2mim][NTf_2][SCN]$. Solid lines have been drawn from eq. (5.4) using coefficients given in Table 5.8.

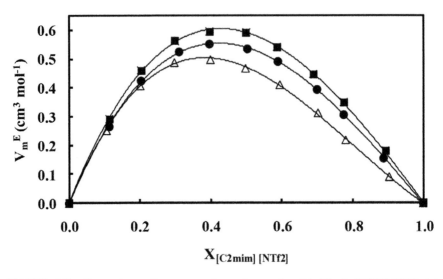

FIGURE 5.3 Comparison of excess molar volumes for (Δ) [C$_2$mim][NTf$_2$][OTf] (•) [C$_2$mim][NTf$_2$][MeSO$_3$] (■) [C$_2$mim][NTf$_2$][SCN] at 298.15K.

with another fluorinated anion [OTf]$^-$ whereas reduce fluorinated interaction when mixed with two nonfluorinated anions [SCN] and [MeSO$_3$]. Clough et al.[23] have demonstrated loss of fluorinated interactions may results in ideal behavior for binary mixtures of [C$_4$mim][MeSO$_4$][NTf$_2$], [C$_4$mim][Me$_2$PO$_4$] [NTf$_2$], and [C$_4$mim][NTf$_2$][Me$_2$PO$_4$].

(4) *Preferential H bonding of anion at H$_2$ position in imidazolium ring*: The preferential H bonding in ILs will change when anions of different type and size are mixed together. The H-bonding effect was explained by Brussel et al.[52] using molecular dynamics simulations and NMR experiments for [C$_2$mim][Cl][SCN]. The Cl$^-$ ions preferentially are attached to H$_2$ position displacing SCN$^-$ ions in mixtures. Hydrogen bond basicity β of anion plays vital role in this effect. Rebelo and coworkers[55,56] explained preferential H bonding between [NH$_4$]$^+$ and [EtSO$_3$]/[EtSO$_4$] over [C$_4$mim] cation in liquid mixtures. [C$_4$mim][Cl][PF$_6$] also reported the preferential H bonding for Cl$^-$. In [C$_4$mim][BF$_4$][PF$_6$], no H bonding have been observed due to same size and H-bond-accepting capacity of these anions.[57] Recently, studies reported by Mattews et al. describes the role of $-\pi$ stacking (interaction between two imidazolium rings) interactions to describe mixing behavior.[24]

5.4 CONCLUSION

Majority of IL mixtures investigated here shows nearly ideal mixing behavior. The investigated mixtures are regarded as ideal solution on the basis of linear dependence of molar volume on common IL involved in DSILs. The slightly higher positive value of excess molar volume is attributed change in size difference of anions, possible cation–anion interactions, and position of anions in bulk mixture.

KEYWORDS

- double-salt ionic liquid
- binary system
- ideal–nonideal behavior
- physicochemical properties
- designer solvent

REFERENCES

1. Earle, M. J.; Seddon, K. R. *Pure Appl. Chem.* **2000**, *72*, 1391–1398.
2. Marsh, K. N.; Deev, A.; Wu, A. C. T.; Tran, E.; Klamt, A. *Korean J. Chem. Eng.* **2002**, *19*, 357–362.
3. Pârvulescu, V. I.; Hardacre, C. *Chem. Rev.* **2007**, *107*, 2615–2665.
4. Welton, T. *Chem. Rev.* **1999**, *99*, 2071–2083.
5. Pârvulescu, V. I.; Hardacre, C. *Chem. Rev.* **2007**, *107*, 2615–2665.
6. Lu, J.; Yan, F.; Texter, J. *Prog. Polym. Sci.* **2009**, *34*, 431–448.
7. Zhao, H.; Malhotra, S. V. *Aldrichim. Acta* **2002**, *35*, 75–83.
8. Sun, P.; Armstrong, D. W. *Anal. Chim. Acta* **2010**, *661*, 1–16.
9. Kocherginsky, N. M.; Yang, Q.; Seelam, L. *Sep. Purif. Technol.* **2007**, *53*, 171–177.
10. Abbott, A. P.; McKenzie, K. J. *Phys. Chem. Chem. Phys.* **2006**, *8*, 4265–4279.
11. Somers, A. E.; Howlett, P. C.; MacFarlane, D. R.; Forsyth, M. *Lubricants* **2013**, *1*, 3–21.
12. Keskin, S.; Kayrak-Talay, D.; Akman, U.; Hortacsu, O. *J. Supercrit. Fluids* **2007**, *43*, 150–180.
13. Plechkova, N. V.; Seddon, K. R. *Chem. Soc. Rev.* **2008**, *37*, 123–150.
14. Malek, N. I.; Ijardar, S. P.; Oswal, S. B. *J. Chem. Eng. Data* **2014**, *59*, 540–55.
15. Malek, N. I.; Singh, A.; Surati, R.; Ijardar, S. P. *J. Chem. Thermodyn.* **2014**, *74*, 103–118.
16. Ijardar, S. P.; Malek, N. I.; *J. Chem. Thermodyn.* **2014**, *71*, 236–248.
17. Vaid, Z. S.; More, U. U.; Gardas, R. L.; Malek, N. I.; Ijardar, S. P. *J. Sol. Chem.* **2015**, *44*, 718–741.

18. Vaid, Z. S.; More, U. U.; Malek, N. I.; Ijardar, S. P. *J. Chem. Thermodyn.* **2015,** *86,* 143–153.
19. Patel, H.; Vaid, Z. S.; More, U. U.; Ijardar, S. P.; Malek, N. I. *J. Chem. Thermodyn.* **2016,** *99,* 40–53.
20. Chatel, G.; Pereira, J. F. B.; Debbeti, C.; Wang, H.; Rogers, R. D. *Green Chem.* **2014,** *16,* 2051–2083.
21. Niedermeyer, H.; Hallett, J. P.; Villar-Garcia, I. J.; Hunt, P. A.; Welton, T. *Chem. Soc. Rev.* **2012,** *41,* 7780–7802.
22. Aparicio, S.; Atilhan, M. *J. Phys. Chem. B* **2012,** *116,* 2626–2637.
23. Clough, M. T.; Crick, C. R.; Grasvick, J.; Hunt, P. A.; Niedermeyer, H.; Welton, T.; Whitaker, O. P. A. *Chem. Sci.* **2015,** *6,* 1101–1114.
24. Matthews, R. P.; Villar-Garcia, I. J.; Weber, C. C.; Griffith, J.; Cameron, F.; Hallett, J. P.; Hunt, P. A.; Welton, T. *Phys. Chem. Chem. Phys.* **2016,** *18,* 8608–8624.
25. Lopes, J. N. C.; Cordeiro, T. C.; Esperanca, J. N. S. S.; Guedes, H. J. R.; Huq, S.; Rebelo, L. P. N.; Seddon, K. R. *J. Phys. Chem. B* **2005,** *109,* 3519–3525.
26. Stoppa, A.; Buchner, R.; Hefter, G. *J. Mol. Liq.* **2010,** *153,* 46–51.
27. Navia, P.; Troncoso, J.; Romani, L. *J. Sol. Chem.* **2008,** *37,* 677–688.
28. Navia, P.; Troncoso, J.; Romani, L. *J. Chem. Eng. Data* **2007,** *52,* 1369–1374.
29. Atilhan, M.; Anaya, B.; Ullah, R.; Costa, L. T.; Aparicio, S. *J. Phys. Chem. C* **2016,** *120,* 17829–17844.
30. Larriba, M.; García, S.; Navarro, P.; García, J.; Rodríguez, F. *J. Chem. Eng. Data* **2012,** *57,* 1318–1325.
31. Seki, S.; Tsuzuki, S.; Hayamizu, K.; Umebayashi, Y.; Serizawa, N.; Takei, K.; Miyashiro, H. *J. Chem. Eng. Data* **2012,** *57,* 2211–2216.
32. Beigi, A. A. M.; Abdouss, M.; Yousefi, M.; Pourmortazavi, S. M.; Vahid, A. *J. Mol. Liq.* **2013,** *177,* 361–368.
33. Seoane, R. G.; Corderi, S.; Gomez, E.; Calvar, N.; Gonzalez, E. J.; Macedo, E. A.; Dominguez, A. *Ind. Eng. Chem. Res.* **2012,** *51,* 2492–2504.
34. Bansal, S.; Kaur, N.; Chaudhary, G.; Mehta, S.; Ahluwalia, A. S. *J. Chem. Eng. Data* **2014,** *59,* 3988–3999.
35. Noda, A.; Hayamizu, K.; Watanabe, M. *J. Phys. Chem. B* **2001,** *105,* 4603–4610.
36. Schreiner, C.; Zugmann, S.; Hartl, R.; Gores, H. J. *J. Chem. Eng. Data* **2010,** *55,* 1784–1788.
37. Vercher, E.; Orchilles, A. V.; Miguel, P. J.; Martinez-Andreu, A. *J. Chem. Eng. Data* **2007,** *52,* 1468–1482.
38. Rodriguez, H.; Brennecke, J. F. *J. Chem. Eng. Data* **2006,** *51,* 2145–2155.
39. Freire, M. G.; Teles, A. R. R.; Rocha, M. A. A.; Schroder, B.; Neves, C. M. S. S.; Carvalho, P. J.; Evtuguin, D. V.; Santos, L. M. N. B. F.; Coutinho, J. A. P. *J. Chem. Eng. Data* **2011,** *56,* 4813–4822.
40. Klomfar, J.; Souckova, M.; Patek, J. *J. Chem. Eng. Data* **2010,** *55,* 4054–4057.
41. Garcia-Miaja, G.; Troncoso, J.; Romani, L. *Fluid Phase Equilib.* **2008,** *274,* 59–67.
42. Montalban, M. G.; Bolivar, C. L.; Banos, F. G. D.; Villora, G. *J. Chem. Eng. Data* **2015,** *60,* 1986–1996.
43. Yusoff, R.; Aroua, M. K.; Shamiri, A.; Ahmady, A.; Jusoh, N. S.; Asmuni, N. F.; Bong, L. C.; Thee, S. H. *J. Chem. Eng. Data* **2013,** *58,* 240–247.
44. Hasse, B.; Lehmann, J.; Assenbaum, D.; Wasserscheid, P.; Leipertz, A.; Froba, A. P. *J. Chem. Eng. Data* **2009,** *54,* 2576–2583.

45. Rabari, D.; Patel, N.; Joshipura, M.; Banerjee, T. *J. Chem. Eng. Data* **2014**, *59*, 571–578.

46. Domanska, U.; Krolikowski, M. *J. Chem. Thermodyn.* **2012**, *54*, 20–27.

47. Anantharaj, R.; Banerjee, T. *Can. J. Chem. Eng.* **2013**, *91*, 245–256.

48. Domanska, U.; Krolikowska, M.; Krolikowski, M. *Fluid Phase Equilib.* **2010**, *294*, 72–83.

49. Krolikowska, M.; Hofman, T. *Thermochim. Acta* **2012**, *530*, 1–6.

50. Widegren, J. A.; Magee, J. W. *J. Chem. Eng. Data* **2007**, *52*, 2331–2338.

51. Malek, N. I., Ijardar, S. P.; Oswal, S. B. *Thermochim. Acta* **2012**, *539*, 71–83.

52. Brussel, M.; Brehm, M.; Pensado, A. S.; Malberg, F.; Ramzan, M.; Stark, A.; Kirchner, B. *Phys. Chem. Chem. Phys.* **2012**, *14*, 13204–13215.

53. Pinto, A. M.; Rodriguez, H.; Colon, Y. J.; Arce Jr., A.; Acree, A.; Soto, A. *Ind. Eng. Chem. Res.* **2013**, *52*, 5975–5984.

54. Castejo'n, H. J.; Lashock, R. J. *J. Mol. Liq.* **2012**, *167*, 1.

55. Pereiro, A. B.; Arau'jo, J. M. M.; Oliveira, F. S.; Bernardes, C. E. S.; Esperança, J. M. S. S.; Canongia Lopes, J. N.; Marrucho, I. M.; Rebelo, L. P. N. *Chem. Commun.* **2012**, *48*, 3656–3658.

56. Oliveira, F. S.; Pereiro, A. B.; Arau'jo, J. M. M.; Bernardes, C. E. S.; Canongia Lopes, J. N.; Todorovic, S.; Feio, G.; Almeida, P. L.; Rebelo, L. P. N.; Marrucho, I. M. *Phys. Chem. Chem. Phys.* **2013**, *15*, 18138–18147.

57. Payal, R. S.; Balasubramanian, S. *Phys. Chem. Chem. Phys.* **2013**, *15*, 21077–21083.

PART II
Polymer Science and Engineering

Part II

Polymer Science and Engineering

CHAPTER 6

UNSATURATED BOND-CONTAINING HETEROCHAIN POLYMERS FOR BIOMEDICAL USE

E. CHKHAIDZE[1*], N. NEPHARIDZE[1], and D. KHARADZE[2]

[1]*Faculty of Chemical Technology and Metallurgy, Department of Chemical and Biological Engineering, Georgian Technical University, 77 Kostava st., Tbilisi 0175, Georgia*

[2]*Ivane Beritashvili Center of Experimental Biomedicine, 14 Gotua st. Tbilisi 0160, Georgia*

Corresponding author. E-mail: ekachkhaidze@yahoo.com

CONTENTS

ABSTRACT

This chapter is devoted to different classes of polymers containing unsaturated bonds (polyesters, polyamides, polyesteramides, polyanhydrides, copolymers of polyesters, polymers modified with unsaturated bonds), which were created for special biomedical purposes. This chapter describes methods of synthesis, properties, and use of these polymers. A significant part of this chapter is dedicated to the works carried out by authors—AABB-type unsaturated polyesteramides based on α-amino acids. Series of new biodegradable unsaturated polyesteramides and copolyesteramides [(1) unsaturated homopolyesteramides (UPEAs) with 100% content of fumaric acid residuals; (2) unsaturated/saturated polyesteramides with <100% content of fumaric acid residuals; (3) unsaturated copolyesteramides with 100% content of fumaric acid residuals] were obtained in the Research Center of Biomedical Polymers and Biomaterials at Georgian Technical University. Fumaric acid was a key monomer by means of which unsaturated bonds were inserted into main chain. For designing of polymers with biodegradation ability—bis-(α-amino acids)- α,ω-alkylene(alkenylene) diesters on the basis of L-leucine and L-phenylalanine and soft bis-electrophiles—di-p-nitrophenyl ethers of dicarbonic acids were used as monomers. The polycondensation was carried out under soft conditions that are necessary to avoid earlier polymerization of double bonds. The chemical transformation of polymers, cross-linking reactions with UV radiation, and properties of UPEAs: solubility, their film-forming ability, thermal properties, and biodegradation were studied. A new family of biodegradable, unsaturated polyesteramides with good material properties is prospective for getting of biodegradable materials with high mechanical strength meant for implantable artificial organs, biodegradable hydrogels, drug-delivery systems, micro-, nanocapsulation, etc.

6.1 INTRODUCTION

Biomedicine is one of the most interesting spheres for use of different classes of polymers containing unsaturated bonds. Aliphatic, unsaturated, biodegradable polyesters were tested in bone tissue repairing surgery—to get bone prostheses in the form of solidifying binding materials (at the expense of double-bond polymerization). Use of unsaturated biodegradable polyesters and polyanhydrides is possible for designing of effective drug delivery systems. Getting of biodegradable hydrogels on the basis of unsaturated polymers is also prospective for systems of controlled release of medicinal preparations.[1]

While a search for biocompatible, stable materials for repair of living tissues (long-acting implants) took place at the initial stage of design of biomedicine materials, in the recent times intensive researches are conducted in the direction of synthesis of biodegradable polymers for different purposes. Among them, advantage is given to biodegradable polymers on the basis of nontoxic building blocks of natural origin.

Implementation of synthesis of biomedical polymers with interesting properties was anticipated on the basis of natural α-amino acids taking into account their role, diversity, and synthesizing abilities, though synthetic polyamino acids—polymers of AB type constructed similarly to natural amino acids—were unable to meet expectations of researchers due to unsatisfactory material properties and high immunogenesity. AABB-type polymers of different classes, where α-amino acids and their derivatives may be used in the form of bis-nucleophilic and bis-electrophilic monomers, turned out to be more prospective. Among researches conducted in this direction,[2–9] lead positions are taken by studies carried out by Research Center of Biomedical Polymers and Biomaterials at Georgian Technical University.[10–12] AABB-type heterochain polymers of different classes are obtained by Georgian researchers on the basis of natural amino acids (AABBP). The latter abbreviation is convenient, since it points at AABB nature of polymers. Obtained polymers are tested in the form of polymer carriers and matrixes for immobilization of medicinal preparations, and enzymes are used in the form of substrates in vitro tests, while some of them are used for study of biodegradation processes in vivo tests, with the purpose of manufacturing the pharmacologically active composites, porous materials, micro-, and nanoparticles.

Further extension of application area for AABB polymers is possible through their functionalization, via insertion of chemically active groups,[13] as well as hydrophobic so-called adhesive groups[14] into basic and lateral chains of polymers. Functional PEAs have the ability of both covalent and noncovalent bonding of medicinal preparations and other bioactive compounds.

Insertion of unsaturated bonds both into basic and lateral chains of macromolecules is one of the most prospective ways of polymers' functionalization.[15,16] Through addition of unsaturated bonds, it is possible to attach other desirable functional groups to macromolecules and to carry out multivarious grafting reactions—copolymerization, hybridization with other unsaturated polymers, synthesis of multifunctional biodegradable hydrogels, etc.

It should be noted that at present, the chemistry and technology of biomedical polymers belong to the area of advanced technologies and biomedical polymers take a leading position among unique polymeric materials.

In the given review, different classes of polymers containing unsaturated bonds (polyesters, polyamides, polyesteramides, polyanhydrides, copolymers of polyesters with acrylates, N-vinilpyrrolidone, etc.) will be considered, which were created for special biomedical purposes.

6.2 POLYESTERS

Aliphatic, unsaturated, biodegradable polyesters are prospective compounds in the form of bone binders, solidifying of which occurs via cross-linking reactions. Selection of polymer structure is possible in such a way that mutually opposed processes of polymer degradation and bone regeneration could last until complete repair of damaged tissue. Poly-1,2-propylene-fumarate is one of such unsaturated biodegradable polymers:

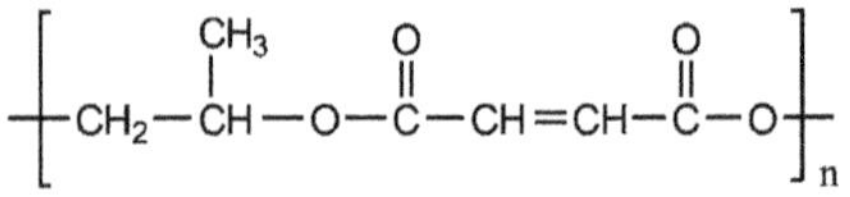

Oligomers of thermo- and polypropylene-fumarate for the first time were obtained by Domb and coworkers,[17] via polycondensation of bis-2-hydroxy-propyl-fumarate or propylene-bis-hydrogenmaleate. Fumaric acid is the Krebs cycle acid, so polymers obtained on its basis are characterized by high biocompatibility. As a result of cross-linking polymers are transformed from plastic state to rigid structures with desirable physical–chemical properties and preservation of biodegradation ability. Initial monomers—bis-2-hydroxypropyl-fumarates (trimers) (PFP) were obtained according to Diagram 6.1, while propylene-bis-hydrogenmaleate (trimer) (MPM)—according to Diagram 6.2.

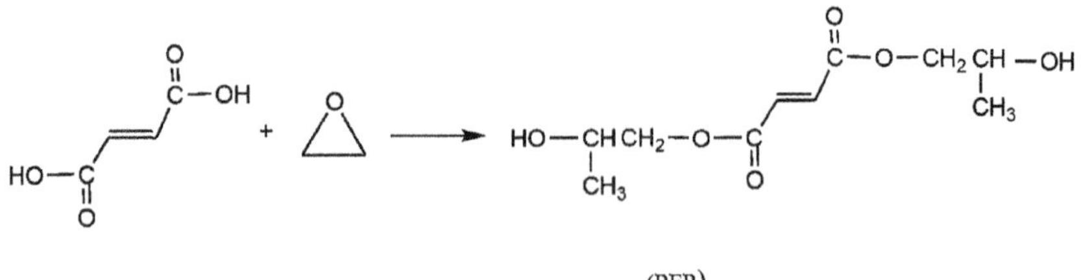

(PFP)

DIAGRAM 6.1

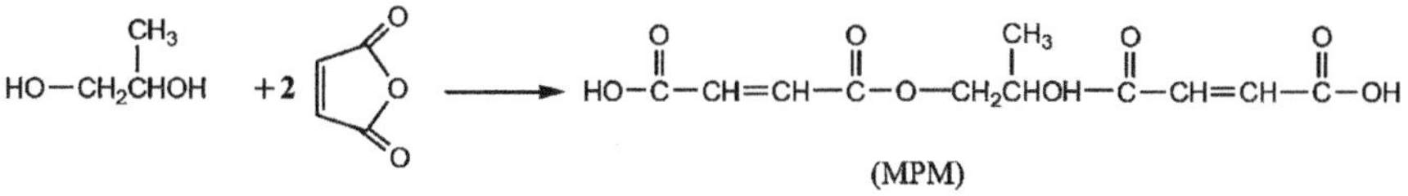

(MPM)

DIAGRAM 6.2

Chains were prolonged through step-by-step transformation of trimers with propylene oxide and maleic anhydride. Properties of obtained polymer are depended on balance between end groups (carboxyl, hydroxyl). Synthesized oligomers were used for preparation of bone cement, which was consisted of oligomer, calcium triphosphate, calcium carbonate, and methyl metacrylate.

Cross-linking reaction was conducted via radical polymerization. Properties of obtained composite depend both on degree of oligomers' polymerization and on the nature of end groups. In case of identical polymerization degree, polymers with carboxyl end groups manifest higher compressive strength in comparison with polymers containing hydroxyl groups.

Poly(propylene fumarate) cross-linked by calcium salts (calcium acetate, calcium hydroxy-apatite) in the form of bioreabsorbed bone cement was tested by Lewandrowski and coworkers[18] on mice. They studied the possibility of pore formation in materials, tissue reaction on implants, etc.

Poly(propylene fumarate)diacrylate was obtained by Timmer and coworkers,[19] who have studied mechanical properties of cross-linked polymers under physiological conditions. Unsaturated polyesters were also obtained[20] via transesterification reaction of diethyl fumarate and 1,4-cyclohexane-dimethanol. Obtained polymers were cross-linked along with N-vinyl-pyrrolidone using radical initiator. Compressive strength and hydrolytic stability of obtained cement is in correlation with polyester structure.

Spassky and coworkers[21–24] have elaborated the method of getting the unsaturated bond-containing polyesters on the basis of maleic and fumaric acids, via interaction of potassium salts of these acids with different α,ω-dibromoalcanes in 1-methyl-2-pyrrolidone (Diagram 6.3).

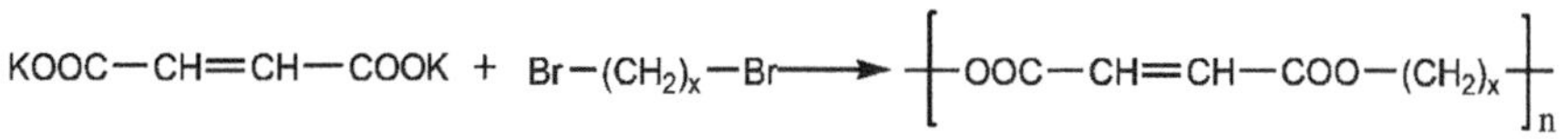

$$KOOC-CH=CH-COOK \ + \ Br-(CH_2)_x-Br \longrightarrow \left[OOC-CH=CH-COO-(CH_2)_x \right]_n$$

X = 1, 2, 3, 4, 5, 6, 8, 10;

DIAGRAM 6.3

Copolyesters were also obtained[25,26] with statistical and multiblock structures on the basis of maleic, fumaric, phthalic, and amber acids.

Djonlagic and coworkers[27] have obtained fumaric acid ring-containing aliphatic, biodegradable copolyesters on the basis of dimethyl-succinate, dimethylfumarate, and 1,4-buthandiol via polycondensation in the melt (Diagram 6.4).

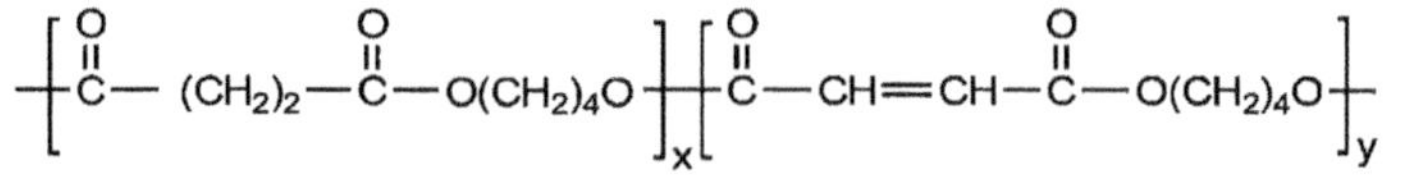

DIAGRAM 6.4

Number of unsaturated bonds in polyesteric chain was varied in the range of 5–20. Crystallinity degree was determined via DSC and X-ray structural analysis. Biodegradation of copolyesters was assessed through enzymatic (lipase) degradation test in buffer solution at 37°C; despite the fact that crystallinity degree of copolyesters reduces with the increase in number of unsaturated bonds, the rate of biodegradation doesn't increase. Authors make conclusion that the rate of biodegradation is depended not only on chemical composition of polymeric chain but also on polymer morphology. The highest biodegradation ability was manifested by copolyesters with 5–10 molar% content of fumarate.

Interesting application area of biodegradable polyesters is represented by medicinal preparation delivery (drug-delivery) systems. Important problem of these systems consists in chemical binding of bioactive compounds with them. Matrixes of medicinal preparation delivery systems, except of bonds with biodegradation ability, contain functional groups, for example, unsaturated bonds, to which covalent attachment of bioactive compounds through their chemical transformation will be possible.

With mentioned end in view, Jerome and coworkers[28] have synthesized polyesters with structure given in Diagram 6.5 on the basis of unsaturated caprolactone—6,7-dihydro-2(3H)-oxyquinone by two mechanisms—(1) via ring opening (with the participation of etheric bonds) and (2) via ring opening through polymerization of metathesis (with the participation of double bonds) [using Schrock (Mo-containing) and Grubbs (Ru-containing) catalysts]. Further transformation of obtained unsaturated polyesters is possible through oxidation of double bonds (epoxidation or transformation into 1,2-diol), bromination, etc. (Diagram 6.5).

In the work,[29] long-chain, symmetric ethers of dicarbonic acids (C_{18}, C_{20}, C_{26}) were obtained via catalytic metathesis condensation of ethers of monobasic unsaturated acids (methyl ethers of 9-decenoic, 10-undecenoic,

13-tetradecenoic acids) by homogenic Grubbs catalyst (Ru dichloride of bis-tricyclohexylphosphine benzylidene), which is dissolved in methylene chloride (Diagram 6.6).

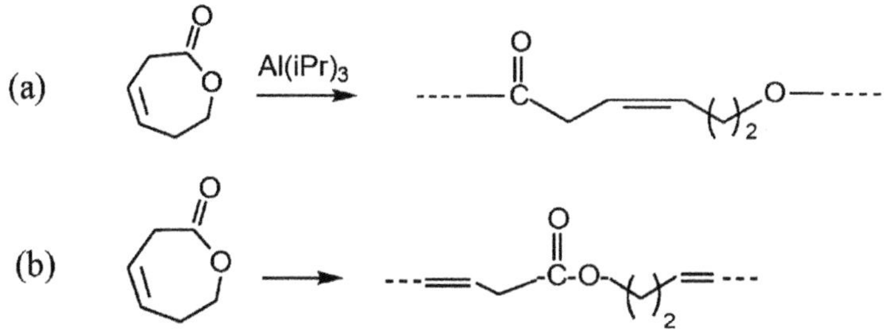

DIAGRAM 6.5

$$H_3C-O-CO-(CH_2)_y-CH{=}CH_2 + H_2C{=}CH-(CH_2)_y-CO-O-CH_3 \xrightarrow[-\ H_2C{=}CH_2]{\text{Grubbs catal.}}$$

$$\longrightarrow H_3C-O-CO-(CH_2)_y-CH{=}CH-(CH_2)_y-CO-O-CH_3$$

where $y = 8, 9, 12$

DIAGRAM 6.6

Epoxidation of unsaturated ethers of dicarbonic acids by H_2O_2/methyl-acetate was also carried out by lipase B immobilized (from *Candida antarctica*) on Novozym 435®. Polyesters were obtained through enzyme (lipase)-catalyzed polycondensation of methyl ethers of unsaturated and epoxidized α,ω-dicarbonic acids with 1,3-propandiol and 1,4-butandiol. Molecular mass of unsaturated and epoxidized polyesters obtained on the basis of 1,3-propandiol have reached 1950–3000, while the mass of polyesters obtained on the basis of 1,4-butandiol is 7900–11,660 that can be explained by higher hydrophobicity of the latter and, respectively, by its higher affinity with catalyst (lipase).

6.3 POLYAMIDES

One of the earlier works in the area of unsaturated polyamides (UPAs) synthesis was conducted by Bader and coworkers.[30] They obtained UPA via polycondensation of dihydrochlorides of L-lysine alkyl ethers with fumaroyl-chloride at the interface (phase boundary) of chloroform/water (Diagram 6.7).

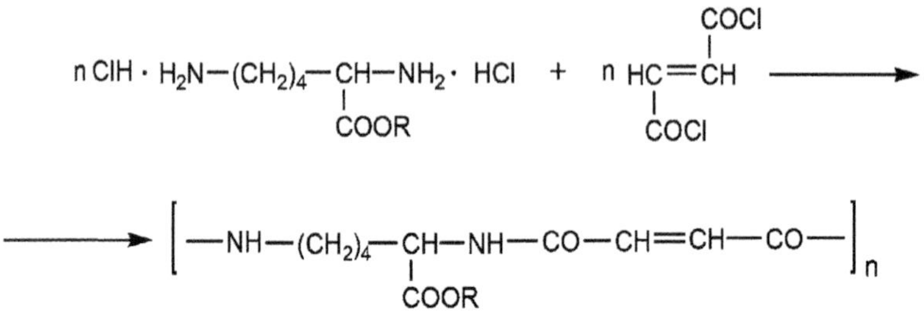

DIAGRAM 6.7

As a result was obtained UPA with 63–75% yield and $M_w = 23{,}000$ molecular mass.

Lenz and coworkers[31] have elaborated the synthesis method of polyamides containing unsaturated bonds and have studied the dependence between structure and rheological properties of polymer. For polycondensation, they have used activated diesters and have studied interaction of bis-(2,4-dinitrophenyl fumarate) and 1,1'-(fumaroyldioxe)-bis-benzotriazol with series of aliphatic and aromatic diamines (in the area of N-methylpyrrolidone, 25°C), according to the following scheme (Diagram 6.8):

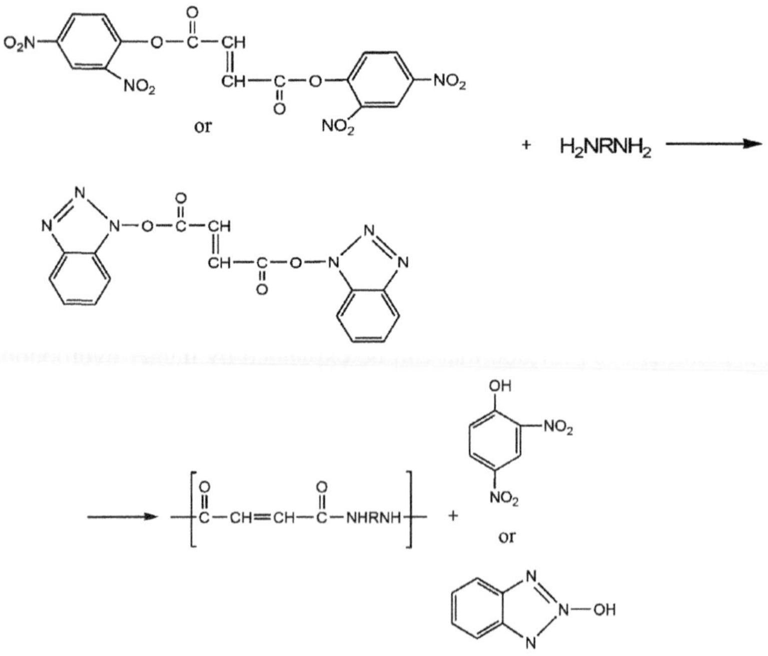

DIAGRAM 6.8

Polymers obtained on the basis of aliphatic diamines were dissolved in trifluoroacetic acid and had clearly expressed melting (softening) temperature. Polymers obtained on the basis of aliphatic diamines were not melted at temperatures lower than 300°C and were dissolved in DMA + LiCl that in general is a good solvent for aromatic polyamides.

6.4   POLYESTERAMIDES

Polyesteramides obtained on the basis of natural amino acids and containing unsaturated bonds should be presumably interesting functional compounds for biomedical use. In the Research Center of Biomedical Polymers and Biomaterials at Georgian Technical University,[10–12,32–34] synthesis of AABB-type heterochain polymers of different classes on the basis of natural amino acids was carried out, which was tested in the form of polymer carriers and matrixes for immobilization of medicinal preparations and enzymes, and for getting the pharmacologically active composites, porous materials, micro-, macroparticles, etc. One of the special places among mentioned materials is taken by polyesteramides, in which are combined positive properties of polyesters and polyamides: biodegradation ability, hydrophilicity, high biocompatibility with tissues, desirable mechanical properties, etc. Further extension of properties and, respectively, application area of these polymers is possible via their functionalization, and insertion of chemically active groups, for example, unsaturated bonds in basic and lateral polymeric chains.

Series of new biodegradable unsaturated polyesteramides (UPEAs) and copolyesteramides (coUPEAs) were obtained in the Research Center of Biomedical Polymers and Biomaterials at Georgian Technical University in collaboration with research group of Cornell University (USA)[35–37] on the basis of unsaturated fumaric acid, saturated aliphatic dicarbonic acids, and α-amino acids—derivatives of L-leucine and L-phenylalanine. For synthesis of targeted polymers, we have used the activated polycondensation method.[10,11] According to this method, polycondensation reaction proceeds under soft conditions that are necessary to avoid earlier polymerization of double bonds. Fumaric acid–unsaturated dicarbonic acid was a key monomer of our work by means of which we inserted unsaturated bonds into basic chain. For designing of polymers with biodegradation ability, we have used bis-(α-amino acids)-α,ω-alkylene(alkenylene)diesters on the basis of L-leucine and L-phenylalanine, while soft bis-electrophiles–di-p-nitrophenyl ethers of dicarbonic acids were used as bis-electrophilic monomers.

To get unsaturated bond-containing functional polymers in the basic chain for the first time, we have synthesized di-*p*-nitro-phenylfumarate–bis-electrophilic monomer, unsaturated active diester of fumaric acid according to two methods: acceptor–catalyst method in the solution and interphase method (Diagram 6.9). Synthesis of fumaric acid and di-*p*-nitrophenyl ether according to interphase method gives us an opportunity of synthesis of high purity monomer with minimum losses. It turned out that one-time recrystallization is sufficient for purification of obtained product to polycondensation purity. In this research, we have also used saturated acids: ethers of amber, adypic and sebacic acids, and di-*p*-nitrophenyl as bis-electrophilic monomers (Diagram 6.9).

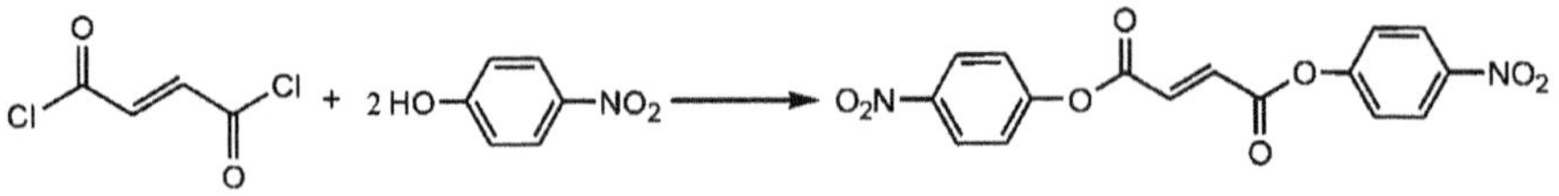

DIAGRAM 6.9

Synthesis of di-*p*-toluene sulphonic acid salts of bis-(α-amino acid)-α,ω-alkylene(alkenylene)diesters was carried out via direct condensation of free α-amino acids with α,ω-diols in the presence of monohydrate of *p*-toluene sulphonic acid (Diagram 6.10). Probable structures of obtained salts were confirmed by FTIR and ¹H NMR spectral and elemental analyses. Salts were obtained with high yield (75–99%) (Diagrams 6.10–6.12).

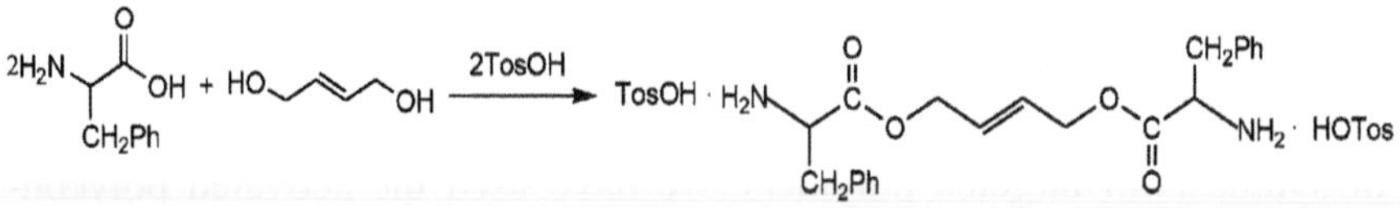

DIAGRAM 6.10

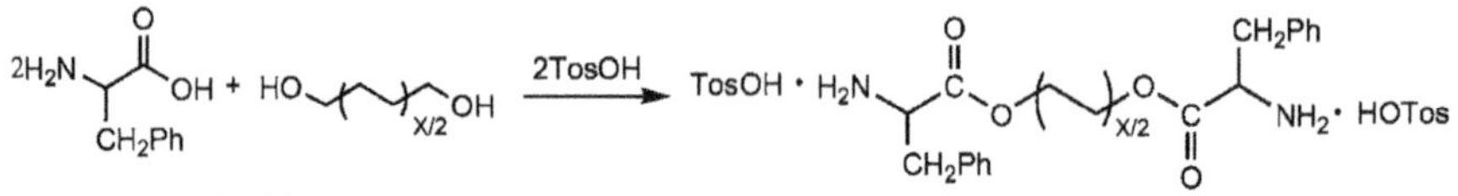

x = 6, 8, 12;

DIAGRAM 6.11

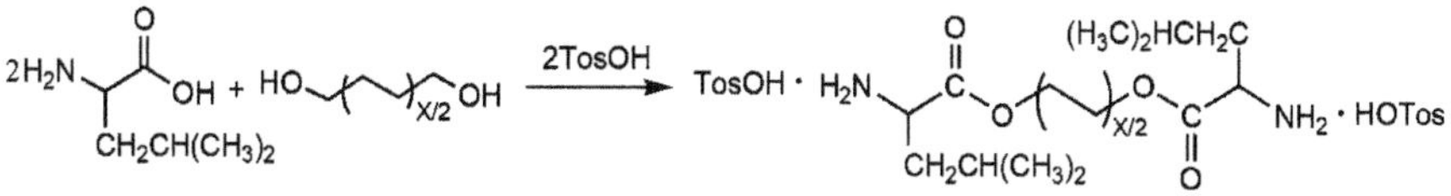

x = 4, 6, 8, 12;

DIAGRAM 6.12

Through polycondensation of above-described monomers, we got: (1) unsaturated homo-polyesteramides with 100% content of fumaric acid residuals; (2) unsaturated/saturated polyesteramides (USPEA) with <100% content of fumaric acid residuals; and (3) coUPEA with 100% content of fumaric acid residuals.

(1) UPEAs on the basis of L-leucine and L-phenylalanine with 100% content of fumaric acid residuals, that is, homo-polyesteramides were obtained via polycondensation of unsaturated di-p-nitrophenyl-fumarate with di-p-toluene sulfonic acid salts of bis-(α-amino acid) α,ω–alkylenediesters of saturated structure or with di-p-toluene sulfonic acid salt of bis-L-phenylala-nine-1,4-butenylenediester (Diagram 6.13). Synthesis was carried out in the solution with equal molarity of monomers (bis-electrophile/bis-nucleophile), in the presence of triethylamine. We have changed reaction temperature, duration, as well as concentrations of solvent and reaction solution according to assigned task and phase state of reaction solution. We have studied the effect of solution nature, reaction temperature, and duration, concentration of reactive substances, length of diol chains, and blocking of end groups on the polycondensation process. Optimal conditions of UPEA synthesis were determined. Structure of obtained polyesteramides was established via NMR, FTIR, and spectral and elemental analyses, their characteristics (molecular masses, molecular mass distribution, and dissolubility) were determined. Properties of obtained homo-polyesteramides were studied.

L-Phenylalanine derivatives give not-cross-linked, soluble polymers, while in case of L-leucine derivatives, formation of cross-linked, insoluble polymers mostly takes place; crucial role of terminal aminogroups in this process is defined. Unsaturated homo-polyesteramides soluble in organic solvents are obtained through their (amino acids) blocking, including those on the basis of L-leucine derivatives—for the first time on the basis of L-leucine are obtained UPEAs soluble in organic solvents with the use of long-chain diols (Diagrams 6.13 and 6.14).

(2) With the aim of getting the UPEA with desirable mechanical prop-erties, containing maximum number of unsaturated bonds and soluble in

organic solvents, we have synthesized unsaturated/saturated polyeste-ramides (USPEA) with <100% content of fumaric acid residuals (Diagram 6.15). During polycondensation, along with *n*-nitrophenyl ethers of fumaric acid, we have used *p*-nitrophenyl ethers of saturated (amber, adypic, sebacic) acids as bis–bis-electrophilic monomers. Regularities of USPEA synthesis were thoroughly studied using the example of polycondensation of *p*-nitro-phenyl ethers of fumaric and sebacic acids and di-*p*-toluene sulfonic acid salt of bis-(L-leucine)-1,6-hexylenediester. The ratio between blocks included into polyesteramides was as follows: 90/10, 80/20, 75/25, 60/40, 40/60, 20/80, respectively; for study of the effect of length of methylene group of diol on properties of unsaturated/saturated polyesteramides, we conducted synthesis of polyesteramides with polycondensation of *p*-nitrophenyl ethers of fumaric and sebacic acids and di-*p*-toluene sulfonic acid salt of bis-(L-leucine)-1,12-dodecaylendiester with blocks' ratio: 40/60, 60/40, 20/80, and 100/0. Optimum conditions of synthesis of copolymers were established, respectively. For studying the effect of end group blocking on molecular mass of obtained unsaturated/saturated polyesteramides, we have determined molecular masses of polyesteramides obtained with and without end group blocking. We obtained USPEA, soluble in organic solvents and containing unsaturated bonds, with high molecular mass (Diagram 6.15).

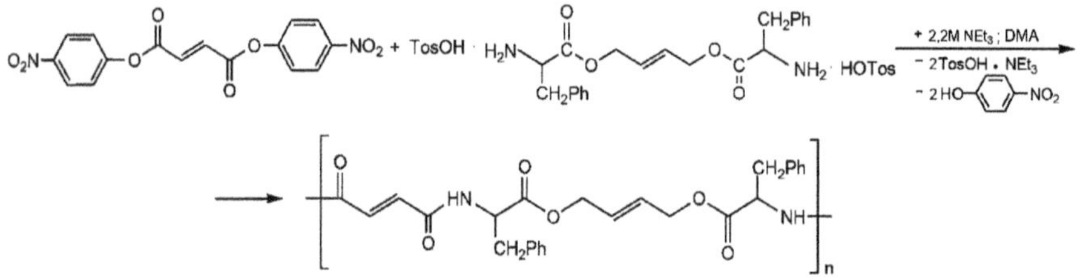

DIAGRAM 6.13

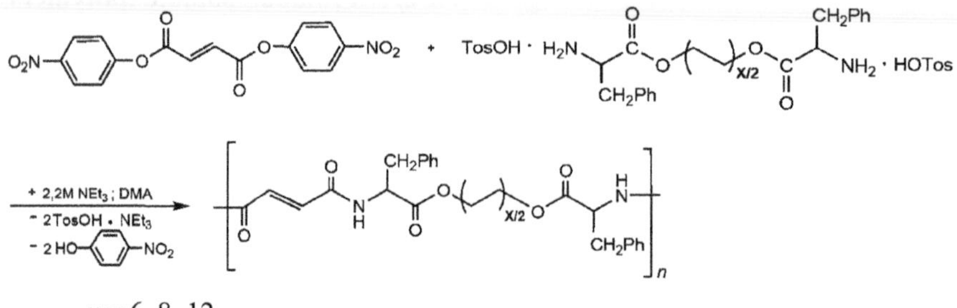

$x = 6, 8, 12$

DIAGRAM 6.14

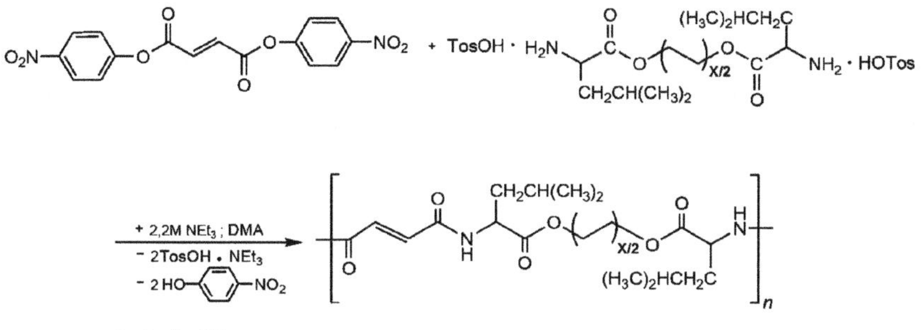

$x = 4, 6, 8, 12$

DIAGRAM 6.15

(3) With the aim of synthesis of coUPEAs with 100% content of fumaric acid residuals in the structures of unsaturated/saturated polymers, we have partially replaced amino acid L-leucine with L-phenylalanine and have studies the effect of the latter on properties of obtained polymers. As a result, we obtained copolymers on the basis of L-leucine and L-phenylalanine, soluble in organic solvents with various ratios of monomers (Diagram 6.16).

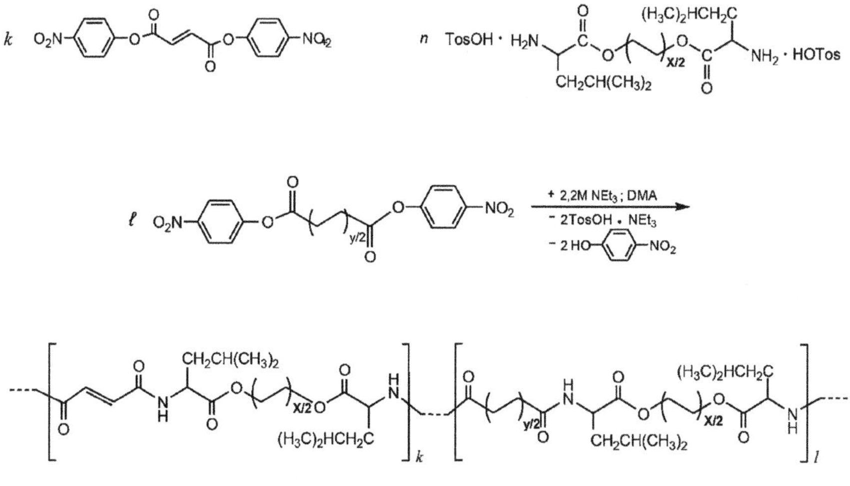

x = 6, 8, 12. y = 2, 4, 8. $k + \ell = n$

DIAGRAM 6.16

We have thoroughly studied properties of UPEAs: solubility in various solvents, their film-forming ability, thermal properties of polymers, and biodegradation ability. We have conducted reactions of chemical transformation of polymers, cross-linking reactions with UV radiation. Relatively high glass transition temperature (T_g) is characteristic for UPEAs in comparison with saturated polyesteramides. Growth in number of methylene groups increases flexibility of macromolecules. One and the same value of glass transition temperature is received at polymers' thermograms, that is, polymers with amorphous, nondomain structure are formed in the synthesis process.

Biodegradation of synthesized polyesteramides (enzymatic hydrolysis) was studied by gravimetric method in vitro tests under close-to-real conditions (pH 7.4; 37°C). Representatives of hydrolases class—α-chymotrypsin and lipase were used as enzymes. We have established dependence of biodegradation ability of polyesteramides on both quantitative and qualitative composition of amino acids, as well as on content of flexible methylene and rigid unsaturated bonds.

Reactions of chemical transformation of unsaturated polymers were conducted using the example of selected sample of unsaturated/saturated polymer by thioglycolic acid, 2-mercaptoethanol, and amino acid—β-alanine. We have established that obtained polyesteramides have an ability of high-strength film forming under UV irradiation (by the example of selected samples) that creates an opportunity of use of obtained polyesteramides with unsaturated structures in biomedicine, for instance, in the form of coating materials with ability of photocrosslinking of blood-vascular stents.

Thus, as a result of carried out research on the basis of natural amino acids, L-phenylalanine and L-leucine synthesized a new family of biodegradable, unsaturated homo-polyesteramides, unsaturated/saturated polyesteramides, copolyesteramides, with good material properties, which are prospective for getting of biodegradable materials with high mechanical strength meant for implantable artificial organs, biodegradable hydrogels, drug-delivery systems, micro-, nanocapsulation, etc.

It should be noted that insertion of double bonds into basic chain of PEA is possible not only with the use of unsaturated diacids but also unsaturated diols.[38–41] Insertion of unsaturated bonds increases rigidity and glass transition temperature of polymers, compared with saturated polymers with similar chemical structure.

Series of unsaturated PEAs was obtained by Chu and coworkers[39–42] using the mixture of diester–diamine monomers on the basis of oligo(ethyleneglicol) and 2-butene-1,4-diol and with the use of mixture of p-nitrophenyl ethers of unsaturated and saturated acids (Diagram 6.17). It is noteworthy that unsaturated bonds were inserted both into diacid and diol rings.

Biodegradable hydrogels, UPEA/PEG-DA, were obtained via UV cross-linking of obtained structures with poly(ethyleneglicol)diacrylate and their properties were studied (Diagram 6.17).

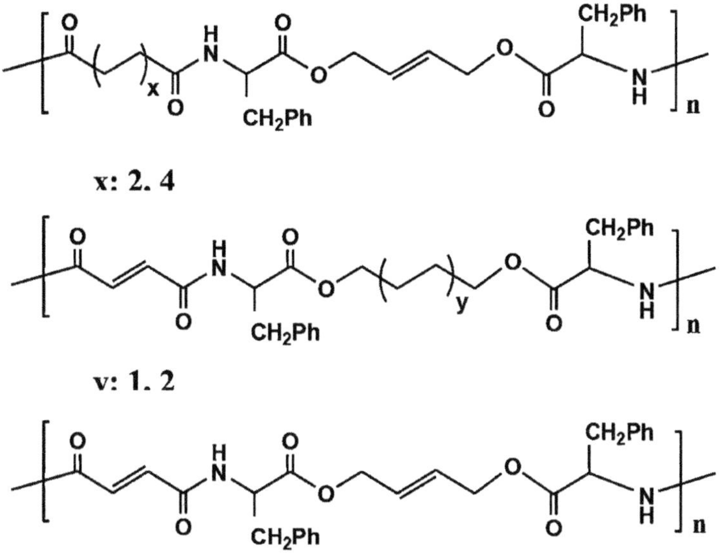

x: 2. 4

v: 1. 2

DIAGRAM 6.17

Properties of PEAs obtained on the basis of amino acids are getting wider in case of use of oligo(ethyleneglicol) instead of common diols.[43] New blocks of poly(ether-ester-amides) by means of etheric bonds can increase hydrophilicity, flexibility, and biodegradation ability of polymers. Materials with wide range of thermal, mechanical, and biological properties were obtained, which can be used in pharmaceutical, biomedical, and tissue engineering.

Biodegradable hydrogels (FPBe-G) were obtained by Chu and coworkers[44] through interaction of polyesteramides (UPEAs) obtained on the basis of fumarate and polyethyleneglycol diacrylate (PEG-DA). Lipophilic medication paklitaxel was inserted into FPBe-G hydrogel and kynetics of its release was studied during 2 months both in pure PBS buffer and in α-chymotrypsine. Correction of preparation release data is possible by pore sizes, molecular mass of cross-linking agent and morphological structure of matrix. Long-term release of paklitaxel during more than 2 months was reached also in case of hydrogels with definite composition.[44]

Series of biodegradable functional polyesteramides were obtained through copolycondensation of *p*-nitrophenyl ethers of dicarbonic acid with monomers (L-phenylalanine and DL-2-allylglycine) received on the basis of amino acids[45] (Diagram 6.18).

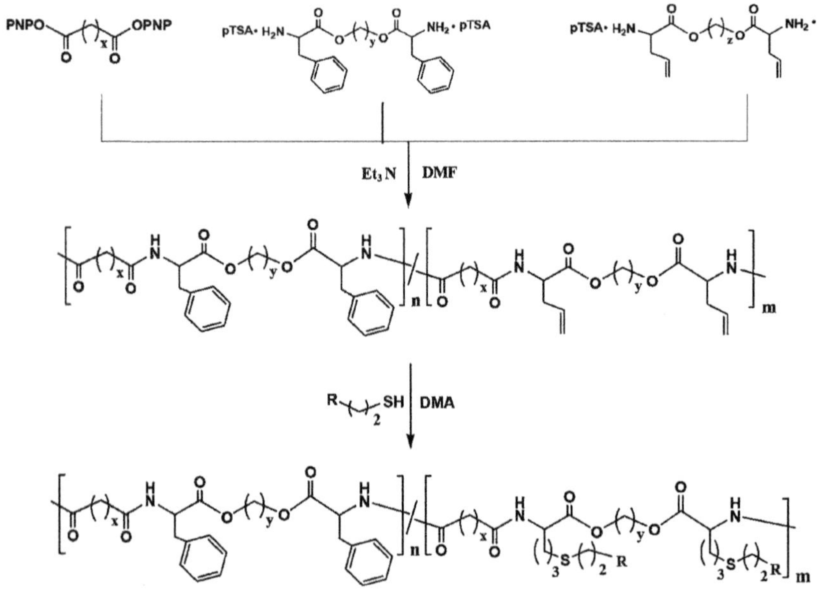

DIAGRAM 6.18

It is noteworthy that glass transition temperature of obtained PEAs reduces with increase of methylene chains both in amino acid and dicarbonic acid segments. Reaction capacity of lateral double bonds of fumaric acid residuals was studied using the method of free-radical attachment.[46–48]

Unsaturated statistical copoly(esteramides) were obtained also through reaction of mixture of phthalic and maleic anhydrides with ε-caprolactam and mixture of ethylene and neopentylene glycols[49] Obtained oligomers (Mn 2100–2600 g/mol) were effectively cross-linked with the use of vynilacetate and benzoyl peroxide–ascorbic acid. New materials have high compressive strength (104.0 MPa) and are hydrolytically degradable. These copolymers potentially can be used as materials for bone fixation. Analogous copolymers were obtained by the similar reaction with the use of 1,6-hexane diamine[50] or glycine[51] instead of ε-caprolactam.

Series of copolyesteramides with positively charged lateral groups of guanidine and lateral double carbon–carbon bonds were obtained similarly by Chu and coworkers via copolycondensation reaction of monomers (L-arginine and DL-2-allylglycine) obtained on the basis of amino acids and p-nitrophenyl ethers of dicarbonic acids[52] (Diagram 6.19).

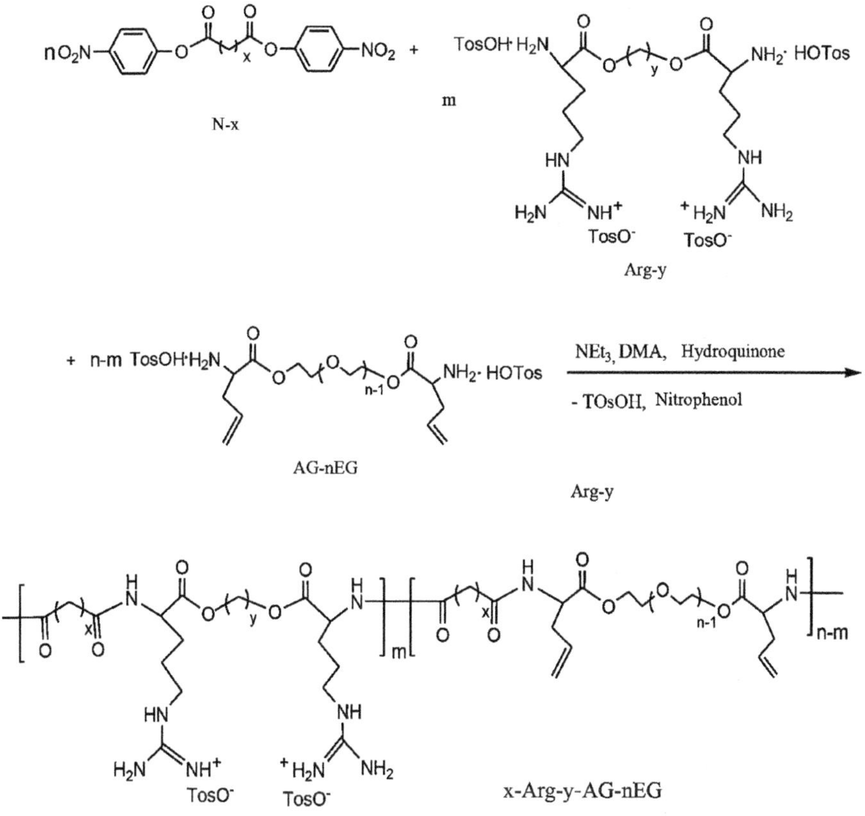

DIAGRAM 6.19

Cationic PEAs are soluble in polar solvents: water, alcohol, DMSO, and DMF. Study of cytotoxicity showed that new PEAs are nontoxical to endothelial cells of bull aorta (during exposition at tested concentration).

Molecular weight of Arg-PEAs was 20,000–60,000 g/mol, with fairly low molecular-mass distribution—less than 1.5.[53] Obtained Arg-PEAs are amorphous material, its T_v varies from 33 to 125°C. Preliminary study of cell morphology and DNA connection showed that a new family of cationic PEAs is nontoxical and more biocompatible than commercial transfection agent (Superfect®), so it can successfully join plasmic DNA. Heavy positive charge of Arg-PEAs, as well as their high solubility is a unique characteristic for gene transfection.

Chu and coworkers[54] have published information about original way of synthesis of unsaturated poly(esteramides) and poly(ether-ester-amides) obtained on the basis of α-amino acids (L-phenylalanine) with lateral amino- and carboxyl groups, to which one can easily graft thiolic compounds in the presence of radical initiators (Diagram 6.20). Mentioned single-step functionalization reaction proceeds with high yield (near to 100%) under soft conditions. This method has advantages compared with methods of selective protection and deprotection of functionalized α-amino acid derivatives.

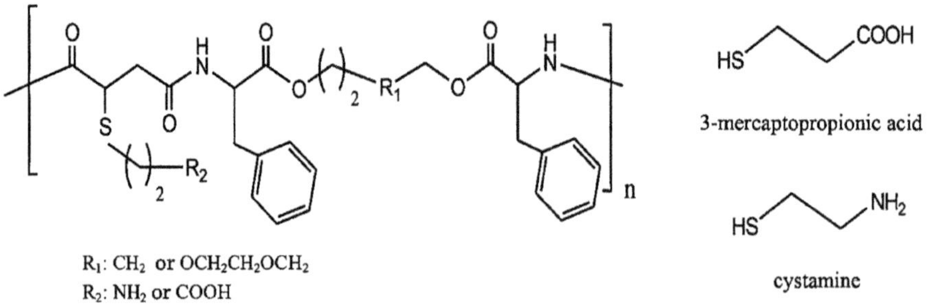

R$_1$: CH$_2$ or OCH$_2$CH$_2$OCH$_2$
R$_2$: NH$_2$ or COOH

3-mercaptopropionic acid

cystamine

DIAGRAM 6.20

6.5 POLYANHYDRIDES

Fast biodegradation ability is an important factor, which predetermines the opportunity of polyanhydrides' use in drug-delivery systems. Insertion of reaction-capable unsaturated bonds into basic chain is expedient, since it was expected that cross-linked polymers would have desirable mechanical properties along with biodegradation ability.

Langer, Domb, and coworkers[55,56] have obtained functional polyanhydrides containing unsaturated bonds via polycondensation of fumaric-, acetylenic-, and stilbene-dicarbonic acids (Diagram 6.21) both in the melt and in the solution. Molecular masses of obtained polymers were more than 30,000.

Polyanhydride obtained on the basis of fumaric acid turned out to be insoluble in a majority of organic solvents. Authors established that fumaric acid copolymers are characterized by better solubility with aliphatic diacids, at that these polymers degrade at a constant speed (rate) under physiological conditions that is prospective, for instance, for creation of controlled drug-release systems that was experimentally confirmed, too. Authors have obtained cross-linked polyanhydrides with high degree of crystallinity and good correlation between rates of drug release and polymer erosion. It was established that polymers degrade within the range of 2–15 days depending on pH and copolymers composition. Obtained polymers are prospective for

designing of systems of fast release of biodegradable medicinal preparations. Cross-linked polyanhydrides also have high-potential use in orthopedics (in the form of bone cement).

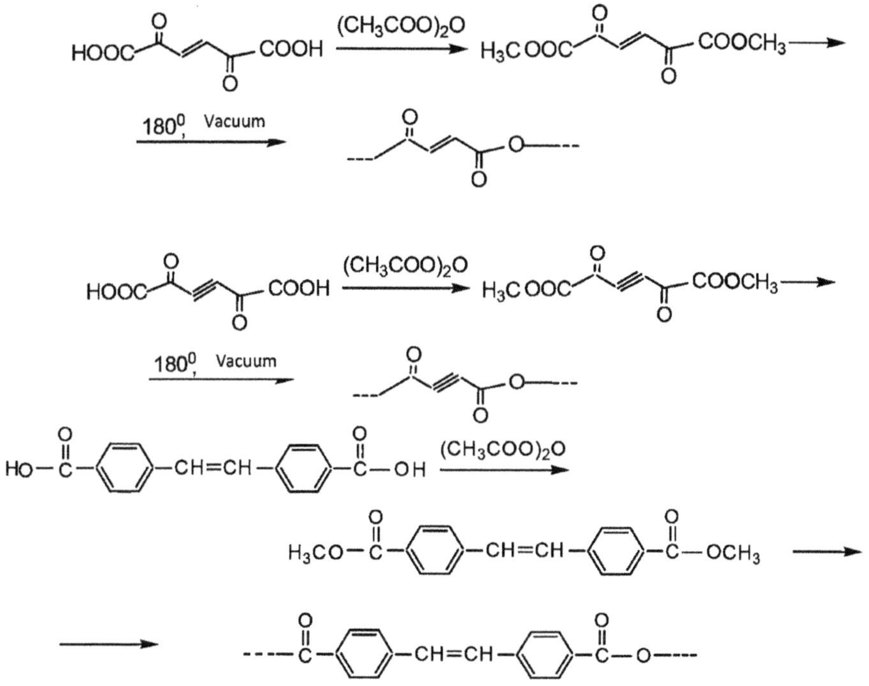

DIAGRAM 6.21

Kricheldorf and coworkers[57] have conducted synthesis of copolymers
of polyesters and polyanhydrides from terephthalic acid and unsaturated
aromatic hydroxy acids. They have used trans-4-hydroxy cinnamic acid and
have obtained cross-linked, insoluble polymers.

Hartmann and coworkers[58] have elaborated the synthesis method of
soluble, photoactive polyanhydrides containing etheric and esteric bonds
on the basis of trans-4-hydroxy cinnamic acid. Ester-linked monomers—
4,4′(octamethylene-dioxe)-di-trans-4-cinnamic acid (3a) were obtained via
interaction of hydroxy acid (1) with 1,8-dibromooctane (2a). 4,4′-(Sebacoyl-
dioxide)-di-trans-cinnamic acid (3b) was obtained through esterification of
(1) hydroxyl group by sebacoylchloride. Acids (3a) and diacids (3) were
transferred into mixed anhydrides via processing with acetic acid anhydride.
Synthesis of polyanhydrides was conducted through homo-polycondensation

of mixed anhydride in the melt, at 160–180°C and in high vacuum. Obtained polyanhydrides are soluble in chloroform, DMF, and DMSO. Cross-linking of polymers was conducted in a photochemical manner via UV irradiation (Diagram 6.22).

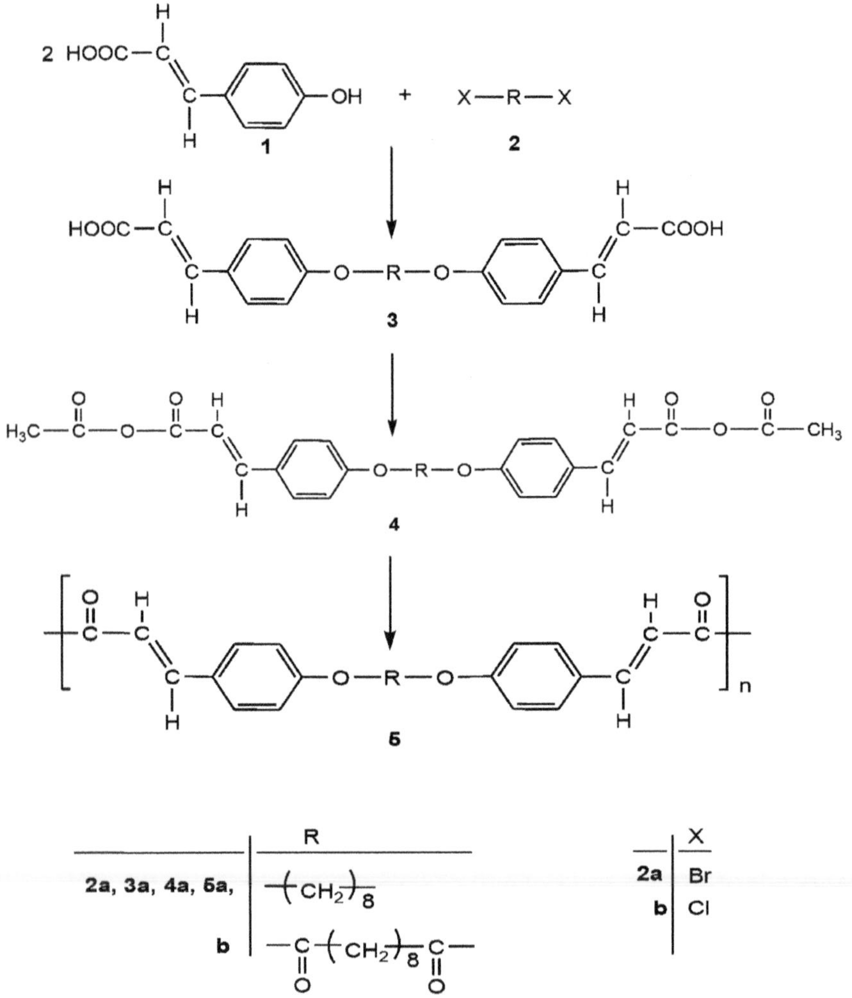

DIAGRAM 6.22

Spiliopoulos and coworkers[59] on the basis of 4-hydroxy cinnamic and 4-hydroxy-benzoic acids have obtained various unsaturated homo- and copolyesters and polyamides. They obtained thermally stable polymers with thermopolymerization ability, characterized them and studied their solubility and water absorption capacity.

6.6   POLYMERS MODIFIED WITH UNSATURATED BONDS

Insertion of unsaturated bonds into polymers is possible in the form of end groups, for ensuing implementation of number of transformation, in particular, cross-linking reactions by their means.

Langer and coworkers[60] have got a new family of polyanhydrides containing terminal methacrylic groups, in which high strength and the ability of controlled degradation are combined (Diagram 6.23).

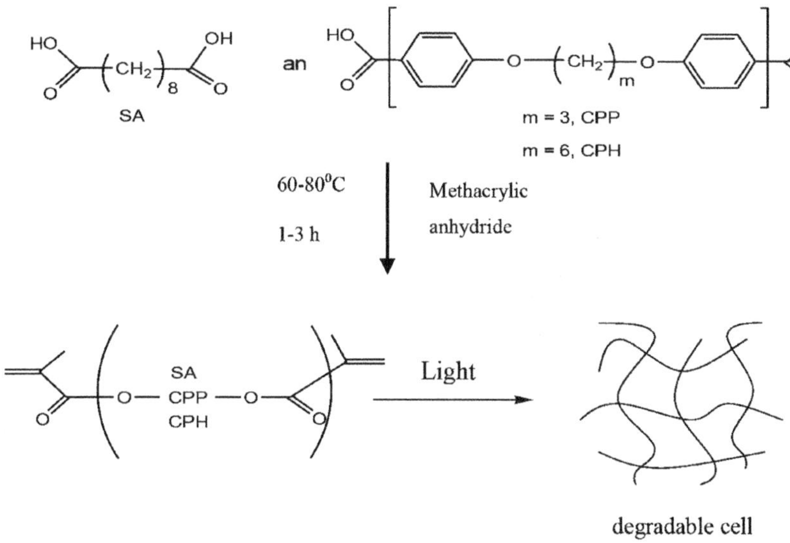

DIAGRAM 6.23

Degradation of cross-linked homopolymers cross-linked on the basis of sebacic acid (SA) and 1,6-bis(p-carboxyphenoxy)-hexane (CPH) was studied in vitro tests. As expected, the network obtained on the basis of SA was completely degraded within 1 week, while only 25% of network obtained on the basis of CPH was resolved (decayed) during 3 months. Polymer networks have basically degraded by surface erosion mechanism and at a constant speed. According to obtained results, almost 1 year was necessary for complete degradation of relatively hydrophobic CPH. At the same time, polymer has kept 90% of strength even in case of 40% mass loss. In vivo study in rats showed that obtained polymers are character- ized by high biocompatibility with soft and bone tissues. Such polymers may find wide application in different fields of medicine—ranging from dentistry to orthopedics.

Langer and coworkers[61] have obtained cross-linked polymer of poly- ether–ester according to Diagram 6.24. At the first stage, they conducted copolymerization of polyethers–polyethylene glycol (PEG) or polypropyl- eneglycol, polytetra-methyleneglycol with cyclic monomer of L-milk acid, then conducted acylation of terminal OH groups by acryloyl chloride, while cross-linked polymers were obtained at the stage of photopolymerization of modified oligomers. Time of in vitro degradation of cross-linked poly- mers at 37°C and 1 N NaOH was changed from 20 to 7 min depending on cross-linking degree and nature of PEG. Obtained materials may be used for coating of various medical devices with the purpose of replacement of nondegradable silicon materials (Diagram 6.24).

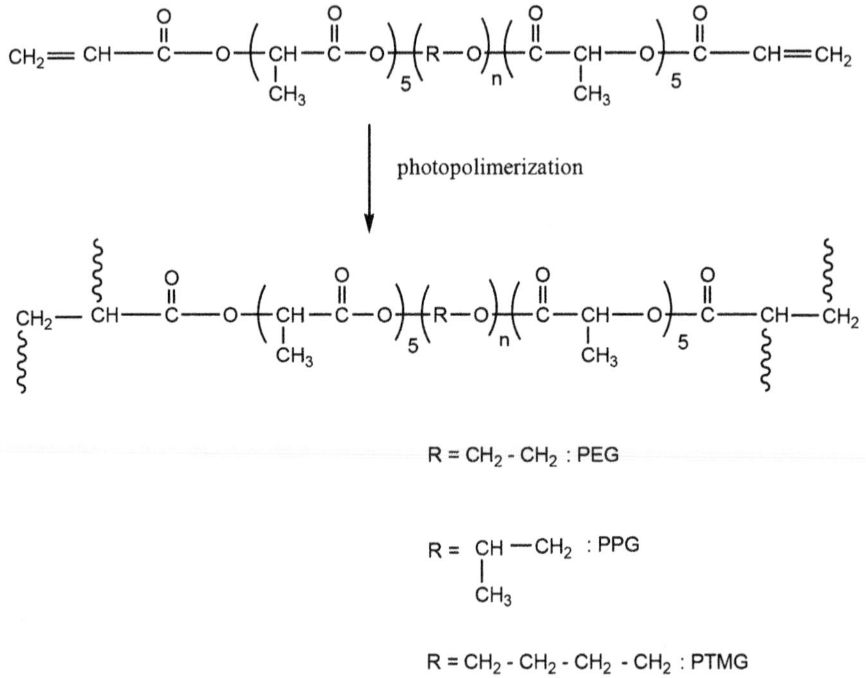

DIAGRAM 6.24

Oligoester containing end unsaturated dicarbonic acids was obtained via interaction of ethyleneglycol lactate diol with maleic anhydride.[62] Polymerization of oligoester was conducted in the melt with toluene-2,4-diisocyanate. New polymers are characterized by porous structure, which is caused by formation of CO_2 (it is biodegradable) at the last stage of synthesis. Even after insertion of isolated C=C double bonds, this polymer remained sufficiently flexible and had the characteristics to form memorization. It is important to note that glass transition temperature was close to human body temperature and met basic requirements related to medical use of polymers.

Dray and coworkers have obtained pyroxamine modified with metacrylic acid and via its copolymerization with 2-hydroxyethyl-methacrylate were obtained hydrogels[63,64] on the surface of which bacteria are not registered and, respectively, biofilm generated by them is not formed. Such hydrogel is prospective for coating of catheter surfaces with the aim of sterility preservation.

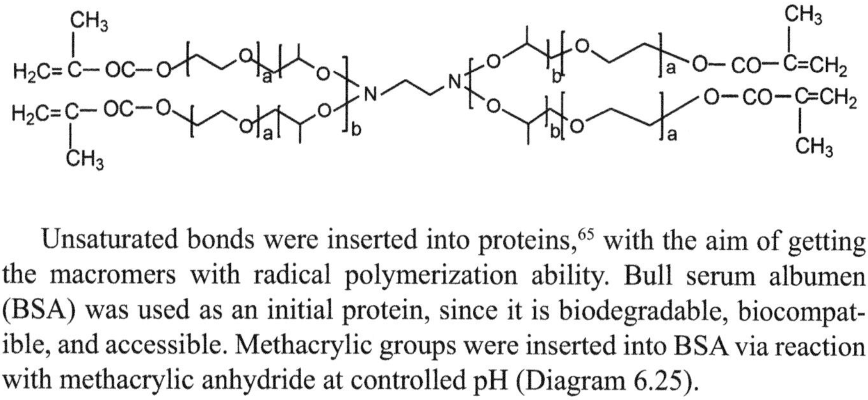

Unsaturated bonds were inserted into proteins,[65] with the aim of getting the macromers with radical polymerization ability. Bull serum albumen (BSA) was used as an initial protein, since it is biodegradable, biocompatible, and accessible. Methacrylic groups were inserted into BSA via reaction with methacrylic anhydride at controlled pH (Diagram 6.25).

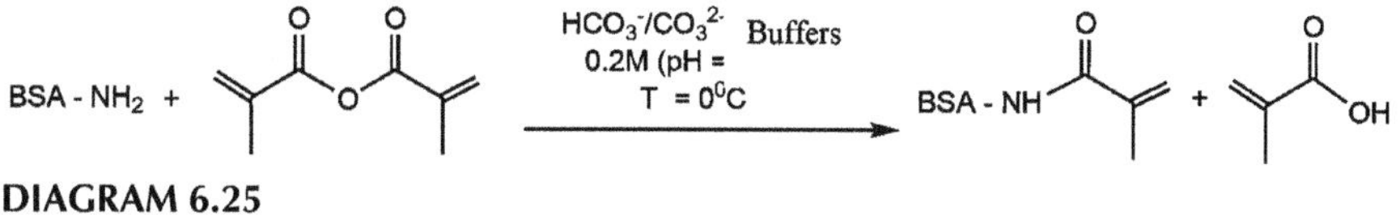

DIAGRAM 6.25

Authors have obtained proteinic hydrogels, which are interesting materials for application in biomedicine and pharmacology.

Brand new class of functional polymers for biomedical purposes—oxirane polymers on the basis of trans- and cis-epoxy-amber acids (Diagram

6.4) is obtained by Georgian chemists,[66–68] and chemical modification reactions of obtained epoxy-polymers with various nucleophilic reagents are studied (Diagram 6.26).

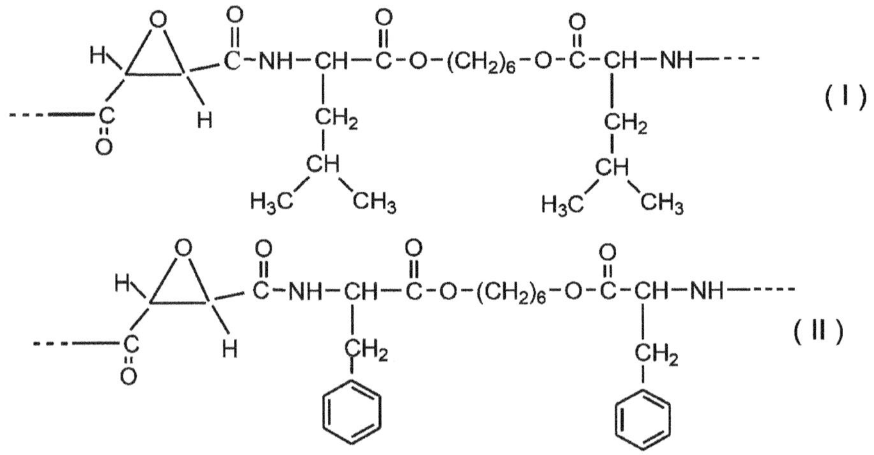

DIAGRAM 6.26

Polyesteramide with lateral hydroxylic and tertiary aminogroups is obtained via interaction (I) with secondary amine (dibutylamine), while at the next stage, acylation of obtained polyesteramide was conducted.

6.7   CONCLUSIONS

Thus, even an above-cited brief review gives us an opportunity to make conclusion that insertion of unsaturated bonds both in basic and in lateral chains of macromolecules makes possible modification of their properties and significant extension of application area. Unsaturated polymers of different classes and three-dimensional (structural) biodegradable systems (networks and hydrogels) obtained on their basis are interesting materials for their use both in engineering and in various fields of medicine—ranging from dentistry to orthopedics: for creation of surgical constructional materials; in the form of binders in tissue repairing surgery, for getting the bone prostheses for implanted artificial organs, drug-delivery systems, gene transfection, cell capsulation, in the form of antibacterial agents, for micro- and nanocapsulation and many other things.

KEYWORDS

- **biomedical polymers**
- **unsaturated polyesteramides**
- **functional polymers**
- **biodegradable polymers**

REFERENCES

1. Guo, K.; Chu, C. C. Synthesis and Characterization of Novel Biodegradable Unsaturated Poly(Ester Amide)/Poly(Ethylene Glycol) Diacrylate Hydrogels. *J. Polym. Sci., A: Polym. Chem.* **2005,** *43,* 3932–3944.
2. Bechaouch, S.; Coutin, B.; Sekiguchi, H. Novel Polyamides from L-Cystine. *Macromol. Rapid Commun.* **1994,** *15,* 125–131.
3. Yu, H. Pseudopoly(Amino Acids): A Study of the Synthesis and Characterization of Polyesters Made from α-L-Amino Acids. Ph.D. Thesis, MIT, 1988.
4. Kohn, J.; Langer, R. Polymerization Reactions Involving the Side Chains of the α-L-Amino Acids. *J. Am. Chem. Soc.* **1987,** *109,* 817–820.
5. Suzuki Sh.; Kondo, T. Disintegration of Poly(*N,N*-tereptaloyl-L-lysine) Microcapsules by Poly(diallyldimethyl Ammonium Chloride). *J. Colloid Interface Sci.* **1978,** *67,* 441–447.
6. Boustta, M.; Huguet, Y.; Vert, M. New Functional Polyamides Derived from Citric Acid and L-Lysine: Synthesis and Characterization. In *Mech. Kinet. Polym. React. Their Use Polym. Synth. Int. Symp. Honor Prof. Pierre Signalt Occas. His 65th Birthday*, Paris, Sept. 9–13, 1990. Prefr. SYMPOL'90: Paris, 1990, p 121.
7. Yasuzava, T.; Yamaguchi, H.; Minoura, Y. Synthesis and Properties of Optically Active Polyurea Interfacial Polyaddition. *J. Polym. Sci., Polym. Chem. Ed.* **1979,** *17,* 3387–3396.
8. Leong, K. W.; Simonte, V.; Langer, R. Synthesis of Polyanhydrides: Melt-Polycondensation, Poly-hydrochlorination and Dehydrative Coupling. *Macromolecules* **1987,** 20, 705–712.
9. Staubi, A.; Mathiowitz, E.; Zucarelli, M.; Langer, R. Characterization of Hydrolytically Degradable Amino Acid Containing Poly(Anhydride-*co*-imides). *Macromolecules* **1991,** *24,* 2283–2290.
10. Katsarava, R. D. Achievements and Problems of Activated Polycondensation. *Adv. Chem.* **1991,** *60,* 1419–1448.
11. Katsarava, R. D.; Synthesis of Heterochain Polymers Using Chemically Activated Monomers (Activated Polycondensation) (Review). High-Molecular Compound. 1983, 31, 1555–1571.
12. Kharadze, D. New Biosimilar Macromolecular Systems on the Basis of Natural Aminoacids. Dissertation for the Scientific Degree of Doctor of Chemical Sciences, Tbilisi, 1998 (in Georgian).

13. Jokhadze, G.; Chu, C. C. Tugushi, D.; Katsarava, R.; Synthesis of Biodegradable Copoly(esteramide) Containing L-Lysine Benzyl Ester Moieties in the Backbones. *Georgian Eng. News* **2006**, *2*, 220–223.

14. Neparidze, N.; Machaidze, M.; Zavradashvili, N.; Mazanashvili, N.; Tabidze, V.; Tugushi, D.; Katsarava, R. Biodegradable Co-polyether-amides with Hydrophobic Side Substituents. *Polym. Med.* **2006**, *2*, 27–33 (in Russian).

15. Lou X., Detrembleur C., Lecomte P., Jerome R. Novel Unsaturated ε−Caprolactone Polymerizable by Ring Opening and Ring Opening Metathesis Mechanisms. *e-Polymers* **2002**, *2*, 505–516.

16. Domb, A. J.; Martinowitz, E.; Ron, E.; Giannos, S.; Langer, R. Polyanhydrides. IV. Unsaturated and Cross-linked Polyanhydrides. *J. Polym. Sci., Polym. Chem.* **1991**, 29, 571–579.

17. Domb, A. J.; Laurencin, C. T.; Israeli, O.; Gerhart, T. N.; Langer, R. The Formation of Propylene Fumarate Oligomers for Use in Bioerodable Bone Cement Composites. *J. Polym. Sci., Polym. Chem.* **1990**, *28* (5), 973–985.

18. Lewandrowski K-U.; Gresser, J. D.; Bondre, S. P.; Silva, A. E.; Wise, D. L.; Trantolo, D. J. Developing Porosity of Poly(propylene glycol-*co*-fumaric acid) Bone Graft Substitutes and The Effect on Osteointegration: A Preliminary Histology Study in Rats. *J. Biomater. Sci., Polym. Ed.* **2000**, *11*, 879–889.

19. Timmer, M. D.; Horch, R. A.; Ambrose, C. G. Mikos, A. G. Effect of Physiological Temperature on the Mechanical Properties and Network Structure of Biodegradable Poly(propylene fumarate)-Based Networks. *J. Biomater. Sci. Polym. Ed.* **2003**, *14*, 369–382.

20. Kharas, G. B.; Scola, A.; McColough, K.; Crawford, A.; Diener, C. A.; Villaseñor, G.; Herrman, J. E.; Passe, L. B.; Watson, K. Synthesis and Characterization of Diethyl Fumarate-1,4-Cyclohexanedimethanol Polyesters for Use in Bioresorbable Bone Cement Composites. *J. Macromol. Sci.* **2006**, *43*, 459–467.

21. Lacoudre, N.; Leborgne, A.; Sepulchre, M.; Spassky, N.; Djonlagic, J.; Jacovic, M. S. Synthese et etude des properietes rheologiques de polyesters stereoreguliers a base d'acide maleique et d'acide fumarique. *Makromol. Chem.* **1986**, *187* (2), 341–350.

22. Jacovic, M. S.; Djonlagic, J.; Sepulchre, M.; Sepulchre, M.-O.; Leborgne, A.; Spassky, N. Synthesis and Rheological Study of Some Maleic and Fumaric Acid Stereoregular Polyesters, 2. Polyesters Derived from 1,4-Dibrombutane or 1,8-Dibromooctane. *Makromol. Chem.* **1988**, *189* (6), 1353–1362.

23. Djonlagic, J.; Sepulchre, M.-O.; Sepulchre, M.; Spassky, N.; Jacovic, M. S. Synthesis and Rheological Study of Some Maleic and Fumaric Acid Stereoregular Polyesters. 3. Synthesis of a Series of Configurationally Pure Aliphatic Polymaleates. *Makromol. Chem.* **1988**, *189* (7), 1485–1492.

24. Djonlagic, J.; Sepulchre, M.-O.; Sepulchre, M.; Spassky, N.; Dunjic, B.; Jacovic, M. S. Synthesis and Rheological Study of Maleic and Fumaric Acid Stereoregular Polyesters. 4. Synthesis of Random, Multi-block and Alternating Unsaturated Copolyesters from Fumaric, Maleic and Phthalic Acid Potassium Salts and 1,4-Dibrombutane. *Makromol. Chem.* **1990**, *191* (7), 1529–1543.

25. Sepulchre, M.-O.; Sepulchre, M.; Spassky, N.; Djonlagic, J.; Jacovic, M. S. Synthesis and Rheological Study of Maleic and Fumaric Acid Stereoregular Polyesters. 6. Influence of Some Parameters on the Polycondensation Reaction of Potassium Maleate and 1,4-Dihalogenobutanes. *Makromol. Chem.* **1991**, *192* (5), 1073–1084.

26. Sepulchre, M.-O.; Moreau, M.; Sepulchre, M.; Djonlagic, J.; Jacovic, M. S. Synthesis and Rheological Study of Maleic and Fumaric Acid Stereoregular Polyesters. 5. Structural Study by ¹H and ¹³C NMR of Random, Multi-block and Alternating Unsaturated Copolyesters from Fumaric, Maleic and Phthalic Acid Potassium Salts and 1,4-Dibrombutane. *Makromol. Chem.* **1990,** *191* (8), 1739–1758.

27. Nikolic, M. S.; Poleti, D.; Djonlagic, J. Synthesis and Charactarization of Biodegradable Poly(Butene Succinate-*co*-Butylene Fumarate)s. *Eur. Polym. J.* **2003,** *39*, 2183–2192.

28. Lou X., Detrembleur C., Lecomte P., Jerome R. Novel Unsaturated ε–Caprolactone Polymerizable by Ring Opening and Ring Opening Metathesis Mechanisms. *e-Polymers* **2002,** 2, 505–516. http://www.e-polymers.org

29. Warwel, S.; Demes, C.; Steinke, G. Polyesters by Lipase-Catalyzed Polycondensation of Unsaturated and Epoxidized Long-Chain α,ω-Dicarboxylic Acid Methyl Esters with Diols. *J. Polym. Sci., A: Polym. Chem.* **2001,** *39*, 1601–1609.

30. Bader, H.; Ruppel, D.; Walich, A. Biologisch abbanbare Polymere fur depootrubereifungen mit kontrollierten Wikstoffabgabe. Patent (ЗаявкаФРГ) N3616320, 1987.

31. Jacovic, M. S.; Djonlagic, J.; Lenz, R. W. Synthesis of Some Fumaric Acid Polyamides from Active Diester Monomers. *Polym. Bull.* **1982,** *8*, 295–301.

32. Katsarava, R.; Beridze, V.; Arabuli, N.; Kharadze, D.; Chu, C. C.; Won, C. Y Amino Acid Based Bioanalogous Polymers. Synthesis and Study of Regular Poly(Ester Amide)s Based on Bis(α-Amino Acid) α,ω-Alkylene Diesters, and Aliphatic Dicarboxylic Acids. *J. Polym. Sci., A: Polym. Chem.* **1999,** 37, 391–407.

33. Markoishvili, K.; Tsitlanadze, G.; Katsarava, L. R.; Morris, J. G.; Sulakvelidze, A. A. Novel Sustained-Release Matrix Based on Biodegradable Poly(Ester Amide)s and Impregnated with Bacteriophages and an Antibiotic Shows Promise in Healing Wounds Infected with Various Pathogenic Bacteria. *Intern. J. Dermatol.* **2002,** *41*, 453–458.

34. Gomurashvili, Z.; Zhang, H.; Da, J.; Jenkins, T. D.; Hughes, J.; Wu, M.; Lambert, L.; Grako, K. A.; DeFife, K. M.; Macpherson, K.; Vassilev, V.; Katsarava, R.; Turnell, W. G. From Drug-Eluting Stents to Biopharmaceuticals: Poly(Ester Amide) a Versatile New Bioabsorbable Biopolymer. In *ACS Symposium Series 977: Polymers for Biomedical Applications*; Mahapatro, A., Kulshrestha, A. S., Eds.; Oxford University Press: Oxford, 2008; pp 10–26.

35. Guo, K.; Chu, C. C.; Chkhaidze, E.; Katsarava, R. Synthesis and Characterization of Novel Biodegradable Unsaturated Poly(Ester Amide)s. *J. Polym. Sci., A: Polym. Chem.* **2005,** *43*, 1463–1477.

36. Chkhaidze, E.; Tugushi, D.; Kharadze, D.; Gomurashvili, Z.; Chu, C. C.; Katsarava, R. New Unsaturated Biodegradable Poly(Ester Amide)s Composed of Fumaric Acid, L-Leucine and α,ω-Alkylene Diols. *J. Macromol. Sci., A: Pure Appl. Chem.* **2011,** *48*, 544–555.

37. Chkhaidze, E. α-Amino Acids Based, Biodegradable, Unsaturated Poly(Ester Amide)s Composed of Fumaric Acid, Synthesis and Transformations. Dissertation for the Ph.D. Degree in Chemistry, Tbilisi, 2008 (in Georgian).

38. Guo, K.; Chu, C. C. Synthesis, Characterization, and Biodegradation of Copolymers of Unsaturated and Saturated Poly(Ester Amide)s. *J. Polym. Sci., A: Polym. Chem.* **2007,** *45*, 1595–1606.

39. Guo, K.; Chu, C. C. Biodegradation of Unsaturated Poly(Ester-Amide)s and Their Hydrogels. *Biomaterials* **2007,** *28*, 3284–3294.

40. Chu, C. C.; Katsarava, R. Elastomeric Functional Biodegradable Copolyester Amides and Copolyester Urethanes. U.S. Patent 6,503,538, January 7, 2003.

41. Guo, K.; Chu, C. C. Copolymers of Unsaturated and Saturated Poly(Ether Ester Amide)s: Synthesis, Characterization, and Biodegradation. *J. Appl. Polym. Sci.* **2008,** *110*, 1858–1869.

42. Guo, K.; Chu, C. C. Synthesis and Characterization of Novel Biodegradable Unsaturated Poly(Ester Amide)/Poly(Ethylene Glycol) Diacrylate Hydrogel. *J. Polym. Sci., A: Polym. Chem.* **2005,** *43*, 3932–3944.

43. Guo, K.; Chu, C. C. Synthesis, Characterization, and Biodegradation of Novel Poly(Ether Ester Amide)s Based on L-Phenylalanine and Oligoethylene Glycol. *Biomacromolecules* **2007,** *8*, 2851–2861.

44. Guo, K.; Chu, C. C. Controlled Release of Paclitaxel from Biodegradable Unsaturated Poly(Ester Amide)s/Poly(Ethylene Glycol) Diacrylate Hydrogels. *J. Biomater. Sci. Polym. Ed.* **2007,** *18*, 489–504.

45. Pang, X.; Chu, C. C. Synthesis, Characterization and Biodegradation of Functionalized Amino Acid-Based Poly(Ester Amide)s. *Biomaterials* **2010,** *31*, 3745–3754.

46. Atkins, K. M.; Lopez, D.; Knight, D. K.; Mequanint, K.; Gillies, E. R. A Versatile Approach for the Syntheses of Poly(Ester Amide)s with Pendant Functional Groups. *J. Polym. Sci., A: Polym. Chem.* **2009,** *47*, 3757–3772.

47. Hoyle, C. E.; Lee, T. Y.; Roper, T. Thiolenes: Chemistry of the Past with Promise for the Future. *J. Polym. Sci., A: Polym. Chem.* **2004,** *42*, 5301–5338.

48. Bantchev, G. B.; Kenar, J. A.; Biresaw, G.; Han Moon, G. Free Radical Addition of Butanethiol to Vegetable Oil Double Bonds. *J. Agric. Food Chem.* **2009,** *57*, 1282–1290.

49. Ai, Y.; Shi, Z.; Guo, W. A New Type of Unsaturated Poly(Ester-Amide): Synthesis and Compressive Strength. *Mater. Des.* **2009,** *30*, 892–895.

50. Ai, Y.; Shi, Z.; Guo, W.; Xie, S. Synthesis and Characterization of a Potential Material as Internal Fixation of Bone Fracture. *Mater. Sci. Eng.* **2009,** *29*, 1001–1005.

51. Ai, Y.; Shi, Z.; Guo, W. Calcium Polyphosphate Fibers/Unsaturated Poly(Ester-Amide) Composites for Bone-Fixation Materials. *Polym. Composite* **2009,** *30*, 1119–1124.

52. Pang, X.; Wu, J.; Reinhart-King, C.; Chu, C. C. Synthesis and Characterization of Functionalized Water Soluble Cationic Poly(Ester Amide)s. *J. Polym. Sci., A: Polym. Chem.* **2010,** *48*, 3758–3766.

53. Song, H.; Chu, C. C. Synthesis and Characterization of a New Family of Cationic Amino Acid-Based Poly(Ester Amide)s and their Biological Properties. *J. Appl. Polym. Sci.* **2012,** *124* (5), 3840–3853.

54. Guo, K.; Chu, C. C. Synthesis of Biodegradable Amino-Acid-Based Poly(Ester Amide)s and Poly(Ether Ester Amide)s with Pendant Functional Groups. *J. Appl. Polym. Sci.* **2010,** *117*, 3386–3394.

55. Domb, A.; Ron, E.; Giannos, S.; Flores, S.; Kim, C.; Dow, R.; Langer, R. In *New Poly(anhydrides): Aliphatic–Aromatic Homopolymers and Unsaturated Polymers*, 14th International Symposium on the Controlled Release of Bioactive Materials; Lee, P. I., Leonharddt, B. A., Eds.; Toronto, 1987, p 138.

56. Domb, A. J.; Martinowitz, E.; Ron, E.; Giannos, S.; Langer, R. Polyanhydrides. IV. Unsaturated and Cross-linked Polyanhydrides. *J. Polym. Sci., Polym. Chem.* **1991,** *29*, 571–579.

57. Kricheldorf, H. R.; Lubbers, D. New polymer syntheses, 46. Thermotropic Poly(Ester-Anhydride)s Derived from Terephthalic Acid and Various Aromatic Hydroxy Acids. *Makromol. Chem. Rapid Commun.* **1990,** *11*, 303–307.

58. Pinther, P.; Hartmann, M.; Wermann, K. Photoreactive Polyanhydrides with Cinnamic Acid Units in the Main Chain. *Makromol. Chem.* **1992,** *193* (10), 2669–2675.

59. Spiliopoulos, I. K.; Mikroyannidis, J. A. Unsaturated Polyamides and Polyesters Prepared from 1,4-Bis(2-carboxyvinyl)benzene and 4-Hydroxycinnamic Acid. *J. Polym. Sci. Polym. Chem.* **1996**, *34*, 2799–2807.
60. Anseth, K. S.; Shastri, V. R.; Langer, R. Photopolymerizable Degradable Polyanhydrides with Osteocompatibility. *Nat. Biotechnol.* **1999**, *17*, 156–159.
61. Kim, B. S.; Hrkach, S. J.; Langer, R. Biodegradable Photo-cross-linked Poly(Ether-Ester) Networks for Lubricious Coatings. *Biomaterials* **2000**, 21, 259–265.
62. Xiao, C.; He, Y. Tailor-Made Unsaturated Poly(Esteramide) Network that Contains Monomeric Lactate Sequences. *Polym. Int.* **2007**, *56*, 816–819.
63. Dray, J. L.; McCoy, C. P. In *Modified Poloxamines for Copolymerization in Novel Anti-adherent Hydrogel Coatings*, ACS Meet. "Polymers in Medicine and Biology", Sonoma Valley, CA, USA, 2005.
64. Ratner, B. D.; Hoffman, A. S.; Schoen, F. J.; Lemons, J. E. *An Introduction to Materials in Medicine* 2nd ed.; Elsevier Academic Press: San Diego, 2002; 100–101.
65. Iemma, F.; Spizzirri, U. G.; Muzzalupo, R.; Puoci, F.; Trombino, S.; Picci, N. Spherical Hydrophilic Microparticles Obtained by the Radical Copolymerisation of Functionalised Bovine Serum Albumin. *Colloid Polym. Sci.* **2004**, *283*, 250–256.
66. Katsarava, R.; Tugushi, D.; Zavradashvili, N.; Jokhadze, G.; Gverdtsiteli, M.; Samkharadze, M. In *Amino Acid Based Epoxy-Poly(Ester Amide)s—a New Class of Biodegradable Functional Polymers: Synthesis and Characterization*, Polymers in Medicine and Biology, Workshop Sponsored by the American Chemical Society, Division of Polymer Chemistry on June 17–20, 2007, Santa Rosa, CA, USA.
67. Zavradashvili, N.; Tugushi, D.; Toidze, P.; Gomurashvili, Z.; Katsarava, R. In *Some Polymer-Analogous Transformations of Amino Acid Based Biodegradable Epoxy-Poly(Ester Amide)s*, Polymers in Medicine and Biology, Workshop Sponsored by the American Chemical Society, Division of Polymer Chemistry on June 17–20, 2007, Santa Rosa, CA, USA.
68. Zavradashvili, N.; Jokhadze, G.; Gverdtsiteli, M.; Otinashvili, G.; Kupatadze, N.; Gomurashvili, Z.; Tugushi, D.; Katsarava, R. Amino Acid Based Epoxy-Poly(Ester Amide)s - a New Class of Functional Biodegradable Polymers: Synthesis and Chemical Transformations. *J. Macromol. Sci. A Pure Appl. Chem.* **2013**, *50* (5), 449–465.

CHAPTER 7

FEATURES OF LANTHANUM EXTRACTION BY INTERGEL SYSTEM BASED ON POLYACRYLIC ACID AND POLY-2-METHYL-5-VINYLPYRIDINE HYDROGELS

T. K. JUMADILOV[1*], R. G. KONDAUROV[1], and L. E. AGIBAYEVA[2]

[1]JSC Institute of Chemical Sciences after A.B. Bekturov, Almaty, Republic of Kazakhstan

[2]Al-Farabi Kazakh National University, Almaty, Republic of Kazakhstan

*Corresponding author. E-mail: jumadilov@mail.ru

CONTENTS

ABSTRACT

Hydrogels mutual activation was studied in aqueous solutions, particularly dependence of swelling coefficient, specific electric conductivity, and pH of aqueous solutions from hydrogels molar ratios in time were studied. Maximum activation of hydrogels occurs at gPAA:gP2M5VP = 5:1 and 4:2 ratio. Lanthanum ions sorption by intergels system polyacrylic acid hydrogel (gPAA)–poly-2-methyl-5-vinylpyridine hydrogel (gP2M5VP) was studied. Maximum sorption degree of the intergel system is observed at gPAA:gP2M5VP = 4:2 ratio, it is 91.09%; this is higher compared with individual hydrogels of gPAA and gP2M5VP (67.71% and 63.65%, respectively). At this hydrogel ratio, polymer-chain total binding degree is 75.83%, which is much higher than polymer-chain binding degree of the initial hydrogels: binding degree of gPAA is 56.50%, of gP2M5VP is 53.00%. Desorption degree lanthanum ions by ethyl alcohol is 80.17%, by nitric acid—92.55%. Obtained results indicate that at gPAA:gP2M5VP = 4:2 ratio in intergel system, there is a significant change of electrochemical, conformational, and sorption properties of the initial macromolecules.

7.1 INTRODUCTION

As a result of previous studies,[1-5] it was found that remote interaction of polymer hydrogels leads to significant changes in their electrochemical and conformational properties. Subsequently, influence of different factors on polymer hydrogels remote interaction in the intergel systems was investigated.[6-12]

At present, technologies of rare earth and other elements concentration and extraction in hydrometallurgy are based on the use of ion-exchange resins. However, the ion exchangers does not have high degree of selective extraction of metals, their distribution coefficient has a low value. In addition, application of ion exchange resins is aimed to selective extraction of one metal, whereas industrial solutions usually contain several valuable components. In this regard, the aim of this chapter is the study sorption capacity relatively to lanthanum as an intergel system based on hydrogels of polyacrylic acid and poly-2-methyl-5-vinylpyridine ions and forecast the possibility of their use for rare earth metals extraction.

7.2 EXPERIMENTAL PART

7.2.1 EQUIPMENT

For measurement of solutions, specific electric conductivity conductometer "MARK-603" (Russia) was used; hydrogen ion concentration was measured on Metrohm 827 pH meter (Switzerland). Samples weight was measured on analytic electronic scales Shimadzu AY220 (Japan). La^{3+} ion concentration in solutions was determined by spectrophotometers SF-46 (Russia) and Perkin Elmer Lambda 35 (USA).

7.2.2 MATERIALS

Studies were carried out in 0.005-M 6-water lanthanum nitrate solution. Polyacrylic acid hydrogels were synthesized in presence of crosslinking agent N,N-methylene-bis-acrylamide and redox system $K_2S_2O_8-Na_2S_2O_3$ in water medium. Synthesized hydrogels were crushed into small dispersions and washed with distilled water until constant conductivity value of aqueous solutions was reached. Poly-2-methyl-5-vinylpyridine (gP2M5VP) hydrogel of Sigma-Aldrich Company (linear polymer crosslinked by divinylbenzene) was used as polybasis.

For investigation task from synthesized hydrogels, an intergel pair polyacrylic acid hydrogel–poly-2-methyl-5-vinylpyridine hydrogel (gPAA–gP2M5VP) was created. Swelling coefficients of hydrogels are $K_{sw(gPAA)}$ = 27.93 g/g and $K_{sw(gP2M5VP)}$ = 3.20 g/g. With gP2M5VP share increase in intergel system, gPAA concentration decreased from 5.96 to 0.97 mmol/L (gPAA:gP2M5VP ratios interval 6:0–1:5). Concentration of gP2M5VP with decrease of gPAA share increased from 1.125 to 6.75 mmol/L (gPAA:gP2M5VP ratios interval 1:5–0:6).

7.2.3 ELECTROCHEMICAL INVESTIGATIONS

Experiments were carried out at room temperature. Studies of intergel system were made in following order: each hydrogel was put in separate glass filters, pores of which are permeable for low molecular ions and molecules but impermeable for hydrogels dispersion. After that, filters with hydrogels were put in glasses with lanthanum nitrate solution. Electric conductivity and pH of over gel liquid were determined in presence of hydrogels in solutions.

7.2.4 DETERMINATION OF HYDROGELS SWELLING

Swelling coefficient was calculated according to the equation:

$$K_{SW} = \frac{m_2 - m_1}{m_1}$$

where m_1 is the weight of dry hydrogel; m_2 is the weight of swollen hydrogel.

7.2.5 LANTHANUM IONS DESORPTION

After sorption each hydrogel separately from another, they were subjected to lanthanum ions desorption in 96% ethanol solution and in 2 M nitric acid.[13]
Desorption degree was calculated according to the following equation:

$$R_{des} = \frac{m_{des}}{m_{sorb}} \times 100$$

7.2.6 METHOD OF LANTHANUM IONS DETERMINATION

Method of lanthanum ions determination in solution is based on the formation of colored complex compound of organic analytic reagent arsenazo III with lanthanum ions.[14]
Extraction (sorption) degree was calculated by equation:

$$\eta = \frac{C_{initial} - C_{residual}}{C_{initial}} \times 100$$

where $C_{initial}$ is the initial concentration of lanthanum in solution (g/L) and $C_{residue}$ is the residual concentration of lanthanum in solution (g/L).
Polymer-chain binding degree was determined by calculations in accordance with equation:

$$\theta = \frac{v_{sorb}}{v} \times 100\%$$

where v_{sorb} is the quantity of polymer links with sorbed lanthanum (mol); v is the total quantity of polymer links (if there are two hydrogels in solution, it is calculated as sum of each polymer hydrogel links) (mol).

7.3   RESULTS AND DISCUSSIONS

7.3.1   STUDY OF gPAA AND gP2 5VP HYDROGELS MUTUAL ACTIVATION IN INTERGEL SYSTEM

Presence of intergel system based on PAA and P2M5VP hydrogels in an aqueous medium provides a variety of different processes which affect on electrochemical equilibrium in solution.

For polyacids, in this case for gPAA, remote interaction with basic hydrogels occurs in three stages:

Stage 1—Ionization, formation of ionic pairs, and dissociation.

Stage 2—After binding of a proton which was released into aqueous medium due to –COOH groups dissociation (dissociation takes place in accordance with the traditional representation[15]: first, there is formation of ionic pairs; after that, ionic pairs partially dissociates to individual ions), by nitrogen atoms equilibrium in solution in reaction ~COOH↔~COO⁻ + H⁺ shifts to right according to the principle of "Le Chatelier" (concentration of H⁺ decreases due to eq. (7.2)).

$$\sim COOH \leftrightarrow \sim COO^- + H^+ \tag{7.1}$$

$$\equiv N + H^+ \rightarrow \equiv NH^+ \tag{7.2}$$

Thus, –COOH groups undergo additional ionization and dissociation.

Stage 3—Internode links are in maximally unfolded state are forced to fold as a result of intramolecular and intermolecular interactions of the type ~COO⁻···H⁺···⁻OOC~ for maintaining of stable condition.

For polybases such as polyvinylpyridine, remote interaction occurs in two stages:

Stage 1—The association of proton cleaved from carboxyl group:

$$\equiv N + H^+ \leftrightarrow \equiv NH^+ \tag{7.3}$$

Stage 2—Interaction of $\equiv NH^+$ with water molecules.

If water medium becomes acidic, then the interaction occurs according to the reaction:

$$\equiv NH^+ + H \cdots OH \leftrightarrow \equiv NH^+ \cdots OH^- + H^+ \tag{7.4}$$

If interaction of vinyl pyridine links with water molecules leads to alkaline medium, then the reaction will be

$$\equiv N + H^+ \cdots OH^- \leftrightarrow \equiv NH^+ + OH^- \qquad (7.5)$$

Result of these interactions is state in which some parts of charged functional groups of hydrogels have no counterions. The concentration of ionized groups, which have no counterions, is in dependence from initial molar ratios of polymer networks and other factors.

Dependence of specific electric conductivity of solutions from hydrogels molar ratio in time is shown in Figure 7.1. During remote interaction, there is an appearance of areas with minimum and maximum conductivity, while this parameter increases for almost all ratios of hydrogels with time. As it can be seen from the figure, minimum values of electric conductivity are observed at gPAA:gP2M5VP = 3:3 ratio for all time of hydrogels remote interaction. Also, low values of conductivity are observed at gPAA:gP2M5VP = 4:2 ratio. The area of maximum electric conductivity is hydrogels ratio 1:5. Maximum values of electric conductivity are achieved after 48 h.

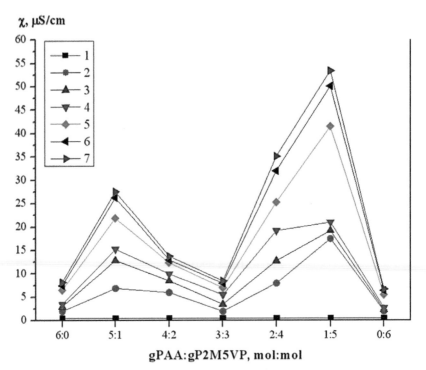

FIGURE 7.1 Dependence of specific electric conductivity from hydrogels molar ratio in intergel system gPAA–gP2M5VP in time in an aqueous medium. Curves' description (hydrogels remote interaction time): 1—0.5 h; 2—1 h; 3—2 h; 4—3 h; 5—6 h; 6—24 h; and 7—48 h.

The minimum conductivity of intergel system gPAA–gP2M5VP is caused by binding of proton which was cleaved from carboxyl group by heteroatom of vinylpyridine. At the time when two hydrogels are put in aqueous medium, both polymers begin to swell due to interaction with water molecules. Carbonyl groups at first ionize and then dissociate to carboxylate anion $-COO^-$ and hydrogen ion (protons) H^+.

Ionization of cationic hydrogel (poly-2-methyl-5-vinylpyridine) in an aqueous medium occurs due to binding of hydrogen ion which were formed during dissociation of carboxylic groups and water molecules into ions H^+ and OH^-. The long-range interaction effect of hydrogels leads to decrease of total content of positively charged ions in an aqueous medium.

High values of electric conductivity, in turn, indicate that for certain ratios of the two hydrogels, dissociation of carboxyl groups prevails over the process of proton association. The reason of this phenomenon is conformational changes of internode links. At certain concentrations, charged NH^+ groups can form intramolecular crosslinks $\geq N\cdots H^+\cdots N\equiv$, which lead to folding of macromolecular globes and to decrease of proton binding.

The dependence of aqueous solutions pH from gPAA:gP2M5VP molar ratios in time are shown in Figure 7.2. With prevalence of polyacid share in solution, decrease of hydrogen ion concentration is observed. Distinct minimum is observed in presence of the polyacid (gPAA:gP2M5VP = 6:0). The maximum pH value at 48 h was observed at gPAA:gP2M5VP = 5:1 ratio. Gradual decrease of pH can be seen with the growth of polybasis share.

The appearance of H^+ ions excess is due to high speed of swelling and $-COOH$ group dissociation and also to an insufficient rate of basic groups swelling and their low concentration. The increase in the OH^- ions content in aqueous medium is associated with low rate of swelling and low concentration of $-COOH$ groups, as well as high speed of swelling and interaction of basic functional groups with H^+ ions. This is possible in case of the second reaction occurrence, in which hydroxyl anions are released in solution. In parallel, there is a third reaction, resulting in a free proton is association with pyridine ring and concentration of positively charged ions in solution dramatically decreases.

Comparing low values of hydrogen ions concentration and high values of electric conductivity in intergel system at ratios of gPAA:gP2M5VP = 4:2 and 3:3, it can be concluded that at these molar ratios, there is maximum activation of hydrogels in intergel system.

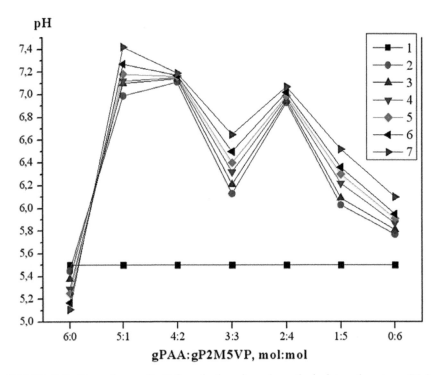

FIGURE 7.2   Dependence of pH from hydrogels molar ratio in intergel system gPAA–gP2M5VP in time in an aqueous medium. Curves' description (hydrogels remote interaction time): 1—0.5 h; 2—1 h; 3—2 h; 4—3 h; 5—6 h; 6—24 h; and 7—48 h.

The concentration of ions in the studied intergel system directly depends on swelling rate and hydrogels concentration in aqueous medium. Swelling rate and deprotonization is in dependence from nature, crosslinking degree, dispersion, and morphology of polymer hydrogels.

Dependencies of swelling coefficients of polyacrylic acid and poly-2-methyl-5-vinylpyridine hydrogels from hydrogels molar ratios in time are shown in Figures 7.3 and 7.4.

Dependence of swelling coefficient of acid hydrogel from hydrogels molar ratio is shown in Figure 7.3. As can be seen with increase of polybasis share, there is an increase of polyacid swelling. Maximum swelling of gPAA is observed at gPAA:gP2M5VP=1:5 ratio at 48 h hydrogels remote interaction.

It should also be noted that at a gPAA:gP2M5VP = 3:3 ratio up to 2 h, there are distinct maximums of polyacid swelling. This phenomenon can be explained that at this polymer, hydrogels ratio ionization of polyacid's links occurs much faster than at other ratios. The area of minimum swelling of polyacrylic acid is observed in presence of only polyacid (gPAA:gP2M5VP = 6:0 ratio) due to the fact that there is no additional dissociation of carboxyl groups as a result of shift of equilibrium to right (toward proton formation).

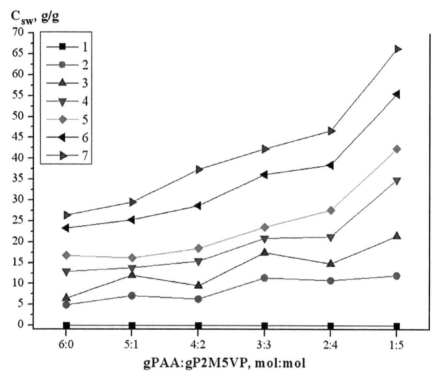

FIGURE 7.3 Dependence of polyacrylic acid hydrogel swelling coefficient from gPAA:gP2M5VP hydrogels molar ratios in time. Curves' description (hydrogels remote interaction time): 1—0.5 h; 2—1 h; 3—2 h; 4—3 h; 5—6 h; 6—24 h; and 7—48 h.

Dependence of poly-2-methyl-5-vinylpyridine swelling coefficient from hydrogels molar ratios in time is presented in Figure 7.4. Minimal swelling of polybasis occurs when hydrogels ratio is 3:3 and 2:4. It occurs due to the fact that macromolecule tries to take the most advantageous form (globule), despite unfolding due to repulsion of same charged units. Swelling of polybasis increases with growth of polyacid share. A distinct maximum of swelling is observed at a ratio 5:1 at 48 h. In a region where polybasis dominates high values of swelling coefficient are caused by additional ionization of pyridine links under the influence of H^+ ions cleaved from $-COOH$ groups.

At intermediate ratios (3:3 and 2:4), concentration of H^+ and OH^- ions has almost similar values in intergel system. As a result, H^+ ions are cleaved from $\equiv N^+H$ and ionization degree of polybasis decreases. Accordingly, there is a contraction of polybasis. With further growth of –COOH groups, ionization degree of nitrogen atoms increases. This is shown in values of poly-2-methyl-5-vinylpyridine swelling coefficient.

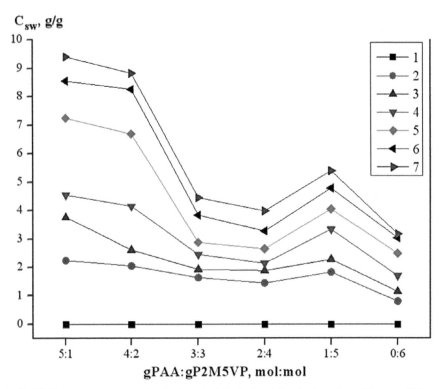

FIGURE 7.4 Dependence of poly-2-methyl-5-vinylpyridine hydrogel swelling coefficient from gPAA:gP2M5VP hydrogels molar ratios in time. Curves' description (hydrogels remote interaction time): 1—0.5 h; 2—1 h; 3—2 h; 4—3 h; 5—6 h; 6—24 h; and 7—48 h.

Comparing values of specific electric conductivity, pH, and swelling coefficient of the two hydrogels, it can be concluded that in result of mutual activation PAA and P2M5VP hydrogels transfer into a highly ionized state. The area of maximum activation of polymer hydrogels is gPAA:gP2M5VP = 5:1 and 4:2 ratios. Highest ionization of gP2M5VP hydrogel occurs at 5:1 ratio. In turn, swelling of gPAA increases significantly when the polybasis prevails in solution (ratio 1:5).

7.3.2   STUDY OF LANTHANUM IONS SORPTION BY INTERGEL SYSTEM gPAA–gP2 5VP

From the results mentioned above, it can be expected that phenomenon of hydrogels mutual activation should be reflected in processes of metals ions extraction. To check this assumption, sorption properties of intergel system gPAA–gP2M5VP in relation to lanthanum ions were studied. There is a change in electrochemical and conformational properties of PAA and P2M5VP hydrogels during extraction of lanthanum ions by intergel system gPAA–gP2M5VP; however, the changes are different comparatively to the situation when the hydrogels interact with each other in an aqueous medium.

Change of lanthanum nitrate electric conductivity in presence of gPAA–gP2M5VP intergel system in time is shown in Figure 7.5. The obtained data show that lanthanum ions sorption causes significant changes of conductivity. As can be seen from figure, electric conductivity decreases for all gPAA:gP2M5VP ratios with time. Minimum values of conductivity are observed at hydrogels ratio 4:2 at 48 h. In solution, there is a presence of ions formed in result of three—stage dissociation of lanthanum nitrate in addition to carboxylate anions and protons.

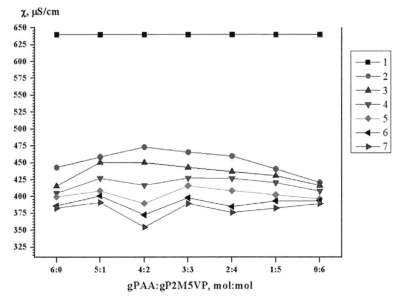

FIGURE 7.5 Dependence of specific electric conductivity from hydrogels molar ratio in time in intergel system gPAA:gP2M5VP in 6-water lanthanum nitrate solution. Curves' description (hydrogels remote interaction time): 1—0.5 h; 2—1 h; 3—2 h; 4—3 h; 5—6 h; 6—24 h; 7—48 h.

In solution, the following chemical reactions occur:

(1) Dissociation of lanthanum nitrate, along with carboxyl groups dissociation:

$$\sim COOH \leftrightarrow \sim COO^- + H^+ \tag{7.6}$$

$$La(NO_3)_3 \cdot 6H_2O \leftrightarrow La^{3+} + 3NO_3^- + 6H_2O \tag{7.7}$$

(2) Sorption of lanthanum ions by polymer hydrogels:

$$3\sim COO^- + La^{3+} \rightarrow \sim COO_3La \tag{7.8}$$

$$3\equiv N + La^{3+} \rightarrow \equiv N^+{}_3La \tag{7.9}$$

These reactions impact on electrochemical equilibrium in solution. Depending on predominance of one of them, there will be changes in values of conductivity.

Figure 7.6 shows dependence of hydrogen ions concentration from polyacrylic acid and poly-2-methyl-5-vinylpyridine hydrogels molar ratios in time. Concentration of hydrogen ions increases with time. Minimum values of pH are observed at ratios gPAA:gP2M5VP = 6:0 and 5:1. This is a result of hydrogel ionization due to formation of coordination bonds with lanthanum ions.

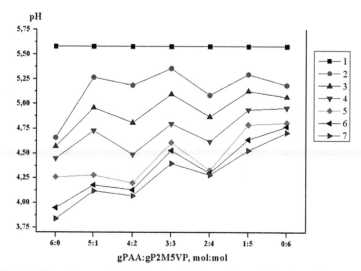

FIGURE 7.6 Dependence of pH from hydrogels molar ratio in time in intergel system gPAA:gP2M5VP in 6-water lanthanum nitrate solution. Curves' description (hydrogels remote interaction time): 1—0.5 h; 2—1 h; 3—2 h; 4—3 h; 5—6 h; 6—24 h; and 7—48 h.

Swelling coefficient of gPAA behavior during lanthanum ion sorption is shown in Figure 7.7. Sharp increase of polyacid swelling is observed with polybasis share increase in solution. Maximum swelling of gPAA occurs at gPAA:gP2M5VP = 1:5 ratio at 1 h of hydrogels remote interaction. There is a gradual decrease of swelling with time. Minimum value of swelling coefficient of gPAA is observed at gPAA:gP2M5VP = 5:1. Due to the fact that there is an occurrence of lanthanum ion sorption, links between node circuits do not have charges, contributing to unfolding of macromolecular globe and as a result, swelling of polyacrylic acid hydrogel decreases.

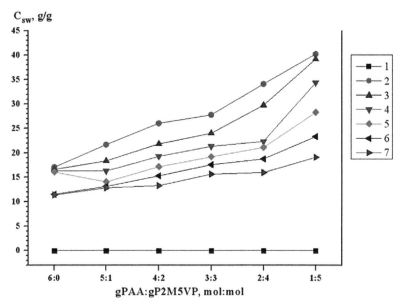

FIGURE 7.7 Dependence of polyacrylic acid hydrogel swelling coefficient in presence of poly-2-methyl-5-vinylpyridine hydrogel in 6-water lanthanum nitrate solution. Curves' description (hydrogels remote interaction time): 1—0.5 h; 2—1 h; 3—2 h; 4—3 h; 5—6 h; 6—24 h; and 7—48 h.

Dependence of swelling coefficient of basic hydrogel poly-2-methyl-5-vinylpyridine from molar ratio gPAA:gP2M5VP in time is shown in Figure 7.8.

Changes of swelling coefficient of base hydrogel are similar to changes in polyacid swelling. First, there is a sharp increase of swelling, maximum value of swelling coefficient is reached at 1 h of hydrogels remote interaction in gPAA:gP2M5VP = 5:1 ratio. Then, there is a consequent gradual decrease of swelling as in the case with polyacrylic acid. The lowest swelling of poly-2-methyl-5-vinylpyridine is observed in presence of only polybasis (gPAA:gP2M5VP = 0:6 ratio) what is associated with the absence of hydrogels mutual activation phenomenon.

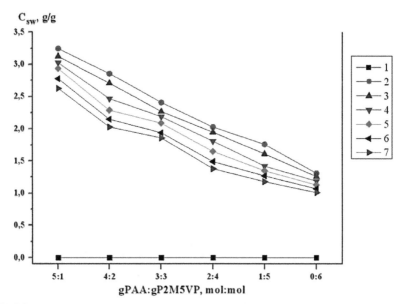

FIGURE 7.8 Dependence of poly-2-methyl-5-vinylpyridine hydrogel swelling coefficient in presence of polyacrylic acid hydrogel in 6-water lanthanum nitrate solution. Curves' description (hydrogels remote interaction time): 1—0.5 h; 2—1 h; 3—2 h; 4—3 h; 5—6 h; 6—24 h; and 7—48 h.

Extraction degree change in dependence of gPAA:gP2M5VP molar ratios is shown in Figure 7.9. As can be seen from the figure, ratios at which there are two hydrogels in solution have higher sorption capacity comparatively to case when solution contains only polyacid or polybasis (ratios 6:0 and 0:6). However, at gPAA:gP2M5VP 5:1 and 4:2 ratios sorption of lanthanum ions is not very intense. A much higher of La^{3+} ions extraction degree is observed at ratios 2:4 and 1:5. Moreover, the maximum values of sorption degree are reached at gPAA:gP2M5VP = 1:5 ratio. Extraction degree at this hydrogels ratio is 90.34%. This is the result of high ionization of gPAA and gP2M5VP due to hydrogels mutual activation in intergel system.

Polymer-chain binding degree is shown in Figure 7.10. Obtained data show that intergel pairs (ratios of gPAA:gP2M5VP from 5:1 to 1:5) have higher binding degree than gPAA and gP2M5VP individual hydrogels (ratio 6:0 and 0:6). Similarly to sorption degree, maximum values of polymer-chain binding degree are observed at gPAA:gP2M5VP=1:5 ratio; it is 73.13%.

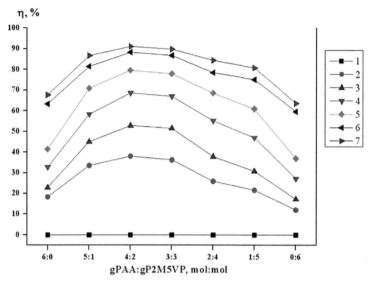

FIGURE 7.9   Dependence of lanthanum ions extraction degree from hydrogels molar ratio in time in intergel system gPAA:gP2M5VP in 6-water lanthanum nitrate solution. Curves' description (hydrogels remote interaction time): 1—0.5 h; 2—1 h; 3—2 h; 4—3 h; 5—6 h; 6—24 h; and 7—48 h.

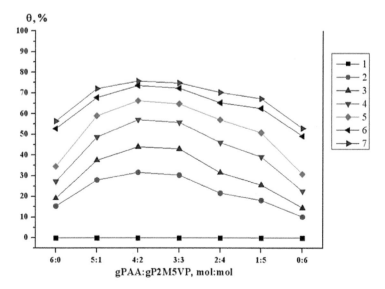

FIGURE 7.10   Dependence of polymer-chain binding degree from the molar ratio of hydrogels in time in intergel system gPAA:gP2M5VP in 6-water lanthanum nitrate solution. Curves' description (hydrogels remote interaction time): 1—0.5 h; 2—1 h; 3—2 h; 4—3 h; 5—6 h; 6—24 h; and 7—48 h.

7.3.3 COMPARISON OF INTERGEL SYSTEM AND INDIVIDUAL HYDROGELS SORPTION ABILITIES

Lanthanum nitrate is present in solution in dissociated state. Dissociation of lanthanum nitrate occurs in three stages, in which dissociation constant of first stage is much higher than in second and third. In this regard, binding of dissociated ions in intergel system occurs according to different mechanisms. Lanthanum nitrate formed in first stage of dissociation is associated according to ionic mechanism. Products of second and third stages are bound due to coordination bond formation.

It should be noted that in intergel system gPAA–gP2M5VP in result of hydrogels mutual activation, their electrochemical and conformational properties are changed significantly. Also there is an increase of extraction degree and polymer-chain binding degree of each hydrogel. Sorption degree of lanthanum ions by gPAA and gP2M5VP individual hydrogels (ratios 6:0 and 0:6) is 67.71% and 63.65%, respectively. As was mentioned above, intergel system extracts 91.09% of lanthanum ions at gPAA:gP2M5VP = 1:5 ratio. Binding degree of gPAA and gP2M5VP individual hydrogels is 56.50% and 53.00%, respectively, when binding degree at gPAA:gP2M5VP = 4:2 ratio is 75.83%. Such increase in sorption properties of polymers is due to the fact that in result of ionization hydrogels undergo various conformational changes in macromolecular structure (e.g., globe unfolding due to repulsion of charged ions on internode links).

7.3.4 STUDY OF LANTHANUM IONS DESORPTION FROM INDIVIDUAL HYDROGELS MATRIX

Desorption kinetics of lanthanum ions ethyl alcohol and nitric acid from the individual hydrogels ratio (gPAA:gP2M5VP = 1:5) are shown in Figures 7.11 and 7.12.

Desorption kinetics of lanthanum ions by ethanol is presented in Figure 7.11. As can be seen, the overwhelming majority is desorbed after 6 h. Then, there is a further increase of lanthanum ions concentration, which becomes negligible after 24 h. The total desorption degree of lanthanum is 80.17% at 48 h.

Lanthanum ions desorption kinetics by nitric acid is shown in Figure 7.12.

The desorption process is similar to desorption with ethanol—the maximum desorption occurs within 6 h. Consequent increase of lanthanum ions concentration in the desorbent reaches final values at 48 h. Desorption degree is 92.55%.

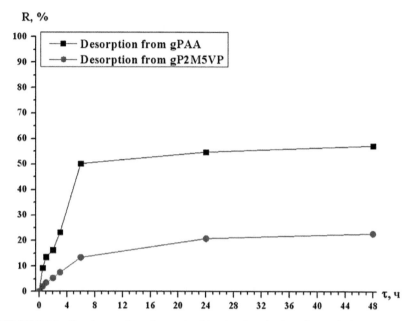

FIGURE 7.11   Lanthanum ions desorption kinetics from hydrogels matrix by ethyl alcohol (96%).

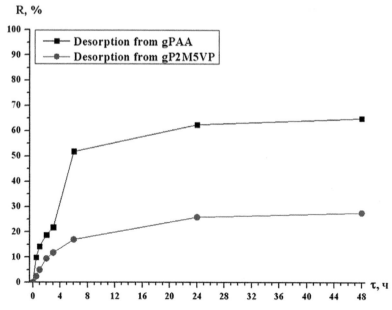

FIGURE 7.12   Lanthanum ions desorption kinetics from hydrogels matrix by 2 M nitric acid.

7.4   CONCLUSION

The obtained results allow us to make the following conclusions:

1.   Result of polymer macromolecules mutual activation is a significant change in electrochemical, conformational properties of hydrogels.
2.   Based on obtained data on specific electric conductivity, pH, and swelling coefficient, it can be concluded that there is an occurrence of lanthanum ions sorption by polymer hydrogels.
3.   Maximum value of lanthanum ions extraction degree (91.09%) is observed at gPAA:gP2M5VP = 4:2 ratio at 48 h of hydrogels remote interaction.
4.   Maximum value polymer-chain binding degree (75.83%) occurs at gPAA:gP2M5VP = 4:2 ratio at 48 h of hydrogels remote interaction.
5.   Desorption degree of lanthanum ions extraction from hydrogels matrix by ethanol is 80.17%. Desorption degree by nitric acid is 92.55%.
6.   Mutual activation of gPAA and gP2M5VP hydrogels in intergel system provides significantly greater sorption degree of lanthanum ions in comparison with individual hydrogels.

ACKNOWLEDGMENT

The work was financially supported by Committee of Science of Ministry of Education and Science of Republic of Kazakhstan.

KEYWORDS

- intergel systems
- polyacrylic acid
- poly-2-methyl-5-vinylpyridine
- hydrogels
- remote interaction
- La^{3+} ions
- sorption
- desorption

REFERENCES

1. Alimbekova, B. T.; Korganbayeva, Zh. K.; Himersen, H.; Kondaurov, R. G.; Jumadilov, T. K. *J. Chem. Chem. Eng.* **2014,** 8 (3), 265.
2. Jumadilov, T. K.; Kaldayeva, S. S.; Kondaurov, R. G.; Erzhan, B.; Erzhet, B. *Proceedings of Symposium ICSP&AM3*, Tbilisi, 2013; p 191.
3. Jumadilov, T. K. *Proceedings of the International Conference of Lithuanian Chemical Society "Chemistry and Chemical Technology"*, Kaunas, 2014; p 226.
4. Alimbekova, B. T.; Jumadilov, T. K.; Korganbayeva, Zh. K.; Erzhan, B.; Erzhet, B. *Bull. d'eurotalent-FIDJIP* **2013,** *5*, 28.
5. Erzhet, B.; Jumadilov, T. K.; Korganbayeva, Zh. K. *Bull. d'eurotalent-FIDJIP* **2013,** *5*, 41.
6. Jumadilov, T. K. *Ind. Kazakhstan* **2011,** *2*, 70.
7. Bekturov, E. A.; Suleimenov, I. E. *Polymer Hydrogels*; Almaty, 1998; p 133.
8. Bekturov, E. A.; Jumadilov, T. K. *News Nat. Acad. Sci. RK. Chem. Ser.* **2009,** *1*, 86.
9. Bekturov, E. A.; Jumadilov, T. K.; Korganbayeva, Zh. K. *KazNU Herald. Chem. Ser.* **2010,** *3* (50), 108.
10. Jumadilov, T.; Shaltykova, D.; Suleimenov, I. *Proceedings of Austrian-Slovenian Polymer Meeting*, Slovenia, 2013; p 51.
11. Jumadilov, T.; Akimov, A.; Eskalieva, G.; Kondaurov, R. *Proceedings of VIII International Scientific-Technical Conference "Advance in Petroleum and Gas Industry and Petrochemistry"* (APGIP-8), LVIV, 2016; p 68.
12. Jumadilov, T. K. *Proceedings of 10th Polyimides & High Performance Polymers Conference*, Montpellier, 2016; p CIV-7.
13. Tereshenkova, A. A.; Statkus, M. A.; Tihomirova, T. I.; Tsizin, G. I. *Mosc. Uni. Herald Chem. Ser.* **2013,** *54* (4), 203.
14. Petruhin, O. M. *Methodology of Physico-chemical Methods of Analysis. Chemistry*, 1987, pp 77–80.
15. Izmailov, N. A. *Electrochemistry of Solutions. Chemistry*, 1976, p 24.

CHAPTER 8

A STUDY ON THE INFLUENCE OF γ-Al$_2$O$_3$ ON POLYVINYL CHLORIDE–POLY(BUTYL METHACRYLATE) NANOCOMPOSITE POLYMER ELECTROLYTES

R. ARUNKUMAR, RAVI SHANKER BABU*, M. USHA RANI, and S. KALAINATHAN

Department of Physics, School of Advanced Sciences, VIT University, Vellore, India

Corresponding author. E-mail: ravina2001@rediffmail.com

CONTENTS

ABSTRACT

Nanocomposite polymer electrolytes (NCPEs) consisting of polyvinyl chloride (PVC), poly(butyl methacrylate) (PBMA), lithium perchlorate (LiClO$_4$), and ethylene carbonate (EC) with various amounts of inorganic filler (γ-Al$_2$O$_3$) have been prepared by solution-casting technique. The prepared nanocomposite films were characterized by X-ray diffraction (XRD), Fourier-transform infrared spectroscopy (FTIR), conductivity studies, dielectric analysis, DC polarization method, and scanning electron microscope analysis. Crystallinity of NCPEs was found to increase beyond 10 wt% of γ-Al$_2$O$_3$ which was confirmed from XRD studies. Complex formation and polymer–salt interaction of NCPEs were identified by FTIR analysis. Conductivity of plasticized polymer electrolytes [PVC (12.5)–PBMA (12.5)–LiClO$_4$ (6)–EC (69)] was observed to be 0.116 m S cm^{-1} at 303 K. On incorporating the inorganic filler, the ionic conductivity of the polymer electrolytes [PVC (12.5)–PBMA (12.5)–LiClO$_4$ (6)–EC (69)–γ-Al$_2$O$_3$ (10)] is increased to 0.711 m S cm^{-1} at 303 K. The temperature-dependent ionic conductivity of NCPEs obeys the Vogel–Tammann–Fulcher relation and frequency-dependent ionic conductivity follows Jonscher's power law. Dielectric constants (ε' and ε'') of different ratios of γ-Al$_2$O$_3$-incorporated PVC–PBMA polymer electrolytes were found to be decreasing with increasing frequency due to existence of electrode polarization. The surface morphology of the NCPEs was confirmed by SEM analysis. Conduction occurs in the NCPEs predominantly due to ions which were confirmed by DC polarization method.

8.1   INTRODUCTION

The present world faces a stringent need in energy sources due to ever increasing essentials of electrical and electronic devices like laptops, cameras, mobile phones, hybrid electric vehicles (EVs), plug-in hybrid EVs, and batteries for EVs.[1] So, to enhance the energy sources, scientist reported on preparation of polymer electrolytes and its application in various electrochemical devices such as batteries, dye-sensitized solar cell, electric double layer capacitor and electrochromic devices, etc. There are several methods to enhance the room-temperature ionic conductivity of solid polymer electrolytes. They are polymer blending, copolymerizing, adding plasticizer, ionic liquids, and inorganic filler.[2] Incorporation of nanosize inorganic filler into the polymer electrolytes is called nanocomposite polymer electrolytes. Nanocomposite polymer electrolytes (NCPEs) have excellent mechanical stability, wide electrochemical stability, high thermal stability, dimensional stability, good interfacial stability, and good compatibility with the electrodes.[3–8]

Generally, high molecular weight aliphatic polymer-based polymer electrolytes have poor mechanical and thermal stability, but the incorporation of inorganic fillers into aliphatic polymers is found to increase both mechanical and thermal stability of the system.[9] The high dielectric constant inorganic filler can enhance the dissociation of salts and decrease the viscosity of composite polymer electrolytes due to ion–ion and ion–polymer interactions. The nanosize ceramic-particle-doped polymer electrolytes have good ionic conductivity and mechanical stability than microsize ceramic particles because nanosize particles have good contact with the electrodes because it has high surface area.[10,11] Usually the ionic conductivity of the polymer electrolyte (gel polymer electrolytes) decreases with increasing the mechanical properties. But in composite polymer electrolytes, the incorporation of inorganic filler was found to increase the mechanical stability and doesn't decrease the ionic conductivity which may be due to the presences of grains in the inorganic filler.[12] The decomposition temperature of inorganic filler-doped polymer electrolytes are high compared to gel polymer electrolytes and it indicates that the composite polymer electrolytes have more thermal stability and are less expensive than the organic additive (plasticizer and ionic liquids)-based polymer electrolytes.

In this chapter, we are focusing on preparation of NCPEs with incorporation of inorganic filler (γ-Al$_2$O$_3$). Dispersion of inorganic filler in the polymer matrix and the formation Lewis acid–base interaction between polymer electrolyte ionic species–polar surface group of inorganic filler are found to increases in the additional sites for ionic migration which improves the ionic conductivity as well as ion transference numbers of the composite polymer electrolytes.[13–15] Further enhancement in interfacial stability between electrodes, decrease in the crystallinity of the polymer matrix, and improved cyclic life improve the physical properties (thermal, dielectric, and mechanical) which are induced by the dispersed inorganic filler.[13–15] Several polymers like polyvinyl chloride (PVC), polyethylene oxide (PEO), polymethyl methacrylate (PMMA), poly(vinylidene fluoride co-hexafluoro propylene) (PVdF–HFP), etc., with inorganic fillers (TiO$_2$, ZrO$_2$, Al$_2$O$_3$, SiO$_2$, and BaTiO$_3$) are found to exhibit enhanced thermal, mechanical, morphological, electrical, and electrochemical properties of the composite polymer matrix.[16]

PVC-based blend polymer electrolytes are extensively studied[9–11] due to its stupendous property of high mechanical and thermal stability. Results based on the studies elucidate that PVC rich phase acts as mechanical stiffener but provides low conductivity. This low conductivity is because of the solid medium where ions find it difficult to penetrate, whereas other combining polymers have high ionic conductivity but poor mechanical stability. PVC is a thermoplastic polymer. Properties dictating the choice of PVC are its (1) good mechanical stability as a result of dipole–dipole interaction between H and Cl atoms,[13] (2) excellent miscibility with other polymers, (3) compatibility with many polymers, and (4) lone pair of electrons at the chlorine atom that helps in solvation of lithium salts.[13] The poor ionic conductivity of PVC-based polymer electrolytes at low frequency hinders the transportation of lithium ions between the electrodes and hence limits their uses in lithium polymer batteries. This problem can be resolved by different methods as discussed earlier. To overcome the disadvantages of PVC electrolytes, PVC blended with other polymers can enhance the electrochemical and mechanical properties. PVC-based blend electrolytes were reported such as PVC–PMMA, PVC–PEMA, PVC–PAN, PVC–PVdF, PVC–P (VdF–HFP), etc.

Methacrylate group (PMMA, PEMA, PPMA, poly(butyl methacrylate) [PBMA])-based electrolytes has special interest in electrochemical device application because of its high transparency, high solvent retention ability, excellent environmental compatibility, good compatibility with most of the polymers, good interfacial stability, and low cost. Furthermore, it has good outdoor weatherability, good resistance to acid, and poor resistance to solvent. Thus, it can be dissolved easily and has good interfacial stability toward the lithium electrodes. It has high ability to solvate the inorganic salt due to the presence of polar functional group in its polymer chain. The presence of lone pair of electron at oxygen atom of PMMA may be to form covalent bond with lithium ions.[18–22] Literature accounts that PVC/PEMA blends have higher stability and a comparative conductivity with that of PVC/PMMA blends.[17] It is hypothesized that PBMA could give higher conductivity compared to PMMA and PEMA, because of its low glass transition temperature and long alkyl chain which could aid conformation of polymer chain ensuring ionic conductivity. PBMA was chosen as the complexing polymer for preparation of polymer blend. PBMA is an amorphous polymer having good dimensional stability with low glass transition temperature (20°C). PBMA-based polymer electrolytes are highly suitable for electrochemical device applications due to its good ionic conductivity at lower temperature itself because of the presence $C=O$ group (electron donation) forming the coordination with cations due to which lithium salt can be easily dissociated in the PBMA matrix which leads to enhancement of the ionic mobility. It has flexible backbone chain due to the presence of bulky ester ($O=C-O-CH_3$) functional group in the polymer matrix[23] which is useful for fabrication of batteries because of flexible-type polymer electrolytes can give required shape, size, and dimension.[23] PBMA acts as a Lewis base (electron donor) and PVC acts as a Lewis acid (electron acceptor).[24]

Aluminum oxide (Al$_2$O$_3$) plays an important role in many places like ceramic, catalytic application, absorbents, abrasives, etc. Al$_2$O$_3$-incorporated composite polymer electrolytes were reported that indicates that Al$_2$O$_3$ mainly influence on the conductivity, mechanical, and thermal stability. Tambelli et al. reported that PEO-based composite polymer electrolytes with addition of α-Al$_2$O$_3$ and γ-Al$_2$O$_3$ which implies that γ-Al$_2$O$_3$-based polymer electrolytes have low activation energy (0.88 eV), average particle size (4700 nm), pore radius (10–250 Å), and good ionic conductivity than α-Al$_2$O$_3$-doped composite polymer electrolytes. The γ-Al$_2$O$_3$-doped polymer electrolytes have enhanced the ionic conductivity up to 25 wt%, but in the case of α-Al$_2$O$_3$, it has decreased the ionic conductivity beyond 5.3 wt% which denotes the blocking effect and variation of γ-Al$_2$O$_3$ because of the particle size and structure.[25–27] So, finally we selected the γ-Al$_2$O$_3$ to prepare the PVC–PBMA composite polymer electrolytes.

In this present communication, the different ratios of inorganic filler (γ-Al$_2$O$_3$)-doped PVC–PBMA blend polymer electrolytes were prepared by solution-casting technique. The structural characters and complex formation of nanocomposite PVC–PBMA polymer electrolytes were confirmed by X-ray diffraction (XRD) studies and Fourier transform infrared (FTIR) analysis, respectively. γ-Al$_2$O$_3$ ratio, temperature, and frequency-dependent ionic conductivity of the NCPEs were studied by AC impedance analysis. γ-Al$_2$O$_3$ ratio and temperature-dependent dielectric constant (ε' and ε'') and dielectric modulus (M' and M'') of NCPEs were analyzed by dielectric studies. Transference number and surface morphology of nanocomposite PVC–PBMA polymer electrolyte membrane were studied by DC polarization techniques and scanning electron microscope analysis, respectively.

8.2 MATERIALS AND METHODS

8.2.1 MATERIALS

PVC with average molecular weight 48,000, PBMA with average molecular weight 337,000, lithium perchlorate (LiClO$_4$), and ethylene carbonate (EC) were procured from Sigma-Aldrich, USA and aluminum oxide (γ-Al$_2$O$_3$) with average particle size (~50 nm) was procured from Alfa-Assar, USA.

8.2.2 ELECTROLYTE PREPARATION

The flowchart for preparation of PVC–PBMA NCPEs is depicted in Figure 8.1. PVC–PBMA NCPEs [PVC (50)–PBMA (50)–LiClO$_4$ (6)–EC (69)–(x) γ-Al$_2$O$_3$] [x = 0 wt% (A1), 5 wt% (A2), 10 wt% (A3), 15 wt% (A4), and 20 wt% (A5)] were prepared by solution-casting technique.

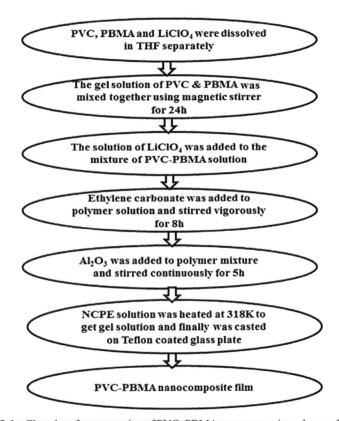

FIGURE 8.1   Flowchart for preparation of PVC–PBMA nanocomposite polymer electrolytes.

8.2.3 XRD STUDIES

XRD studies is a subserve tool to analyze the structural characters of polymers, salts, and NCPEs. Structural characters of nanocomposite PVC–PBMA polymer electrolytes were investigated by XRD studies using BRUKER X-ray diffractometer with Cu Kα radiation in the range of 10°–80° at steps of 0.02°/s.

8.2.4 FTIR ANALYSIS

FTIR spectroscopy is an important tool to investigate the interactions between polymer, lithium salt, and inorganic filler (γ-Al$_2$O$_3$), based on functional group analysis. Functional group, presence of some new peaks, absence of pristine peaks and complex formation of NCPE were confirmed by FTIR analysis using SHIMADZU IR AFFINITY spectrometer under transmittance mode from 4000 to 400 cm^{-1}.

8.2.5 CONDUCTIVITY STUDIES

Impedance analysis is a stupendous tool to investigate the conduction mechanism of NCPEs and how the ionic conductivity is varied due to incorporation of inorganic filler (γ-Al$_2$O$_3$). Ionic conductivity (depends on γ-Al$_2$O$_3$ ratio, temperature, and frequency) of nanocomposite PVC–PBMA polymer electrolyte membranes were studied by AC impedance analysis by using HIOKI 3532-50 LCR Hi TESTER meter at different temperatures (303–373 K) in the frequency range 50 Hz–5 MHz.

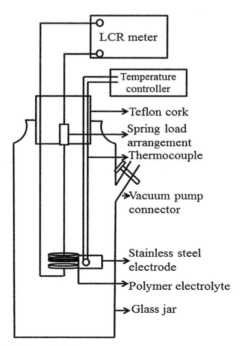

FIGURE 8.2   Conductivity cell setup.

The specially designed conductivity cell setup is depicted in Figure 8.2. The circular NCPE film (1.5 cm diameter) was placed between the stainless steel electrodes (blocking electrode) with a spring load arrangement to ensure good contact between the sample and electrodes. The stainless steel electrodes were connected with LCR meter and temperature was measured by temperature controller. During the conductivity measurement, the sample was kept in vacuum condition with using vacuum creator to avoid contact of the sample with atmosphere. The variation in ionic conductivity of the NCPEs with increasing temperature is recorded in the range of 303–373 K.

As per Ohm's law

$$V = IR; \; R = \frac{V}{I} \tag{8.1}$$

$$R = \rho \frac{L}{A} \; (\rho = 1/\sigma)$$

$$R = \frac{1}{\sigma} \frac{L}{A}$$

$$\sigma = \frac{L}{RA}; \; (R = R_b)$$

The ionic conductivity of NCPE were calculated using the relation

$$\sigma = \frac{L}{R_b A} \tag{8.2}$$

where L is the thickness of the sample measured with using peacock meter, A is the area of the film $(A = nr^2)$, and R_b is the bulk resistance obtained from intercept on X-axis in Cole–Cole plot.

8.2.6 DIELECTRIC STUDIES

Dielectric behavior (ε', ε'', M', and M'') of nanocomposite PVC–PBMA polymer electrolyte membranes were studied by dielectric studies by using HIOKI 3532-50 LCR Hi TESTER meter at different temperatures (303–403 K) in the frequency range 50 Hz–5 MHz. Dielectric behavior (ε', ε'', M', and M'') of polymer matrix are calculated by using the following relations. The complex dielectric constant is

$$\varepsilon^* = \varepsilon' - i\varepsilon'' \tag{8.3}$$

Real part of dielectric constant

$$\varepsilon' = \frac{Cd}{\varepsilon_o A} \tag{8.4}$$

Imaginary part of dielectric constant

$$\varepsilon'' = \frac{\sigma}{\omega\varepsilon_o} \tag{8.5}$$

The complex dielectric modulus

$$M^* = M' - iM'' \tag{8.6}$$

Real part of dielectric modulus

$$M' = \frac{\varepsilon'}{(\varepsilon')_2 + (\varepsilon'')_2} \tag{8.7}$$

Imaginary part of dielectric modulus

$$M'' = \frac{\varepsilon''}{(\varepsilon')_2 + (\varepsilon'')_2} \tag{8.8}$$

where d is the thickness; A is the area of the film; C is the parallel capacitance; σ is the conductivity of the composite polymer electrolytes; ε_o is the permittivity of free space (8.854×10^{-12} F m^{-1}); and ω is the angular frequency ($\omega = 2\pi f$).

8.2.7 DC POLARIZATION METHOD

The transference number measurement setup is shown in Figure 8.3. Transference number of nanocomposite PVC–PBMA polymer electrolyte membrane was calculated by DC polarization techniques by using 6485 Picoammeter, DC power, supply, and conductivity cell. DC voltage (fixed at 1.5 V) is applied across the polymer electrolyte membrane through stain less steel electrodes (blocking electrode) with using DC power supply. DC current is monitored in 6485 Picoammeter in the range of µA with respect to time. During the conductivity measurements, the sample was kept in vacuum condition to avoid contact of the sample with atmosphere. The nanocomposite PVC–PBMA polymer electrolytes to be used in the rechargeable lithium battery applications and the charge carriers occur must be in ions. This present technique confirms the conduction because of the occurrences of ions or electrons. The ionic conductivity should be close to one and electronic conductivity should be negligible or close to zero.

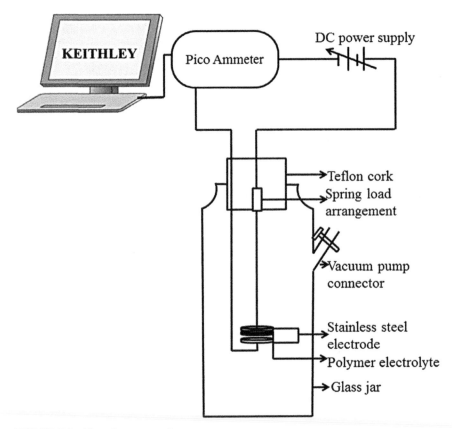

FIGURE 8.3   Transference number measurement setup.

The transference number of PVC–PBMA polymer electrolyte membrane has been calculated by using following relations:

The ionic transference number

$$t_+ = \frac{I_i - I_f}{I_i} \qquad (8.9)$$

The electronic transference number

$$t_- = \frac{I_f}{I_i} \tag{8.10}$$

where I_i is the initial current and I_f is the final current.

8.2.8 SEM ANALYSIS

The surface morphology of nanocomposite PVC–PBMA polymer electrolytes were identified from scanning electron micrographs obtained using Carl Zeiss EVO/185H, UK.

8.3 RESULTS AND DISCUSSION

8.3.1 STRUCTURAL ANALYSIS

XRD pattern of pristine (PVC, PBMA, LiClO$_4$, and γ-Al$_2$O$_3$) and nanocomposite PVC–PBMA-blended polymer electrolytes are shown in Figure 8.4a–i. XRD patterns of pristine PVC, PBMA, LiClO$_4$, and γ-Al$_2$O$_3$ (Fig. 8.4a–d) showing the absence of sharp crystalline peaks (Fig. 8.4a and b) reveal the amorphous nature of pure PVC and PBMA. The sharp diffraction peaks (Fig. 8.4c and d) at $2\theta = 24°$, $28°$, $31°$, $34°$, $40°$, $49°$, $50°$, and $54°$ of LiClO$_4$ and $2\theta = 37°$, $39°$, $42°$, $45°$, and $67°$ of γ-Al$_2$O$_3$ indicate the crystalline nature of LiClO$_4$ and γ-Al$_2$O$_3$ used. The disappearance of sharp peaks of LiClO$_4$ in the NCPE indicate that LiClO$_4$ is complexed in the polymer matrix.[28] Figure 8.4e–i shows a broad halo which mentions that nanocomposite PVC–PBMA polymer electrolytes with different proportions of γ-Al$_2$O$_3$ are amorphous in nature. It is observed that intensity of the broad halo is decreased till 10% of γ-Al$_2$O$_3$ (Fig. 8.4g), and on further addition, broadening of the halo is decreased and also intensity of that halo is increased which denotes the diminution of amorphous phase at higher concentrations of γ-Al$_2$O$_3$. The peak at $67°$ (Fig. 8.4d) of γ-Al$_2$O$_3$ is found to reappear above 10 wt% of γ-Al$_2$O$_3$ indicating the recurrence of crystalline nature of CPE. The prepared NCPEs with 10% of γ-Al$_2$O$_3$ exhibits high ionic conductivity when compared to other films as shown in Figure 8.4.

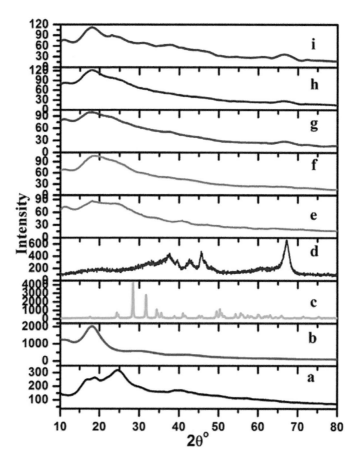

FIGURE 8.4   XRD pattern for (a) PVC, (b) PBMA, (c) LiClO₄, (d) γ-Al₂O₃, and (e–i) complex PVC (12.5)–PBMA (12.5)–LiClO₄ (6)–EC (69)–*X* γ-Al₂O₃ (*X* = 0, 5, 10, 15, and 20 wt%), respectively.[29] (Reprinted with permission from Arunkumar, R.; Babu, R. S.; Usha Rani, M. Investigation on Al2O3 doped PVC–PBMA blend polymer electrolytes, J. Mater. Sci.: Mater. Electron. DOI:10.1007/s10854-016-5924-0. © 2017, Springer.)

8.3.2   FUNCTIONAL GROUP ANALYSIS

The FTIR spectra of pure PVC, PBMA, LiClO₄, EC, and nanocomposite PVC–PBMA-blended polymer electrolytes with different concentrations of γ-Al₂O₃ are shown in Figure 8.5a–i. The bands are assigned for pristine PVC; symmetric C–H stretching [ν(C–H)] at 2920 cm⁻¹, CH₂ symmetric stretching [ν(CH₂)] at 2848 cm⁻¹, symmetric C=C stretching [ν(C=C)] at 1630 cm⁻¹, CH rocking [ρ(C–H)] at 1260 cm⁻¹, *trans*-CH wagging [ω(CH)] at 989 cm⁻¹, symmetric C–Cl stretching [ν_s(C–Cl)] at 831 cm⁻¹, and γ(C–Cl) stretching [γ-ν(C–H)] at 607 cm⁻¹. The group frequencies are assigned for pure PBMA, C–H stretching [ν(C–H)] at 2958 cm⁻¹, antisymmetric C–H stretching [ν$_{as}$(C–H)] at 2873 cm⁻¹, symmetric C=O stretching [ν(C=O)] at 1720 cm⁻¹, CH₂ twisting [τ(CH₂)] at 1382 cm⁻¹, C–O stretching [ν(C–O)] at 1265 cm⁻¹, C–O–C antisymmetric stretching [ν$_{as}$(C–O–C)] at 1141 cm⁻¹, CH₂ twisting [τ(CH₂)] at 1062 cm⁻¹, C–O stretching [ν(C–O)] at 963 cm⁻¹, and CH₂ rocking [ρ(CH₂)] at 748 cm⁻¹. The vibrational bands at 1070, 939, and 630 cm⁻¹ (Fig. 8.5c) correspond to the ClO₄⁻ anion of LiClO₄ salt. The vibrational bands of pristine EC (Fig. 8.5d) appearing at 2931, 1068, and 717 cm⁻¹ are assigned to CH₂ scissoring [δ(CH₂)], C–O symmetric stretching [ν(C–O)], and CH₂ rocking [ρ(CH₂)], respectively.

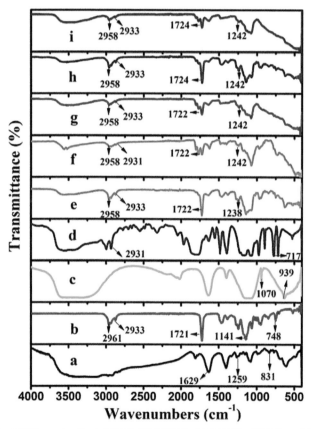

FIGURE 8.5 FTIR spectra for (a) PVC, (b) PBMA, (c) LiClO₄, (d) EC, and (e–i) complex PVC (12.5)–PBMA (12.5)–LiClO₄ (6)–EC (69)–X γ-Al₂O₃ (X = 0, 5, 10, 15, and 20 wt%), respectively.[29] (Reprinted with permission from Arunkumar, R.; Babu, R. S.; Usha Rani, M. Investigation on Al2O3 doped PVC–PBMA blend polymer electrolytes, J. Mater. Sci.: Mater. Electron. DOI:10.1007/s10854-016-5924-0. © 2017, Springer.)

The C–H stretching at 2920 cm⁻¹, C=C symmetric stretching at 1630 cm⁻¹, CH rocking at 1260 cm⁻¹, and C–Cl stretching at 831 cm⁻¹ of pristine PVC are found to be shifted in the complexes around 2931–2933, 1625–1631, 1238–1242, and 842–844 cm⁻¹, respectively. C–H stretching at 2958 cm⁻¹, CH antisymmetric stretching at 2873 cm⁻¹, C=O stretching at 1720 cm⁻¹, and CH_2 rocking at 748 cm⁻¹ of pristine PBMA are shifted to the complexes around 2958, 2870–2877, 1722–1724, and 721–748 cm⁻¹, respectively. CH_2 scissoring of pristine EC at 2931 cm⁻¹ is shifted to slightly higher frequency region at 2933 cm⁻¹ in all the complexes and CH_2 rocking at 717 cm⁻¹ of EC is shifted to 721, 727, 745, and 748 cm⁻¹ in the complexes corresponding to the film A1–A4 and it is absent in A5. The disappearances of some pristine peaks in the complexes at 2848 cm⁻¹ [CH_2 antisymmetric stretching], 1412 cm⁻¹ [CH bending], 1265 cm⁻¹ [C–H rocking], 1061 cm⁻¹ [CH bending], 989 cm⁻¹ [*trans*-C–H wagging], and 607 cm⁻¹ [symmetric C–Cl stretching] and appearance of some new peaks in the complexes at 1770, 1475, 1176, 970, and 435 cm⁻¹ denote the confirmation of complex formations (Table 8.1).[30]

TABLE 8.1   Assignments of Vibrational Modes of Pristine PVC, PBMA, and PVC–PBMA Complexes.

Assignments	PVC	PBMA	PVC–PBMA complexes
ν(C–H)		2958	2958
ν(C–H)	2920	–	2931–2933
$ν_{as}(CH_2)$		2873	2870–2877
$ν_s(CH_2)$	2848	–	–
$ν_s(C=O)$		1720	1722–1724
ν(C=C)	1630	–	1625–1631
$τ(CH_2)$		1382	–
ν(C–O)		1265	–
ρ(C–H)	1260	–	1238–1242
$τ(CH_2)$		1062	–
ω(CH	989	–	–
ν(C–O)		963	–
$ν_s(C–Cl)$	831	–	842–844
$ρ(CH_2)$	–	748	721–748
γ-ν(C–H)	607	–	–

8.3.3   CONDUCTIVITY STUDIES

8.3.3.1   γ-Al$_2$O$_3$ RATIO-DEPENDENT IONIC CONDUCTIVITY

The ionic conductivity of nanocomposite PVC–PBMA-blended polymer electrolytes as a function of filler concentrations are shown in Figure 8.6. The ionic conductivity of NCPE bereft of γ-Al$_2$O$_3$ (film A1) is obtained as 0.116×10^{-3} S cm^{-1} at 303 K, while the γ-Al$_2$O$_3$ incorporated into the polymer electrolytes which increases with its conductivity to 0.712×10^{-3} S cm^{-1} at 303 K for film A3 (with 10% γ-Al$_2$O$_3$). According to Ratner et al., ionic conduction in the polymer electrolytes occurs easily in amorphous phase and it has been accomplished by the incorporation of nanocomposite. The addition of nanocomposite into polymer electrolytes promotes the formation of Lewis acid–base interaction between the nanocomposite–lithium salts (ClO$_4^-$) segments of polymer, dispersion of γ-Al$_2$O$_3$ stimulates to the augments of the amorphous phase and diminution of crystallinity to produce

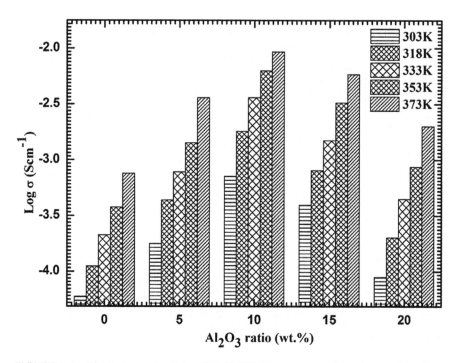

FIGURE 8.6   The ionic conductivity of PVC–PBMA nanocomposite polymer electrolytes dependence on γ-Al$_2$O$_3$ ratio.

the liberty ways for ionic migrations in the NCPE.[31] The conductivity of the polymer matrix can also be viewed based on the following equation:

$$\sigma = \Sigma_i \, n_i Z_i e \mu_i \qquad (8.11)$$

where n_i is the number of charge carrier density (cm^{-3}), Z_i is the valency of charge carriers, e is the elementary electron charge, and μ_i is the ionic mobility (cm^2 V^{-1} s^{-1}) of the polymer matrix. Ionic conductivity is found to increase with increase in filler concentration due to the specific interaction of the ceramic surfaces promoting fast ion transport, surpassing the dilution effect. At higher filler content, the dilution effect predominates resulting in the lowering of conductivity. γ-Al$_2$O$_3$ ratio-dependent ionic conductivity of NCPEs is given in Table 8.2.

8.3.3.2   TEMPERATURE-DEPENDENT IONIC CONDUCTIVITY

Temperature-dependent ionic conductivity of nanocomposite PVC–PBMA electrolytes is shown in Figure 8.7. The ionic conductivity of NCPE is found to increase with increase in temperature due to the effortless expansion of

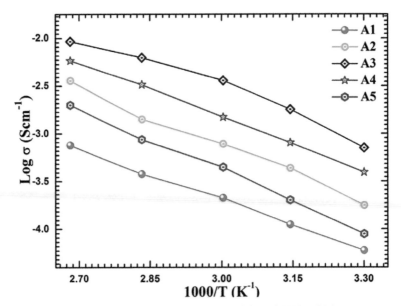

FIGURE 8.7   Temperature-dependent ionic conductivity of PVC–PBMA nanocomposite polymer electrolytes.[29] (Reprinted with permission from Arunkumar, R.; Babu, R. S.; Usha Rani, M. Investigation on Al2O3 doped PVC–PBMA blend polymer electrolytes, J. Mater. Sci.: Mater. Electron. DOI:10.1007/s10854-016-5924-0. © 2017, Springer.)

polymer electrolytes producing more free volume for segmental motion. In that free volume, ionic transport occurs easily and hence enhancing the ionic conductivity of NCPE at higher temperatures. Presence of nonlinearity line in the temperature-dependent ionic conductivity of the nanocomposite PVC–PBMA polymer electrolytes plot obeys the Vogel–Tammann–Fulcher (VTF) relation and it confirms the ionic conductivity which occurs due to migration of ions in a viscous matrix.[32] The VTF relation is

$$\sigma = AT^{-1/2}\exp(\frac{-B}{T-T_0})$$
(8.12)

where A is the pre-exponential factor to the number of charge carriers, T_0 is the temperature at which the conductance tends to zero, T is the absolute temperature, and B is the activation energy. Ionic conductivity for 10 wt% of γ-Al$_2$O$_3$-doped PVC–PBMA polymer electrolyte is found to be 0.712 m S cm^{-1} at 303 K and it increases to 9.265 m S cm^{-1} at 373 K. Ionic conductivities of NCPE are varied two orders of magnitude with increase in temperature (Table 8.2).

TABLE 8.2 Ionic Conductivity for γ-Al$_2$O$_3$-Incorporated PVC–PBMA Nanocomposite Polymer Electrolytes.[29]

Sample code	γ-Al$_2$O$_3$ ratio (wt%)	Ionic conductivity 10^{-3} (S cm^{-1})				
		303 K	318 K	333 K	353 K	373 K
A1	0	0.116	0.189	0.214	0.371	0.742
A2	5	0.178	0.435	0.781	1.415	3.597
A3	10	0.712	1.801	3.611	6.271	9.265
A4	15	0.393	0.805	1.491	3.261	5.823
A5	20	0.089	0.202	0.445	0.867	1.998

(Reprinted with permission from Arunkumar, R.; Babu, R. S.; Usha Rani, M. Investigation on Al2O3 doped PVC–PBMA blend polymer electrolytes, J. Mater. Sci.: Mater. Electron. DOI:10.1007/s10854-016-5924-0. © 2017, Springer.)

8.3.3.3 FREQUENCY-DEPENDENT IONIC CONDUCTIVITY

Frequency-dependent ionic conductivity for 10 wt% of γ-Al$_2$O$_3$-incorporated PVC–PBMA composite polymer electrolytes at different temperatures (303–373 K) is depicted in Figure 8.8. Figure 8.8 implies that ionic conductivity is found to increase with increase in frequency and temperature. Typically, frequency-dependent ionic conductivity shows three different regions: low, intermediate, and high-frequency regions. The conductivity of the nanocomposite PVC–PBMA polymer electrolytes has poor ionic conductivity at low frequency region due to presence of space charge polarization effect at the blocking electrodes. The drop-in conductivity at low frequency may be due to drop-in mobile ions as a result of accumulation of charges at electrode electrolyte interface.[33] At intermediate frequency, the conductivity is found to be virtually frequency independent and the extrapolation of the plot to zero frequency that gives the value of DC conductivity at all temperatures. The consequence of high frequency leads to high mobility of charge carriers resulting in the increased conductivity.[34,35]

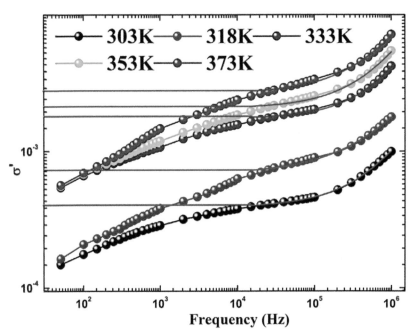

FIGURE 8.8   The frequency-dependent ionic conductivity for 10 wt% γ-Al₂O₃-doped PVC–PBMA composite polymer electrolytes at different temperatures.

AC conductivity of the PVC–PBMA blend polymer electrolytes obeys the Jonscher's power law[36]

$$\sigma_{AC} = \sigma_{DC} + A\omega_n \qquad (8.13)$$

where σ_{DC} is the DC conductivity, A is the pre-exponential factor, ω is the angular frequency ($\omega = 2nf$) and n is the fractional exponent. The values of σ_{DC}, A, and n for 10 wt% of γ-Al₂O₃-doped PVC–PBMA polymer electrolytes at different temperatures have been calculated by using (Origin 8.5) nonlinear curve fitting relation, $y = a + b\,x^c$. The values of σ_{DC}, A, and n are tabulated in Table 8.3. The DC conductivity of the polymer matrix varies from 3.873×10^{-4} to 2.671×10^{-3} S cm^{-1} at 303–373 K. Conductivity increases with increasing temperature may be due to the increased mobility of ion increases because of increases segmental motion of the polymer chain and ion concentration increases with increase in the electrolyte temperature which may produce more free volume in the polymer matrix. The exponential factor (n) varies from 1.010 to 0.775 with change in temperature, which confirms that nanocomposite PVC–PBMA polymer electrolytes are predominantly an ionic conductor.[37]

TABLE 8.3 The Values of σ_{DC}, A, and n for PVC (12.5)–PBMA (12.5)–LiClO$_4$ (6)–EC (69)–γ-Al$_2$O$_3$ (10) Complex Polymer Electrolytes at Different Temperatures.

Temperature (K)	$\sigma_{DC} \times 10^{-3}$ (S cm^{-1})	A	n
303	0.3873	2.067×10^{-9}	0.912
318	0.6885	2.431×10^{-8}	0.775
333	1.750	2.070×10^{-9}	1.010
353	2.080	6.766×10^{-9}	0.947
373	2.671	1.865×10^{-8}	0.894

8.3.4 DIELECTRIC STUDIES

8.3.4.1 REAL AND IMAGINARY PARTS OF DIELECTRIC CONSTANT

The real and imaginary parts of the dielectric constant (ε' and ε'') for different concentrations of γ-Al$_2$O$_3$-incorporated PVC–PBMA composite polymer electrolytes with the function of frequency are depicted in Figures 8.9a–e and 8.10a–e, respectively. Both real (ε') and imaginary (ε'') parts of dielectric constant of composite polymer electrolytes increase with the increase in the filler concentration up to 10 wt% of γ-Al$_2$O$_3$ due to higher charge carrier density.[38] Furthermore, they are decreased at higher concentration of γ-Al$_2$O$_3$; the decrease in the dielectric behavior of composite polymer electrolytes for higher concentration of γ-Al$_2$O$_3$ (15 and 20 wt%) is due to the decrease in the charge carrier density compared to 10 wt% of γ-Al$_2$O$_3$ concentration.[38] The real part of dielectric constant of PVC–PBMA composite polymer electrolytes was found to increase from 110,242 (A1) to 758,267 (A3) at 373 K; beyond that it decreases to 174,424 (A5) at 373 K.

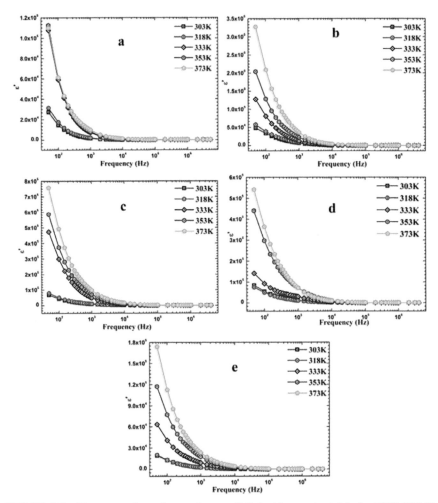

FIGURE 8.9 Frequency-dependent real part of dielectric constant (ε') for PVC–PBMA nanocomposite polymer electrolytes (a) A1, (b) A2, (c) A3, (d) A4, and (e) A5.

The real (ε') and imaginary (ε'') parts of dielectric constant for all γ-Al$_2$O$_3$ ratio-doped PVC–PBMA composite polymer electrolytes at different temperatures are maximum at low frequency which decreases gradually with increasing frequency and tends almost to zero at higher frequency. This signifies the existence of electrode polarization effect.[39] The real and imaginary parts of dielectric constant of the composite polymer electrolytes increases with increasing temperature (303–373 K) due to the increase in the flexibility as well as segmental motion of the polymer chain. The ionic conductivity of the composite polymer electrolytes increases with the increase in temperature signifying the thermal agitation. Due to the thermal activation, high value of dielectric constant composite polymer electrolytes favors in the higher ionic conductivity. This indicates that the ionic mobility of charge carriers increases with increasing temperature due to the dissociation in lithium salts, which influences in the transportation of ions between the electrodes.[40] The temperature-dependent dielectric constant (ε') of 10 wt% of γ-Al$_2$O$_3$-incorporated PVC–PBMA electrolytes are found to be increasing with increase in the temperature from 67,522 to 758,267 at 303–373 K.

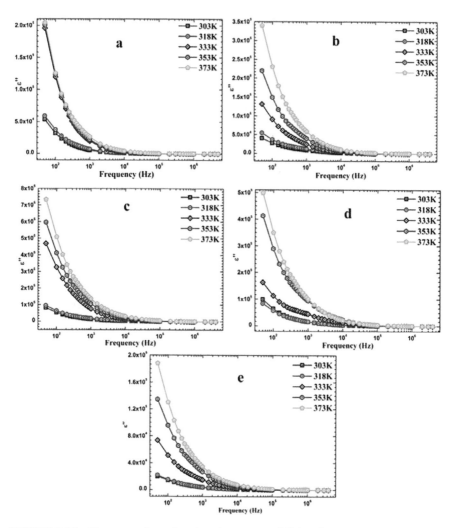

FIGURE 8.10 Frequency-dependent imaginary part of dielectric constant (ε'') for PVC–PBMA nanocomposite polymer electrolytes (a) A1, (b) A2, (c) A3, (d) A4, and (e) A5.

8.3.4.2   REAL AND IMAGINARY PARTS OF DIELECTRIC MODULUS

The real (M') and imaginary (M'') modulus for different ratios γ-Al$_2$O$_3$-doped PVC–PBMA composite polymer electrolytes with the function of frequency are shown in Figures 8.11a–e and 8.12a–e, respectively. Dielectric modulus of both real and imaginary parts decreases with increase in γ-Al$_2$O$_3$ ratio up to 10 wt% beyond that the ratio starts to increase at higher ratio of γ-Al$_2$O$_3$ (15 and 20 wt%) may be accumulation of charge carriers. The presence of long tail at low frequency for both real and imaginary modulus for NCPEs

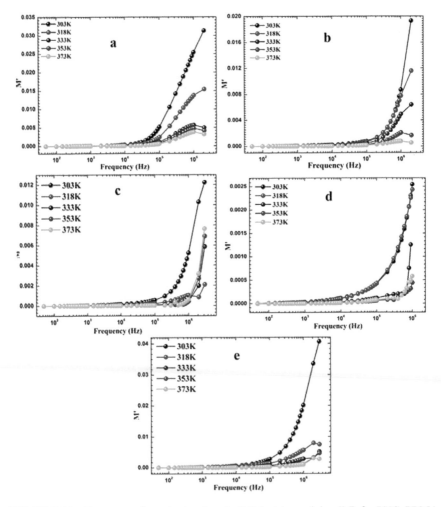

FIGURE 8.11   Frequency-dependent real part of dielectric modulus (M') for PVC–PBMA nanocomposite polymer electrolytes (a) A1, (b) A2, (c) A3, (d) A4, and (e) A5.

could be because of large capacitance associated with the electrode–electrolyte interface. The peak intensity for both real and imaginary modulus is high for lower temperature at higher frequency regions. The peak intensity of the dielectric modulus decrease with increasing temperature could be due to the presence of plurality of relaxation mechanism.[41–43] Among these different ratios of γ-Al₂O₃-doped PVC–PBMA composite polymer, electrolyte modulus (M' and M'') spectrum illustrates that the film A3 has low intensity modulus and it signifies the high ionic conductivity than other films.

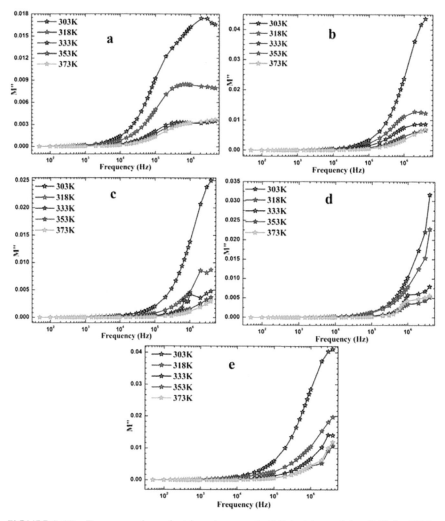

FIGURE 8.12 Frequency-dependent imaginary part of dielectric modulus (M'') for PVC–PBMA nanocomposite polymer electrolytes (a) A1, (b) A2, (c) A3, (d) A4, and (e) A5.

8.3.5   TRANSFERENCE NUMBER MEASUREMENT

DC polarization method is used to calculate the transference number of PVC–PBMA composite polymer electrolyte. The ionic (t_+) and electronic transference number (t_-) of different concentrations of γ-Al$_2$O$_3$-incorporated PVC–PBMA polymer electrolytes have been calculated and values are shown in Table 8.4. The plot for polarization current versus time for best room temperature ionic conductivity film A3 [PVC (50)–PBMA (50)–LiClO$_4$ (6)–EC (69)–γ-Al$_2$O$_3$ (10)] is depicted in Figure 8.13. The ionic and electronic transference numbers (t_+) of PVC–PBMA NCPEs are found to be in the range of 0.88–0.95 and 0.04–0.11, respectively. Ionic transference number is too high compared to electronic transference number. This implies that charge transport occurs in this polymer electrolyte membrane predominantly due to ions.[44,45]

TABLE 8.4   Transport Parameters of Nanocomposite PVC–PBMA Polymer Electrolytes at 303 K.

Sample code	Current (µA)		Transference number	
	I_i	I_f	t_+	t_-
A1	18.3	2.1	0.88	0.114
A2	20.8	1.7	0.91	0.086
A3	24.2	1.2	0.95	0.049
A4	21.6	1.4	0.93	0.064
A5	19.7	1.8	0.90	0.091

8.3.6   MORPHOLOGY ANALYSIS

Scanning electron micrographs of composite PVC–PBMA–LiClO$_4$–EC–X (γ-Al$_2$O$_3$) (X = 0, 10, and 20 wt%) polymer electrolyte membranes at 2000× magnifications is shown in Figure 8.14a–c. Polymer electrolyte with 0 wt% of γ-Al$_2$O$_3$ (Fig. 8.14a) shows smooth and ununiform-sized pores helps in trapping plasticizer ensuing conductivity. The appearance of pores in polymer electrolytes may be due to evaporation of solvent under vacuum. The presence of pores in the polymer matrix can assists in ionic migration between cathode to anode during discharging and vice versa at charging. SEM images of composite polymer electrolyte membranes with 10 wt% of γ-Al$_2$O$_3$ are shown in Figure 8.14b. The smooth edged

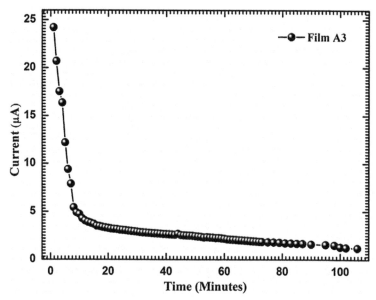

FIGURE 8.13 The polarization current versus time for 10 wt% of γ-Al$_2$O$_3$-doped PVC–PBMA polymer electrolytes (A3) at room temperature (303 K).

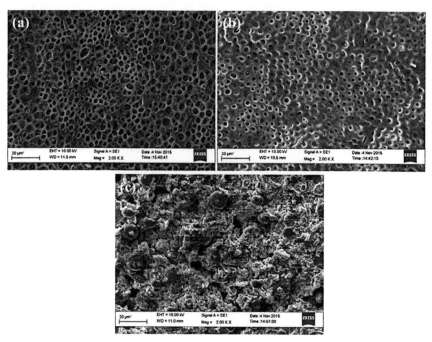

FIGURE 8.14 SEM images for PVC–PBMA nanocomposite polymer electrolyte: (a) filler free membrane (A1), (b) 10 wt% γ-Al$_2$O$_3$, and (c) 20 wt% γ-Al$_2$O$_3$.

and increased uniform pore size could help in higher entrapment of plasticizer may be responsible for the increased ionic conductivity. Figure 8.14c shows polymer electrolyte membrane with 20 wt% of γ-Al$_2$O$_3$. It is evident from the figure that at higher concentration of γ-Al$_2$O$_3$, the pores are found to be distorted. This may be due to the particles occupying the pores or interaction between particles. This aggregation of γ-Al$_2$O$_3$ content (20 wt%) blocks the ionic transportation between the electrodes leading to decreased ionic conductivity.[46] No phase separated morphology is obtained in the films indicating the complete miscibility and good compatibility between polymers, salt, and plasticizer.

8.4 CONCLUSION

In this chapter, we have discussed the influence of γ-Al$_2$O$_3$ on PVC–PBMA composite polymer electrolytes. PVC–PBMA polymer electrolytes with addition of different ratios of γ-Al$_2$O$_3$ were prepared by solution-casting technique. Structural characters and complex formation of NCPE are confirmed by XRD and FTIR analysis, respectively. XRD studies revealed the amorphous phase present in all the NCPE and which ensures ionic conduction. The shift, presence, and absence of peaks in NCPE when comparing with those of pristine constituents lead to the conclusion of the occurrence of polymer–plasticizer, polymer–polymer, and polymer–salt interaction. This has been ascertained from FTIR studies. Among the different ratios of γ-Al$_2$O$_3$, Film A3 (10 wt% of γ-Al$_2$O$_3$) has best ionic conductivity which is found to be 0.712 m S cm^{-1} at 303 K and the obtained nonlinearity in temperature dependence of ionic conductivity could be explained based on the free volume model and obeys Vogel–Tammann–Fulcher relation. NCPEs film A3 [PVC (50)–PBMA (50)–LiClO$_4$ (6)–EC (69)–γ-Al$_2$O$_3$ (10)] has high dielectric constant [758,267 (50 Hz) at 100 K]. The conduction occurs in the nanocomposite PVC–PBMA polymer electrolytes which are predominantly due to ions was confirmed by DC polarization method. Finally, we conclude that 10 wt% γ-Al$_2$O$_3$-incorporated PVC–PBMA NCPEs is suitable for lithium ion battery applications because of its high amorphicity, ionic conductivity, and dielectric constant than its contemporaries.

KEYWORDS

- **nanocomposites**
- **poly(butyl methacrylate)**
- **structural analysis**
- **DC polarization technique**
- **AC impedance analysis**

REFERENCES

1. Bernhard, R.; Latini, A.; Panero, S.; Scrosati, B.; Hassoun, J. Poly(Ethylenglycol) Dimethylether–Lithium Bis(trifluoromethanesulfonyl) Imide, PEG500DME–LiTFSI, as High Viscosity Electrolyte for Lithium Ion Batteries. *J. Power Sources* **2013**, *226*, 329–333.

2. Marcinek, M.; Syzdek, J.; Marczewski, M.; Piszcz, M.; Niedzicki, L.; Kalita, M.; Plewa-Marczewska, A.; Bitner, A.; Wieczorek, P.; Trzeciak, T.; Kasprzyk, M.; Łężak, P.; Zukowska, Z.; Zalewska, A.; Wieczorek, W. Electrolytes for Li-Ion Transport—Review. *Solid State Ion.* **2015**, *276*, 107–126.

3. Rahman, I. A.; Padavettan, V. Synthesis of Silica Nanoparticles by Sol–Gel: Size-Dependent Properties, Surface Modification, and Applications in Silica–Polymer Nanocomposites—A Review. *J. Nanomater.* **2012**, *2012*, 8.

4. Scrosati, B.; Croce, F.; Panero, S. Progress in Lithium Polymer Battery R&D. *J. Power Sources* **2001**, *100*, 93–100.

5. Stephan, A. M. In *Recent Advances in Polymer Nanocomposites*; Thomas, S., Zaikov, G. E., Valsaraj, S. V., Eds.; Brill NV: Leiden, Netherlands, 2009.

6. Ramanavicius, A. In *Progress in Polymers Nanocomposites Research*; Thomas, S., Zaikov, G. E., Eds.; Nova Publishers: Hauppauge, NY, 2008.

7. Zaikov, G. E.; Thomas, S. In *Polymer Nanocomposite Research Advances*; Thomas, S., Zaikov, G. E., Eds.; Nova Publishers: Hauppauge, NY, 2008.

8. Subramaniam, R. T.; Rajantharan, R. S. In *Characterization of Polymer Blends*; Thomas, S., Grohens, Y., Jyotishkumar, P., Eds.; Wiley-VCH Verlag GmbH & Co. KGaA: Weinheim, Germany, 2015.

9. Bruce, P. G.; Scrosati, B.; Tarascon, J. M. Nanomaterials for Rechargeable Lithium Batteries. *Angew. Chem. Int. Ed.* **2008**, *47*, 2930–2946.

10. Lim, C. S.; Ramesh, S.; Majid, S. R. The Effect of Antimony Trioxide on Poly(Vinyl Alcohol)–Lithium Perchlorate Based Polymer Electrolytes. *Ceram. Int.* **2013**, *39*, 745–752.

11. Awadhia, A.; Patel, S. K.; Agrawal, S. L. *Prog. Cryst. Growth Charact. Mater.* **2006**, *52*, 61.

12. Quartarone, E.; Mustarelli, P.; Magistris, A. PEO-Based Composite Polymer Electrolytes. *Solid State Ion.* **1998**, *110*, 1–14.

13. Panero, S.; Scrosati, B.; Sumathipala, H. H.; Wieczorek, W. Dual-Composite Polymer Electrolytes with Enhanced Transport Properties. *J. Power Sources* **2007,** *167,* 510–514.

14. Prasanth, R.; Shubha, N.; Hng, H. H.; Srinivasan, M. Effect of Nano-Clay on Ionic Conductivity and Electrochemical Properties of Poly(Vinylidene Fluoride) Based Nanocomposite Porous Polymer Membranes and their Application as Polymer Electrolyte in Lithium Ion Batteries. *Eur. Polym. J.* **2013,** *49,* 307–318.

15. Croce, F.; Appetecchi, G. B.; Persi, L.; Scrosati, B. Nanocomposite Polymer Electrolytes for Lithium Batteries. *Nature* **1998,** *394,* 456–458.

16. Croce, F.; Sacchetti, S.; Scrosati, B. Advanced, High-Performance Composite Polymer Electrolytes for Lithium Batteries. *J. Power Sources* **2006,** *161,* 560–564.

17. Han, H. S.; Kang, H. R.; Kim, S. W.; Kim, H. T. Phase-Separated Polymer Electrolyte Based on Poly(Vinyl Chloride)/Poly(Ethyl Methacrylate) Blend. *J. Power Sources* **2002,** *112,* 461–468.

18. Rajendran, S.; Babu, R. S. Ionic Conduction Behavior in PVC–PEG Blend Polymer Electrolytes upon the Addition of TiO_2. *Ionics* **2009,** *15,* 61–66.

19. Mendoza, N.; Paraguay-Delgado, F.; Hechavarría, L.; Nicho, M. E.; Hu, H. Nanostructured Polyethylene Glycol–Titanium Oxide Composites as Solvent-Free Viscous Electrolytes for Electrochromic Devices. *Solar Energy Mater. Solar Cells* **2011,** *95,* 2478–2484.

20. Panero, S.; Scrosati, B.; Baret, M.; Cecchini, B.; Masetti, E. Electrochromic Windows Based on Polyaniline, Tungsten Oxide and Gel Electrolytes. *Solar Energy Mater. Solar Cells* **1995,** *39,* 239–246.

21. Vondrák, J.; Reiter, J.; Velická, J.; Sedlaříková, M. PMMA-Based Aprotic Gel Electrolytes. *Solid State Ionics* **2004,** *170,* 79–82.

22. Agnihotry, S. A.; Ahmad, S.; Gupta, D.; Ahmad, S. Composite Gel Electrolytes Based on Poly(Methylmethacrylate) and Hydrophilic Fumed Silica. *Electrochim. Acta* **2004,** *49,* 2343–2349.

23. Sengwa, R. J.; Choudhary, S. Dielectric Properties and Fluctuating Relaxation Processes of Poly(Methyl Methacrylate) Based Polymeric Nanocomposite Electrolytes. *J. Phys. Chem. Solids* **2014,** *75,* 765–774.

24. Bhattacharyya, R.; Roy, N.; Chakraborty, D. Mechanical, Thermomechanical, and Morphology of PVC/PBMA Blends and Full IPNs. *J. Appl. Polym. Sci.* **2006,** *99,* 2033–2038.

25. Tambelli, C. C.; Bloise, A. C.; Rosa´rio, A. V.; Pereira, E. C.; Magon, C. J.; Donoso, J. P. Characterisation of PEO–Al_2O_3 Composite Polymer Electrolytes. *Electrochim. Acta* **2002,** *47,* 1677–1682.

26. Nunes-Pereira, J.; Costa, C. M.; Lanceros-Mendez, S. Polymer Composites and Blends for Battery Separators: State of the Art, Challenges and Future Trends. *J. Power Sources* **2015,** *281,* 378–398.

27. Siqi, B. L.; Qingbai, T.; Chuangsheng, J.; Xu, C.; Shengliang, C.; Yan, L. X. Preparation and Characterization of PEO–PMMA Polymer Composite Electrolytes Doped with Nano-Al_2O_3. *Electrochim. Acta* **2015,** *169,* 334–341.

28. Rajendran, S.; Babu, R. S.; Usha Rani, M. *Bull. Mater. Sci.* **2012,** *34,* 1525–1530.

29. Arunkumar, R.; Babu, R. S.; Usha Rani, M. Investigation on Al2O3 doped PVC–PBMA blend polymer electrolytes. *J. Mater. Sci.: Mater. Electron.* DOI:10.1007/s10854-016-5924-0.

30. Vien, D. L.; Colthup, N. B.; Fateley, W. G.; Grasselli, J. G. *Infrared and Raman Characteristic Frequencies of Organic Molecules*; Academic Press: New York, 1991; p 85.

31. Rhoo, H.-J.; Kim, H.-T.; Park, J.-K.; Hwang, T.-S. *Elecrochim. Acta* **1997**, *42*, 1571–1579.

32. Capiglia, C.; Saito, Y.; Yamamoto, H.; Kageyama, H.; Mustarelli, P. Transport Properties and Microstructure of Gel Polymer Electrolytes. *Electrochim. Acta* **2000**, *45*, 1341–1345.

33. Gogulamurali, N.; Suthanthiraraj, S. A., Maruthamuthu, P. In *Solid State Ionics: Materials and Applications*; Chowdari, B. V. R., Chandra, S., Singh, S., Srivastava, P. C., Eds.; World Scientific: Singapore, 1992.

34. Venkateswarlu, M.; Satyanarayana, N. AC Conductivity Studies of Silver Based Fast Ion Conducting Glassy Materials for Solid State Batteries. *Mater. Sci. Eng.: B* **1998**, *54*, 189–195.

35. Polu, A. R.; Kumar, R.; Rhee, H. W. Magnesium Ion Conducting Solid Polymer Blend Electrolyte Based on Biodegradable Polymers and Application in Solid-State Batteries. *Ionics* **2015**, *21*, 125–132.

36. Sengwa, R. J.; Dhatarwal, P.; Choudhary, S. Effects of Plasticizer and Nanofiller on the Dielectric Dispersion and Relaxation Behaviour of Polymer Blend Based Solid Polymer Electrolytes. *Curr. Appl. Phys.* **2015**, *15*, 135–143.

37. Das, A.; Thakur, A. K.; Kumar, K. Conductivity Scaling and Near-Constant Loss Behavior in Ion Conducting Polymer Blend. *Solid State Ionics* **2014**, *268*, 185–190.

38. Kumar, D. A.; Selvasekarapandian, S.; Baskaran, R.; Savitha, T.; Nithya, H. Thermal, Vibrational and AC Impedance Studies on Proton Conducting Polymer Electrolytes Based on Poly(Vinyl Acetate). *J. Non-Cryst. Solids* **2012**, *358*, 531–536.

39. Tripathi, N.; Thakur, A. K.; Shukla, A.; Marx, D. T. Ion Transport Study in Polymer–Nanocomposite Films by Dielectric Spectroscopy and Conductivity Scaling. *Phys. B: Condens. Matter* **2015**, *468*, 50–56.

40. Tripathi, S. K.; Gupta, A.; Kumari, M. Studies on Electrical Conductivity and Dielectric Behaviour of PVdF–HFP–PMMA–NaI Polymer Blend Electrolyte. *Bull. Mater. Sci.* **2012**, *35*, 969–975.

41. Mishra, R.; Baskaran, N.; Ramakrishnan, P. A.; Rao, K. J. Lithium Ion Conduction in Extreme Polymer in Salt Regime. *Solid State Ionics* **1998**, *112*, 261–273.

42. Nithya, H.; Selvasekarapandian, S.; Kumar, D. A.; Sakunthala, A.; Hema, M.; Christopherselvin, P.; Sanjeeviraja, C. Thermal and Dielectric Studies of Polymer Electrolyte Based on P(ECH–EO). *Mater. Chem. Phys.* **2011**, *126*, 404–408.

43. Ramesh, S.; Arof, A. K. Ionic Conductivity Studies of Plasticized Poly(Vinyl Chloride) Polymer Electrolytes. *Mater. Sci. Eng.: B* **2001**, *85*, 11–15.

44. Sikkanthar, S.; Karthikeyan, S.; Selvasekarapandian, S.; Arunkumar, D.; Nithya, H.; Junichi, K. Structural, Electrical Conductivity, and Transport Analysis of PAN–NH$_4$Cl Polymer Electrolyte System. *Ionics* **2016**, 1–10.

45. Asmara, S. N.; Kufian, M. Z.; Majid, S. R.; Arof, A. K. Preparation and Characterization of Magnesium Ion Gel Polymer Electrolytes for Application in Electrical Double Layer Capacitors. *Electrochim. Acta* **2011**, *57*, 91–97.

46. Lim, C. S.; Teoh, K. H.; Liew, C. W.; Ramesh, S. Capacitive Behavior Studies on Electrical Double Layer Capacitor Using Poly(Vinyl Alcohol)–Lithium Perchlorate Based Polymer Electrolyte Incorporated with TiO$_2$. *Mater. Chem. Phys.* **2014**, *143*, 661–667.

PART III

Physical Chemistry for the Life Sciences

CHAPTER 9

GLOBAL WATER CRISIS, THE VAST INNOVATIONS, AND THE VISION FOR THE FUTURE

SUKANCHAN PALIT[1,2*]

[1]*Department of Chemical Engineering, University of Petroleum and Energy Studies, Energy Acres, Post Office Bidholi via Premnagar, Dehradun 248007, India*

[2]*43, Judges Bagan, Post Office Haridevpur, Kolkata 700082, India*

E-mail: sukanchan68@gmail.com; sukanchan92@gmail.com

CONTENTS

ABSTRACT

Mankind and human scientific endeavor are today on the path of immense scientific regeneration and vast scientific vision. Environmental engineering science and water science and technology are in the similar manner surpassing vast and versatile scientific frontiers. Today, global water status stands in the midst of deep scientific comprehension and vast scientific imagination. In this chapter, the author deeply comprehends the scientific progress in water technology and gives a vast glimpse on the recent innovations and the visionary technologies in water science and water purification. With assuring cadence and deep introspection, the author delves deep into the next-generation scientific innovations and scientific vision with the prime objective of furtherance of science and engineering. Global water crisis today is in the critical juncture of deep scientific catastrophe and vast scientific vision. The world today witnesses the burning issues of climate change, loss of ecological biodiversity, environmental disasters, and depletion of fossil fuel resources. In such a situation, engineering science and technology needs to be envisioned and readdressed with the passage of scientific history and visionary timeframe. In this chapter, the author pointedly focuses on traditional and nontraditional environmental engineering techniques such as advanced oxidation processes and membrane separation processes. Scientific innovation, deep scientific instinct, and the vast scientific needs of the human society are the torchbearers toward a newer era in the field of environmental engineering science and water technology. Global water shortage and research and development initiatives in environmental engineering science are today leading a long and visionary way in the true emancipation of global environmental sustainability. This chapter gives a wide glimpse on the vast scientific potential and the scientific success of environmental engineering tools mainly nontraditional techniques with a sole objective of furtherance of science and engineering.

9.1 INTRODUCTION

Science and technology are witnessing drastic and dramatic challenges in our present day human civilization. Environmental engineering science is moving from one visionary paradigm toward another. Environmental catastrophes, global climate change, loss of ecological biodiversity, and global water shortage have urged scientists and engineers to move toward newer vision and newer innovation. The vision and the challenge of technology and engineering science are immense and path-breaking today. In this chapter, the author pointedly focuses on the immense importance of innovations and new technologies in water science and technology and the needs of water purification, drinking water treatment, and industrial wastewater treatment. Technological advancements, scientific vision, and the needs of human society will all lead a long and effective way in the true emancipation of environmental engineering science today. This chapter is an eye-opener to the world of challenges in environmental engineering science and water purification. Modern science today stands in the midst of deep scientific introspection and vast technological vision. The needs of modern science are the visionary world of provision of basic human needs such as food, water, shelter, and electricity. Environmental engineering paradigm is in the midst of deep vision and scientific fortitude. As human civilization moves forward toward a newer visionary era, environmental engineering science and water technology assume immense importance. In this chapter, the author pointedly and instinctively touches upon the grave concerns of global water shortage and global water hiatus and the immediate need of innovations and newer technologies. This chapter will open new avenues and new windows of vision and fortitude in the field of water science and technology in decades to come. The vision of science in global water research and development is immense and groundbreaking. Technology and engineering science have few answers to the scientific intricacies of arsenic and heavy metal groundwater remediation in developing and developed economies. Here needs the importance of environmental engineering tools such as membrane science and advanced oxidation processes (AOPs).

9.2 THE AIM AND THE OBJECTIVE OF THIS STUDY

Science and technology are huge colossus today with a vast vision of their own. Environmental engineering science and environmental protection in the similar manner stand in between vast scientific vision and scientific ingenuity. The aim and objective of this study is to articulate the visionary world of innovations in water science and technology and raise the awareness of global water catastrophe. In this chapter the author deeply elucidates the scientific success, the scientific potential, and the vast scientific imagination in the path toward innovations in environmental protection and water purification. The author in this chapter deeply comprehends the need of technology and engineering science in tackling global water shortage and global water hiatus. The author in this chapter pointedly focuses on human scientific fortitude and deep human scientific farsightedness in grappling the worldwide problem of drinking water shortage, water purification, and industrial wastewater treatment. Engineering science and technology are today retrogressive with the passage of scientific history and visionary time-frame. Environmental engineering is in a state of immense scientific crisis and scientific acuity. This chapter opens up new thoughts and new ideas in the field of chemical process engineering, environmental engineering, biological sciences, and applied geology in the tackling with the enigma of arsenic and heavy metal groundwater remediation. Mankind's immense scientific divination and prowess, the vast technological profundity, and the immediate need of pure drinking water will all lead a long and visionary way in the true emancipation of water purification and drinking water treatment today. Industrial wastewater treatment is another cornerstone of this well-researched chapter. Industrial wastewater treatment is a burning issue of human civilization and human scientific endeavor today. Man's immense scientific journey, mankind's vast scientific vision, and the research forays in environmental engineering are the veritable torchbearers toward a newer era in environmental protection. The purpose and the aim of this study is to proclaim the importance of water purification technologies in the sole objective of scientific advancement.

9.3   GLOBAL WATER SHORTAGE AND THE SCIENTIFIC DOCTRINE OF ENVIRONMENTAL SCIENCE

Global water shortage and global water hiatus are the scientific blunders of human civilization today. South Asia today stands in the midst of deep crisis and immense scientific fortitude. The scientific doctrine of environmental engineering science today needs to be envisioned and reorganized. Arsenic and heavy metal groundwater contamination is a burning issue in present day human civilization. Mankind stands in the midst of deep scientific devastation and in the similar manner vast scientific vision. Environmental engineering science thus needs to be streamlined with newer vision and newer innovations as science and technology surges forward. Global water shortage and groundwater heavy metal contamination are the burning issues and are veritable scientific enigma. Technology, engineering, and science have practically no answers to the marauding and monstrous issue of arsenic and heavy metal groundwater remediation. Scientific and academic rigor are today at the helm in every research and development initiatives in water purification and water technology. The vision and challenge of science are immensely retrogressive at this crucial juncture of scientific history and time.

9.4 THE VISION OF ENVIRONMENTAL ENGINEERING TECHNIQUES

The vision of environmental engineering tools is today surpassing vast and versatile scientific frontiers. Human scientific imagination and scientific candor are in a state of immense devastation. Environmental engineering and chemical process engineering tools such as membrane science and AOPs are the utmost need of the hour as water purification and drinking water treatment stand in the midst of scientific upheaval. This vision of water purification, drinking water treatment, and industrial wastewater treatment needs to be readdressed and revamped if global water crisis needs to be eliminated. The author in this chapter reiterates the success of application of membrane science in water purification with the sole objective of furtherance of science and engineering. Desalination and water treatment are today linked by an unsevered umbilical cord. Human scientific progress in membrane science and water technology needs to be equally addressed and envisioned as science and engineering move from one visionary paradigm to another.

9.5 THE STATUS OF GLOBAL WATER CRISIS

The status of global water crisis is grave and needs immediate attention. South Asia, particularly Bangladesh and India, is in the threshold of major environmental crisis. Arsenic groundwater contamination is challenging the scientific firmament of Bangladesh and India. Drinking water contamination is at its helm in South Asia today. Water purification paradigm needs to be envisioned and restructured with the passage of scientific history and time. The state of global water crisis today stands in the midst of deep hope and scientific optimism. Scientific grit, vast scientific vision, and the world of scientific challenges will all lead a long and effective way in the true emancipation of water science and water technology. Newer technology, newer innovation, and newer vision are the pivots of scientific endeavor in water purification today. The author pointedly focuses on human scientific progress in membrane science, AOPs, and nontraditional environmental engineering tools. Technology and engineering science are highly advanced today as human civilization moves from one environmental crisis to another.

The answers to these environmental catastrophes are the new technologies and newer innovations. In this chapter, the author repeatedly proclaims the importance of membrane science, AOPs, and desalination in the scientific emancipation of water purification technologies. Mankind needs to be alert to the disasters of climate change today. Climate change and loss of ecological biodiversity are challenging the scientific firmament today. These areas of scientific endeavor need to be streamlines as human civilization moves forward.

9.6 TRADITIONAL ENVIRONMENTAL ENGINEERING TOOLS

Traditional environmental engineering tool was the cornerstone of scientific research pursuit previously. Sedimentation, flocculation, and activated sludge treatment are the conventional areas of industrial wastewater treatment. Industrial wastewater treatment and water purification thus stands in the midst of vast scientific vision and vast scientific cognizance. The immediate need of the hour is the scientific forays into nontraditional environmental engineering techniques such as membrane science and advanced oxidation techniques. Human scientific progress and strides of human civilization today need to be envisioned and restructured as environmental engineering science moves toward a newer visionary realm. Today, water science and technology and water purification tools are vastly latent and need to be streamlined as human scientific progress surges forward. Science and engineering are two huge colossi today with a definite and purposeful vision of their own. In this chapter, the author reiterates with immense cadence and scientific candor the success of innovation and the need of technological profundity in the furtherance of environmental protection and environmental engineering. Mankind's immense scientific prowess, the needs of human society, and the world of challenges in environmental protection and environmental engineering will all lead a long and visionary way in the true emancipation of environmental sustainability today. Sustainable development and infrastructural prowess are the pillars of human progress today. Environmental and energy sustainability are the need of the hour.

9.7 NONTRADITIONAL ENVIRONMENTAL ENGINEERING TOOLS

Nontraditional environmental engineering tools are today the cornerstones of environmental engineering scientific endeavor. Membrane science is revolutionizing the scientific fabric of environmental engineering and chemical process engineering. Mankind's immense scientific vision is in a state of immense catastrophe as environmental disasters redefine the vast history of science and technology. AOP is another branch of nontraditional environmental engineering scientific endeavor. Human scientific pursuit's immense prowess, the vast technological profundity, and the world of scientific validation will all lead a long, effective, and visionary way in the true realization of environmental sustainability and environmental protection. Today is the world of space science, nuclear technology, and renewable energy technology. Man's scientific vision is witnessing immense challenges as energy crisis and environmental engineering disasters devastate the vast and wide scientific firmament. Ozonation or ozone oxidation is a branch of AOP which today is surpassing vast scientific frontiers. In this chapter, the author deeply ponders on the immense scientific and technological acuity and farsightedness in the application of AOPs, integrated oxidation processes, and ozonation. The challenge and vision of nontraditional environmental engineering techniques are slowly unfolding today. Scientific progress and emancipation of technology and engineering science in today's world are surpassing vast and versatile frontiers. Today the world is faced with the issues of groundwater heavy metal and arsenic contamination. Science and engineering have practically no answers to the enigmatic and monstrous issue of groundwater remediation. Here comes the necessity of innovation and deep scientific vision.

9.8   THE SCIENTIFIC DOCTRINE OF AOPS AND THE VISION FOR THE FUTURE

Human scientific vision and the vast scientific doctrine in AOPs are today surpassing scientific frontiers. Today, AOPs are visionary scientific endeavor in the field of water purification and industrial wastewater treatment. Nonconventional advanced oxidation techniques are challenging the scientific landscape today. Ozone oxidation, sonochemical industrial wastewater treatment, and electro-Fenton treatment are the branches of environmental engineering tools which have immense scientific potential and replete with scientific profundity. The world of science and technology are faced with immense scientific travails and deep scientific barriers. In this chapter, the author deeply pronounces the need of scientific innovation, the need of scientific ingenuity, and the vast scientific cognizance in the field of both traditional and nontraditional environmental engineering applications. Today, the world of environmental engineering and chemical process engineering are huge colossus with a definite and purposeful vision of their own. Human civilization's immense scientific prowess is at a state of immense disaster as frequent environmental disasters destroy the scientific fabric of deep intellect and precision. Zero-discharge norms and global environmental regulations are scarcely followed by developing and developed countries around the world. Here comes the importance of innovation and vision. AOPs and integrated oxidation processes will with immense lucidity and clarity open a new avenue of research pursuit in years to come.

9.9 SIGNIFICANT SCIENTIFIC ENDEAVOR IN AOPS

Technology and engineering science are advancing at a rapid pace moving from one visionary paradigm to another. Today, the challenge and the vision of nontraditional environmental engineering techniques such as AOPs are immense and path-breaking. Human scientific regeneration, human scientific cognizance, and the vast scientific fervor of environmental engineering will go a long and effective way in the true realization of environmental protection and water purification. Research pursuit in AOPs and integrated AOPs are changing the entire scientific landscape and the vast vision of global scientific regeneration. Ozonation or ozone oxidation needs to be restructured and envisioned as human scientific progress, science and engineering move toward visionary directions. Today is the world of energy and environmental sustainability. Sustainable development, infrastructural development, and water purification are all linked by visionary cords. The fetters and scientific travails of environmental sustainability and environmental protection need to be ameliorated.

Munter[1] discussed, with deep and cogent insight, current status and prospects in AOPs. The paper provides a deep overview of theoretical basis, efficiency, economics, laboratory and pilot plant testing, design, and modeling of different AOPs (combinations of ozone and hydrogen peroxide with UV radiation and catalysts). Hazardous organic wastes from industrial, military, and commercial operations represent one of the greatest challenges in environmental engineering scientific endeavor.[1] Technological vision, vast scientific motivation, and profundity are the veritable pillars of this paper. Conventional incineration has immense disadvantages as environmental engineering science surges forward. The AOPs have proceeded along the two visionary routes: (1) oxidation with oxygen in temperature

ranges intermediate between ambient conditions and those found in incin-
erators and (2) the use of high-energy oxidants such as ozone and hydrogen
peroxide and/or photons that are able to generate highly reactive intermediates—the hydroxyl radicals.[1] In 1987, scientists from USA defined AOPs as
"near ambient temperature and pressure water treatment processes which
involve the generation of hydroxyl radicals in sufficient quantity to effect
water purification."[1] Scientific assertiveness, deep scientific vision, and
vast scientific farsightedness are the pillars of this well-researched paper.
The hydroxyl radical is a powerful, nonselective chemical oxidant, which
acts very rapidly with most organic compounds.[1] Technology of AOP is
deeply addressed in this paper. The attack by the OH radical, in the presence of oxygen, initiates a complex cascade of oxidative reactions leading to
mineralization of the organic compound. The exact routes of these reactions
are still unclear. Scientific clarity and deep profundity are the hallmark of
this research endeavor. The routes are—chlorinated organic compounds—
oxidized first to intermediates, such as aldehydes and carboxylic acids, and
finally to carbon dioxide, water, and chloride ion.[1] In this paper, the author
deeply elucidates with cogent farsightedness the scientific success of non-
photochemical and photochemical methods. These are ozonation, ozone
+ hydrogen peroxide, ozone + catalyst, Fenton system, ozone/ultraviolet,
hydrogen peroxide/ultraviolet, ozone/hydrogen peroxide/ultraviolet, photo-
Fenton, and photocatalytic oxidation.[1] The author in this paper addressed
and thoroughly reviewed various AOPs with the sole objective of furtherance of science and engineering.

Sharma et al.[2] deeply elucidate with deep and cogent insight AOPs for
wastewater treatment. Today water purification paradigm stands in the midst
of deep scientific vision and vast scientific cognizance. AOPs constitute
promising technologies for the treatment of wastewaters containing non-
easily removable organic compounds. All AOPs are designed to produce
hydroxyl radicals.[2] Technology and engineering science of AOPs are today
surpassing vast and versatile scientific boundaries. It is the hydroxyl radicals
that act with high efficiency to destroy organic compounds. Scientific ingenuity and vast scientific clarity and lucidity are the utmost need of the hour.
This paper presents a general review of efficient AOPs developed to decol-
orize and degrade organic pollutants for environmental protection purposes.[2]
The deep scientific ardor in application of AOP in industrial wastewater
treatment is thoroughly addressed in this paper.[2]

Gilmour[3] deeply discussed in his doctoral thesis application perspectives
in water treatment using AOPs. The challenge, the vision, and the targets of

AOP are deeply discussed in this thesis. AOPs using hydroxyl radicals and other oxidative radical species are being studied extensively for removing of organic pollutants from industrial waste streams. Scientific validation, the vast scientific fervor, and the barriers in application of AOP are discussed in deep details here. This study focuses on the evaluation of upstream processing and downstream posttreatment analysis of AOP techniques.[3] The function of pilot-scale immobilized photocatalytic reactor is described in details in this paper. Technology and engineering science of nonconventional environmental engineering tools are gaining immense importance as human scientific research pursuit treads forward. This is a vast and versatile eye-opener toward scientific emancipation of environmental science. Trace concentrations of numerous organic compounds (emerging contaminants and endocrine disrupting compounds) such as pharmaceuticals and personal care products (PPCPs) including prescription drugs and biologics, nutraceuticals, fragrances, sunscreen agents, and numerous others are reported in industrial wastewater.[3] There is a growing health and environmental concern for PPCP.[3] Thus, this research work addresses the techniques to eliminate these organic pollutants in wastewater.[3]

Oller et al.[4] with vast and versatile vision discussed combination of AOPs and biological treatments for wastewater decontamination. Technological vision, vast scientific ardor, and scientific prowess are the pillars of scientific advancements today. Vast and wide scientific ingenuity is the pivot of this entire paper. Nowadays there is a continuously evergrowing worldwide concern for development of alternative water reuse technologies, basically focusing on agriculture and industry. In this vast context, AOPs are considered a highly competitive water treatment technology for the removal of those organic and inorganic pollutants not treatable by conventional techniques due to their high chemical stability and low biodegradability.[4] Scientific vision and vast scientific profundity are the immediate need of the hour as environmental disasters devastate the scientific firmament. Although chemical oxidation for complete mineralization is usually expensive, its combination with a biological treatment is vastly and effectively reported to reduce operating costs. The hallmarks of this treatise are the deep introspection in degradation kinetics and the reactor modeling of the combined process. Human civilization's immense scientific prowess, the needs of human society, and the futuristic vision will all lead a long and effective way in the true emancipation of environmental engineering and industrial wastewater treatment.[4]

9.10 THE VAST SCIENTIFIC VISION OF MEMBRANE SCIENCE

Membrane science is a branch of novel separation process in chemical process engineering. The vision of modern science and modern scientific endeavor is vast and versatile. Today, membrane science and water technology are the two opposite sides of the visionary coin. Human civilization stands today in the midst of deep scientific introspection and vast comprehension. Water purification and desalination in the similar vein are the visionary aim and objectives of scientific endeavor today. Technology and engineering science need to be readdressed and reenvisioned as civilization surges forward. Membrane science scientific doctrine involves reverse osmosis, nanofiltration, ultrafiltration, microfiltration, electrodialysis, and some other effective water treatment processes. Human scientific vision and human scientific regeneration in the field of membrane science and desalination are of highest order and global research and development initiatives surge forward. Technology needs to be rebuilt and water technology needs to be given utmost importance as human civilization treads forward toward a newer paradigm and a newer scientific era. Food processing technology in developing nations is at a state of immense scientific distress. This area of engineering science needs to be overhauled with the application of membrane science.

9.11 SIGNIFICANT SCIENTIFIC RESEARCH PURSUIT IN MEMBRANE SCIENCE

Membrane science today is in the path of immense vision and scientific rejuvenation. Global water shortage and lack of clean drinking water are challenging the scientific vision of environmental engineering science and water technology. Mankind's immense scientific determination, scientific girth, scientific prowess, and the utmost need of environmental protection will all lead a long and effective way in the true emancipation of environmental sustainability. Energy and environmental sustainability are the pillars of human civilization and human scientific endeavor today. Water technology and environmental sustainability are the two opposite sides of the visionary coin. Membrane science and environmental engineering techniques are thus the utmost needs of human society today.

Palit[5] discussed with vast foresight in a far-reaching review advanced environmental engineering separation processes, environmental analysis, and application of nanotechnology. The challenge and vision of nanotechnology applications and nanofiltration applications are deeply addressed in this chapter.[5] Challenges, barriers, and difficulties are the focal points in environmental engineering applications today. This treatise, with immense insight, delineates the success of environmental separation processes and the efficiency of advanced environmental analysis. Technological advancements are today in the path of rejuvenation and vision. This treatise delineates with deep and cogent insight the success of environmental separation processes, mainly membrane separation processes and tertiary treatment tools such as AOPs and integrated AOPs.[5]

Palit[6] discussed lucidly, with deep and cogent insight, application of nanotechnology, nanofiltration, and drinking and wastewater treatment. The world of environmental engineering science and water technology is moving at a faster pace toward newer visionary paradigm. Water process engineering and chemical process engineering are the hallmarks of effective scientific endeavor in water purification technology today.[6] In this treatise, the author delineates with cogent insight and deep introspection the application and importance of nanotechnology in drinking water treatment and industrial wastewater treatment. The author with immense vision targets the recent scientific endeavor in nanofiltration in drinking water treatment and industrial wastewater treatment. Global drinking water crisis is at a devastating peril. The success of membrane science and the vast futuristic vision will all lead a long and visionary way in the true realization of both water purification and environmental sustainability. This paper redrafts and revisits the entire gamut of water purification technologies and opens new windows of innovation and scientific instinct in decades to come.[6]

9.12   OTHER ENVIRONMENTAL ENGINEERING TOOLS

Environmental engineering science and chemical process engineering are today in the path of immense validation and scientific profundity. Industrial wastewater treatment, drinking water treatment, and desalination are the innovations of tomorrow. Primary, secondary, and tertiary wastewater treatments are changing the face of scientific endeavor. Desalination is the technology of utmost need in many developing and developed nations throughout the world. The world of water purification and environmental engineering today stands in the midst of a deep crisis with the growing concern for arsenic and heavy metal contamination of groundwater and drinking water. This scientific paradigm needs to be veritably streamlined and reenvisioned. Science of groundwater remediation in the similar manner needs to be reenvisioned and reframed, if the human civilization needs to be survived. Environmental engineering tools are the pillars of human civilization and human scientific genre. This paper veritably proclaims the success and scientific revelation of traditional and nontraditional environmental engineering tools. The world of science needs to be overhauled and restructured if the human planet needs to be saved.

9.13 MODERN SCIENCE: DIFFICULTIES, CHALLENGES, AND OPPORTUNITIES

Modern science and engineering endeavor are today veritably the necessities of human civilization. The difficulties, challenges, and opportunities in application of engineering science in water technology are immense and path-breaking. Human scientific research pursuit in environmental protection, the needs of human survival, and the wide vision of environmental engineering will lead a long and effective way in the true scientific understanding of environmental sustainability. Sustainable development and infrastructural development as regards to energy and environment are the prime objectives of any scientific endeavor today. Modern science today stands in the midst of deep scientific vision and unending scientific introspection. The opportunities of modern science and its immense visionary research pursuit are vast and versatile. Today, the world of science and engineering is in the process of newer regeneration as space technology and nuclear science research are changing the vast scientific firmament of vision and determination. Technology of nanoscience needs to be revived and restructured with the passage of scientific history and time. Nanoscience and nanotechnology are the opposite sides of the visionary coin of environmental sustainability. Sustainable development whether it is energy or environment is the engineering marvel of modern science. The author deeply comprehends the utmost need of modern science in the refurbishment of environmental engineering science and environmental protection today. Human scientific endeavor and the needs of human society and human advancement are today linked by an unsevered umbilical cord. The difficulties, challenges, and opportunities in modern science are immense and need to be readdressed and reemphasized with the passage of scientific history. Human scientific judgment, the vast scientific truth, and the difficult scientific barriers are the torchbearers toward a newer era of scientific emancipation and deep scientific validation. Desalination, water purification, industrial wastewater treatment, and drinking water treatment are today replete with scientific vision and vast scientific overhauling. The grit and determination of modern science will all lead a long and visionary way in the true emancipation of environmental protection and environmental sustainability. Science of environmental protection today is equally ensconced with vision and scientific fortitude. The targets of modern science should be the need of provision of pure drinking water. Today, mankind's scientific prowess and vast stature are in a dismal state. Today's scientific endeavor should lead human civilization toward a newer era in the field of water technology.

9.14   ENVIRONMENTAL SUSTAINABILITY AND THE PURSUIT OF SCIENCE

Environmental sustainability and water purification are two visionary avenues of scientific research pursuit today. The visionary words of Dr. Gro Harlem Brundtland, former Prime Minister of Norway, on the science of "sustainability" need to be readdressed and reemphasized with the progress of human civilization. Global water research and development initiatives, the needs of human scientific advancements, and the futuristic vision of environmental protection will all lead a long and effective way in the true emancipation of environmental sustainability today. Drinking water treatment and industrial wastewater treatment are the veritable needs of human society today. Arsenic and heavy metal poisoning of groundwater in developing and developed nations throughout the world is a burning and deeply enigmatic issue. Technology and engineering science have few answers to the scientific enigma of arsenic groundwater remediation. Scientists and engineers around the world are today faced with tremendous academic rigor and vast scientific vision. Human scientific progress thus needs to be streamlined and reorganized as water science and water technology surge forward toward a newer era. Today, the science of "sustainability" is a huge colossus with a vast vision of its own. Energy sustainability and energy security are the other sides of the visionary coin. The pursuit of modern science in water purification will today lead a long and visionary path in the true emancipation of environmental sustainability. Technology and engineering science thus need to be redrafted and re-envisaged as human civilization and human scientific research pursuit surge forward. The burning issue of arsenic and heavy metal contamination needs to be vehemently addressed as science and engineering moves toward a newer visionary eon.

Human scientific stature, scientific genre, and the vast scientific ingenuity are the necessities of scientific research pursuit in water technology today. The challenge and the vision of water technology and industrial wastewater treatment should be the targets of science in both developed and developing nations around the world. In this chapter, the author repeatedly stresses on the scientific success, the vast scientific fortitude, and the imminent scientific needs of water purification technology and industrial wastewater treatment technologies in the path toward emancipation of science and engineering. Thus, the scientific vision of environmental sustainability will be veritably realized.

9.15 WATER PURIFICATION, DRINKING WATER TREATMENT, AND INDUSTRIAL WASTEWATER TREATMENT

Technological and scientific advancements in drinking water treatment and industrial wastewater treatment are the visionary avenues of scientific emancipation today. Human scientific progress and the human scientific genre are the pillars of human civilization and mankind today. The needs of water purification are immense and groundbreaking today. Every nation around the world is veritably struggling in dealing with the burning issues of groundwater remediation and drinking water treatment. Arsenic and heavy metal groundwater contamination are a bane to human civilization today. South Asian countries such as Bangladesh and India are in the throes of an immense environmental disaster of untold proportions. Science, engineering, and technology have veritably no answers to global research forays in groundwater remediation. Zero discharge norms and environmental regulations are hardly followed in developing and developed countries around the world. The success of human civilization and human scientific endeavor is thus at deep stake. Human scientific forbearance and human scientific ingenuity need to be restructured and overhauled as science and engineering of environmental protection surge forward. The whole world is today in the threshold of a newer era of challenges and immense vision. In a similar vein, environmental engineering science and water purification technologies need to be restructured and revamped with the passage of scientific history and time. In this chapter, the author repeatedly pronounces the needs of innovations in water purification technologies in the path toward scientific emancipation and engineering innovations.

9.16 ARSENIC AND HEAVY METAL CONTAMINATION OF GROUNDWATER AND THE VISION FOR THE FUTURE

Arsenic and heavy metal contamination of groundwater is a deep curse to human scientific vision and human scientific research pursuit today. Bangladesh and India are today in a state of immense distress as regards successful implementation of environmental sustainability. Answers to groundwater remediation are few, but the scientific vision is immensely resounding. Technology and engineering science needs to be redrafted and reenvisioned, if arsenic groundwater contamination needs to be ameliorated. Human scientific progress and human scientific paradigm are in a state of immense distress and turmoil. Provision of clean drinking water and the huge domain of industrial wastewater treatment in developed and developing economies need to be vehemently addressed and vastly reenvisioned. Arsenic groundwater contamination in Bangladesh is world's largest environmental disaster. Mankind and human scientific profundity need to be reenvisioned if arsenic or heavy metal contamination is to be totally erased. Groundwater exploration should be the science and engineering of tomorrow. In the similar vein, environmental sustainability should be the pillar of all environmental engineering research and development initiatives. Chemical process engineering and basic chemical engineering operations need to be overhauled and reenvisioned in the similar manner if arsenic groundwater remediation needs to be enhanced and water science and technology emancipated. Scientific acuity, deep vision, and the vast needs of the human society are the torchbearers toward a newer era in cross-boundary research in groundwater remediation. In this chapter, the author repeatedly proclaims the success, the needs, and the profundity in research innovations in groundwater exploration and groundwater remediation. This chapter will surely open out windows of scientific invention and deep scientific instincts in environmental engineering in decades to come.

9.17 THE SCIENCE OF GROUNDWATER REMEDIATION

Groundwater remediation and groundwater exploration are today areas of interdisciplinary research. Applied geology, biological sciences, chemical process engineering, and environmental engineering science are today molded into one science in the research pursuit in arsenic and heavy metal groundwater remediation. Today, the challenge and vision of water technology and environmental chemistry are far-reaching. Technology of arsenic groundwater remediation in South Asia has few positive answers. The challenge lies in the hands of environmental engineering science and biological sciences. Human scientific pursuit is dismal as science and engineering of groundwater decontamination moves toward a newer paradigm. Global water research and development initiatives should be in the path of a new beginning. Civil society challenges and governmental policies are the torchbearers toward a newer eon in the field of drinking water treatment, wastewater treatment, and water purification.

9.18 SIGNIFICANT SCIENTIFIC RESEARCH PURSUIT IN GROUNDWATER REMEDIATION

Research pursuit in heavy metal groundwater remediation is witnessing immense challenges today. The whole world is awaiting newer innovation and a complete scientific revival in arsenic groundwater decontamination. Interdisciplinary research endeavor is the pillar and mainstay of arsenic groundwater remediation. Human scientific challenges are veritably in the state of distress and vision.

Hashim et al. (2011)[7] discussed, with deep and cogent insight, remediation technologies for heavy metal contaminated groundwater. Scientific validation in environmental management, deep scientific fervor, and the needs of human society will all lead a long and visionary way in the true emancipation of environmental engineering science and sustainable development. In this research work, 35 approaches for groundwater treatment are reviewed and classified under three categories which are chemical, biological, and physico-chemical treatment processes. The authors with deep conscience discussed sources, chemical property, and speciation of heavy metals in groundwater.[7] The challenges of science and engineering in this paper are the technologies for treatment of heavy metal contaminated groundwater which are chemical treatment technologies, in-situ treatment by using reductants, reduction by dithionite, reduction by using iron-based technologies, soil washing, in-situ soil flushing, in-situ chelate flushing, biosorption treatment technologies, and enhanced biorestoration.[7] Human scientific regeneration and deep scientific acuity are the pillars of this well-researched paper.

Global water crisis and significant technological innovations are the two opposite sides of the visionary coin. Mankind today stands in the midst of deep scientific introspection and vast vision. The author in this treatise deeply comprehends the vast and versatile technological innovations and its imminent needs to human scientific research pursuit.

9.19 FUTURE RECOMMENDATIONS AND FUTURE FLOW OF SCIENTIFIC THOUGHTS

Science, technology, and engineering science are today moving at a fast pace surpassing one visionary frontier over another. Today, the world of environmental engineering science stands in the midst of deep scientific introspection and vast scientific divination.[8,9] Climate change, loss of ecological biodiversity, and frequent environmental disasters are veritably challenging the scientific firmament in today's present day human civilization. The challenges and the vision of science are immense and far-reaching. Future of human civilization and human scientific endeavor today lies in the hands of environmental engineers and environmental scientists. In this chapter, the author deeply comprehends and poignantly depicts the scientific success, the scientific adroitness, and the vision behind traditional and nontraditional environmental engineering tools. The success and the vision of this chapter go beyond scientific imagination and scientific excellence.[10,11] Human scientific progress in environmental protection is at a state of immense scientific distress and in the similar manner at a state of scientific regeneration. Future research trends and future recommendations should be targeted toward newer innovation and a newer visionary scientific era. Science of groundwater remediation and heavy metal groundwater decontamination is the enigmatic and puzzling scientific issue today. Technology and engineering science have few answers to the burning issue of arsenic and heavy metal groundwater contamination in developing and developed economies throughout the world. Human scientific progress and human scientific paradigm are in the state of immense scientific distress and replete with immense failures.[12,13] Future recommendations and future flow of thoughts should be targeted toward newer innovations and newer techniques. Environmental engineering science today needs to be veritably overhauled and reorganized as science treads forward in a newer century. Novel separation processes, advanced oxidation techniques, integrated AOPs, desalination, and sonochemistry are the visionary endeavor of tomorrow. The challenge and the vision of science are immense and far-reaching. Water purification and industrial wastewater treatment should involve the abovementioned processes. Membrane science and other novel separation processes are the utmost needs of water treatment today. In this chapter, the author reiterates the success of novel separation processes and desalination in provision of pure drinking water to the teeming millions in developing and developed economies around the world. Scientific vision, deep scientific candor, and vast scientific ingenuity stand as the futuristic targets of water purification and industrial wastewater treatment.

9.20   FUTURE RESEARCH TRENDS IN WATER PURIFICATION

Water purification and the vision of science and technology are the needs of human society. Future research trends in water purification need to be streamlined toward novel separation processes and nontraditional environmental engineering techniques. Human scientific paradigm today stands in the midst of deep scientific fortitude and vision. Scientific clarity, scientific acuity, and deep scientific vision are the utmost need of research and development initiatives today. Future trends in water purification should be directed toward more innovations in novel separation processes, traditional and nontraditional environmental engineering techniques. Science has no answers to the environmental disasters and global climate change today. Human scientific ingenuity is in a state of immense distress. In such critical juncture of human scientific history and time, validation of science is of immense necessity. Scientific motivation in water science, the vast technological vision, and the wide world of science will surely open up new vistas of candor and scientific vigor in decades to come. Today, membrane science and AOPs have immense potential in solving environmental engineering problems. The state of environment is immensely dismal. Future research trends in global water research and development forays should focus toward technologies and innovations in tackling climate change, loss of ecological biodiversity, and frequent environmental catastrophes. In this chapter, the author deeply elucidates the success of innovations in traditional and nontraditional environmental engineering paradigm and the vast necessities of science and engineering in confronting environmental engineering disasters.

9.21   CONCLUSION AND SCIENTIFIC PERSPECTIVES

Science and technology are two huge colossi with a definite and purposeful vision of their own. Academic and scientific rigor in global water research and development initiatives need to be targeted with immense scientific might and vision today. The future of science and engineering in today's global scientific order is bright and path-breaking. Innovations, scientific vision, and vast scientific candor are the pillars of research pursuit today. Developed and developing nations around the world are giving immense and widespread importance to research forays in water technology and water purification technologies. In this chapter, the author deeply reiterates the scientific success of traditional and nontraditional environmental engineering tools with the sole objective of furtherance of science and engineering of environmental protection. Today, environmental engineering is of utmost need to human society and human scientific endeavor. Technology and engineering science today needs to be reemphasized and reorganized as the human planet witnesses immense scientific travails and unending environmental disasters. Climate change and loss of ecological biodiversity are the pivots of scientific endeavor and scientific genre today. Future scientific perspectives in global water initiatives and environmental science are facing uphill tasks today as mankind trudges forward toward newer knowledge dimensions. Membrane science, desalination, and AOPs are the necessities of human science, human acuity, and human innovation. The author in this chapter repeatedly pronounces the success of human scientific research pursuit in both traditional and nontraditional environmental engineering techniques and gives a vast glimpse on the difficulties, challenges, and opportunities in furtherance of science and engineering today. The future perspectives in environmental engineering, chemical process engineering, and nanotechnology are wide, vast, and versatile. Today, every branch of scientific endeavor is linked with research pursuit in nanotechnology. Nanotechnology in today's human planet is the engineering marvel of today. This chapter will surely go a long and visionary way in the true emancipation of nanotechnology applications in environmental engineering science and chemical process engineering. The challenges and the vision of science are far-reaching and need to be reorganized. Nanotechnology and its applications in environmental engineering and chemical process engineering are the needs of human civilization today. The author in this chapter repeatedly urges the scientific community and the vast technological domain to reorganize themselves toward a newer era in water technology and environmental engineering. Human scientific vision will then be truly revisited.

KEYWORDS

- environment
- engineering
- water
- oxidation
- vision
- techniques
- membranes

REFERENCES

1. Munter, R. Advanced Oxidation Processes: Current Status and Prospects. *Proc. Estonian Acad. Sci. Chem.* **2001,** *50* (2), 59–80.
2. Sharma, S.; Ruparelia, J. P.; Patel, M. L. A In *General Review on Advanced Oxidation Processes for Wastewater Treatment,* International Conference on Current Trends in Technology, NUICONE, Ahmedabad, India, Dec 8–10, 2011, pp 1–7.
3. Gilmour, C. R. Water Treatment Using Advanced Oxidation Processes: Application Perspectives. Electronic Thesis and Dissertation Repository, Master of Engineering Thesis, University of Western Ontario, Canada, 2012.
4. Oller, I.; Malato, S.; Sanchez-Perez, J. A. Combination of Advanced Oxidation Processes and Biological Treatments for Wastewater Decontamination: A Review. *Sci. Total Environ.* **2011,** *409,* 4141–4166.
5. Palit, S. Advanced Environmental Engineering Separation Processes, Environmental Analysis And Application Of Nanotechnology: A Far-Reaching Review, Chapter 14. In *Advanced Environmental Analysis: Application of Nanomaterials;* Chaudhery, M. H., Boris, K., Eds.; Royal Society of Chemistry: United Kingdom, 2017; Vol. 1, pp 377–416.
6. Palit, S. Application of Nanotechnology, Nanofiltration and Drinking and Wastewater Treatment- A Vision for the Future, Chapter 17, In *Water Purification*; Grumezescu, A. M., Ed.; Academic Press: USA, 2017; pp 587–620.
7. Hashim, M. A.; Mukhopadhyay, S.; Sahu, J. N.; Sengupta, B. Remediation Technologies for Heavy Metal Contaminated Groundwater. *J. Environ. Manage.* **2011,** *92,* 2355–2388.
8. Palit, S. Nanofiltration and Ultrafiltration: The Next Generation Environmental Engineering Tool and A Vision For The Future. *Int. J. Chem. Tech. Res.* **2016,** *9* (5), 848–856.
9. Palit, S. Filtration: Frontiers of the Engineering and Science of Nanofiltration: A Far-Reaching Review. In *CRC Concise Encyclopedia of Nanotechnology*; Ortiz-Mendez, U., Kharissova, O. V., Kharisov, B. I., Eds.; Taylor and Francis: UK, 2016; pp 205–214.
10. Palit, S. Advanced Oxidation Processes, Nanofiltration, and Application of Bubble Column Reactor. In *Nanomaterials for Environmental Protection*; Kharisov, B. I., Kharissova, O. V., Rasika Dias, H. V., Eds.; Wiley: USA, 2015; pp 207–215.
11. Shannon, M. A.; Bohn, P. W.; Elimelech, M.; Georgiadis, J. A.; Marinas, B. J. *Science and Technology for Water Purification in the Coming Decades*; Nature Publishing Group: UK, 2008; pp 301–310.
12. Cheryan, M. *Ultrafiltration and Microfiltration Handbook*; Technomic Publishing Company Inc.: USA, 1998.
13. Mathioulakis, E.; Bellesiotis, V.; Delyannis, E. Desalination by Using Alternative Energy: Review and State-of-the-Art. *Desalination* **2007,** *203,* 346–365.

CHAPTER 10

THE VISION OF WATER PURIFICATION AND INNOVATIVE TECHNOLOGIES: A BRIEF REVIEW

SUKANCHAN PALIT[1,2*]

[1]*Department of Chemical Engineering, University of Petroleum and Energy Studies, Bidholi via Premnagar, Dehradun 248007, Uttarakhand, India*

[2]*43, Judges Bagan, Haridevpur, Kolkata 700082, India*

E-mail: sukanchan68@gmail.com, sukanchan92@gmail.com

CONTENTS

ABSTRACT

Water science and water technology today are witnessing drastic and dramatic challenges. The immense scientific prowess and scientific potential of environmental engineering tools are delineated in details in this chapter with a clear vision toward furtherance of science and engineering. Global water issues have changed the vision of research and development initiatives and water purification technologies are being re-envisioned as human civilization moves toward a newer era of scientific regeneration and wide scientific vision. In this treatise, the author pointedly focuses on the different environmental engineering tools in water purification. The immense challenge and the vision of science are the forerunners toward a greater emancipation of environmental sustainability today. This treatise also discusses lucidly the conventional and nonconventional techniques of environmental engineering. Novel separation tools and membrane-separation processes are the hallmarks of this writing endeavor. The science of water purification is highly advanced yet many features of environmental engineering paradigm still needs to be understood. This treatise reviews the varied conventional and nonconventional environmental engineering tools such as membrane science and advanced oxidation processes with a primary goal toward the furtherance of science and engineering. Human civilization and human scientific endeavor are witnessing paradigmatic changes as regards environment. The status of environment and ecology today stands in the midst of deep peril. The author of this chapter rigorously points out the immense scientific potential of environmental engineering pursuit in solving global water issues.

10.1   INTRODUCTION

Water science and water technology are today the frontiers of science and technology. Water purification stands today in the critical midst of scientific introspection and vision. Human mankind and human scientific research pursuit in a similar vein are deeply challenged and are facing the immense test of our times. Technology of water science needs to be re-envisioned and restructured as mankind moves from one decade to another. This century is a period of immense scientific rejuvenation. Environmental engineering science and chemical process engineering are in a state of immense scientific comprehension as global water challenges and water issues derail the development of human society. Technology is baffled and scientific vision derailed as human scientific research pursuit gains newer heights. The world of engineering science and water purification science are passing through difficult scientific terrains as global water concerns challenge the wide scientific fabric. Water catastrophe is a bane to human civilization. The only hope and scientific determination lies in the hands of conventional and nonconventional environmental engineering tools. The success of human scientific endeavor in water purification is slowly evolving toward visionary directions and newer innovations. In this treatise, the author pointedly focuses on the various environmental engineering tools and the immense scientific potential and scientific vision behind it.

10.2 THE AIM AND OBJECTIVE OF THIS STUDY

Scientific vision of global water challenges and groundwater remediation are the primary objectives of this study. Water purification concepts encompass these widely relevant areas.[3] Water technology is more advanced today, yet the problems of arsenic and heavy metal contamination are till today unsolved. This treatise explores these areas. The question of environmental sustainability is of immense importance in the road toward success in realization of environmental engineering techniques today. In this century, science stands as a huge colossus with a vast and versatile vision of its own. Technology and engineering science of water science are witnessing immense restructuring and surpassing visionary scientific frontiers.[3] The author in this chapter rigorously points out the different water purification techniques, conventional as well as nonconventional, with the sole aim of furtherance of science and technology. Global water research and development initiatives are experiencing immense paradigmatic changes with the passage of scientific history, scientific vision, and time. This treatise gives a wide view of the different scientific genres behind the success of environmental sustainability and environmental engineering science. Sustainability, whether it is energy or environmental, is of utmost importance in the road toward success of human civilization and human scientific research pursuit. Environmental technology and water purification are moving toward a wide world of scientific justification and scientific candor as global water issues plunges the scientific domain to a world of deep peril. The aim of this study is clear, vast, and versatile. Scientific boundaries of water purification are witnessing a newer dawn and a newer visionary future as human civilization treads toward scientific destiny in this century.[3]

10.3   WATER PURIFICATION AND THE VISION FOR THE FUTURE

Water purification, industrial wastewater treatment, and drinking water treatment are the domains of human scientific endeavor which are deeply challenged and needs to be re-envisioned and re-envisaged with the passage of scientific history, scientific vision, and time. The vision for the future in the field of water purification and water science and technology are challenged and needs to be re-envisaged as human mankind passes through newer scientific trials.[3] Global water shortage is the deepening crisis of today. In this chapter, the author pointedly focuses on the different water treatment procedures and the vast and versatile world of drinking water treatment. The other side of the visionary coin is the evergrowing concerns of arsenic and heavy-metal groundwater contamination. Technology and engineering science are veritably baffled at this vexing global water issue. The challenge of science, the visionary roads toward future and the futuristic vision are all the forerunners toward a newer era in the field of water purification today (www.google.com; www.wikipedia.com).

Global water research and development initiatives are in a state of immense scientific rejuvenation and deep scientific introspection.[1] Water shortage and groundwater contamination with arsenic and heavy metals are a bane to human civilization and toward scientific progress. Technology of water science needs to be revamped as growing concerns for environmental protection derail human growth and the progress of human civilization. The visions for the future in scientific endeavor in environmental engineering science and the holistic world of environmental protection are slowly evolving into new dimensions. Global water purification status today stands between intense scientific introspection and deep crisis. Human mankind needs to delve deep into the crossroads of scientific validation and scientific vision. Technology of water science and engineering needs to be redefined and re-envisaged as groundwater heavy metal contamination devastates the scientific landscape.[1] The success of scientific endeavor needs to be revamped as human civilization witnesses the immense whirlpool of scientific validation and scientific vision. In this treatise, the author reiterates the wide visionary success of innovations and new technologies of drinking water treatment and industrial wastewater treatment. Chemical process engineering and environmental engineering today stands in the crossroads of vision and scientific contemplation. In such a crucial juncture of scientific history and time, vision and validation of science are of primary importance (www.google.com; www.wikipedia.com).[1,3]

10.3.1 SCIENTIFIC VISION, SCIENTIFIC TRUTH, AND DEEP SCIENTIFIC UNDERSTANDING OF INDUSTRIAL WASTEWATER TREATMENT

Scientific vision in the field of water technology and environmental engineering science is in a state of immense distress today. Growing concerns of environmental disasters and groundwater heavy metal contamination have urged the scientific domain to gear forward toward newer innovation and a visionary journey in scientific research pursuit. Technology is baffled and science has few answers to the immense environmental catastrophes facing human mankind today. Human civilization's immense scientific girth and human scientific endeavor need to be re-envisioned and re-addressed with the passage of scientific decade and time. A deep scientific understanding and scientific cognizance is needed as environmental engineering science enters a newer era of scientific regeneration. Water is an immense asset and a necessity to human progress and human civilization. The challenge is immense and goes beyond scientific imagination. Sustainable water management is the other side of the visionary coin today. Scientific challenges in sustainability are beyond scientific vision and scientific profundity today. The success is short lived as climate change and environmental disasters derails the progress of human civilization and wide scientific rigor. Industrial wastewater treatment needs to be reorganized in a similar vein with the sole aim of furtherance of science and engineering (www.google.com; www.wikipedia.com).[1]

Environmental and energy sustainability are the cornerstones of a nation's advancement whether it is developed or developing. The challenge and vision of a nation's advancement are equally dependent upon provision of basic human needs such as water. Water is a veritable and primary component of a nation's economic growth. The scientific truth needs to be re-envisioned as industrial wastewater treatment and water pollution control reaches newer and visionary heights. The primary concern for today's human civilization is the provision of basic human needs such as water and electricity. At such a crucial juncture of human scientific research pursuit, scientific validation, and scientific vision is of utmost and veritable importance.[1]

10.4 THE PRESENT STATUS OF GLOBAL INDUSTRIAL WATER POLLUTION CONTROL

Global industrial water pollution control status is grim and needs to re-envisaged and readdressed as science and engineering ushers in a revolutionary century. Scientific domain is slowly gaining visionary height as scientific needs for full-scale environmental engineering research assumes immense importance. Sustainable water management in its broadest perspective encompasses ecosystems protection and restoration, integrated water resources protection as well as infrastructure development, operation, and maintenance. These areas are totally neglected in both developed and developing nations. Environmental engineering science needs to be widely revamped as industrial pollution is the utmost need of the hour.[1,3]

10.4.1   GLOBAL POPULATION, WATER SUPPLIES, AND THE VISION FOR THE FUTURE

Global population and water supplies are interlinked today as human civilization marches forward toward a newer era. Technology needs to be re-enshrined and revamped as scientific and academic rigor of water technology ushers in a newer scientific domain. Water scarcity is a bane to human advancement. The veritable success, the wide vision, and the targets of science will go a long way in true realization of environmental engineering science and environmental sustainability today. Sustainable development is of utmost importance to the progress of human civilization today. Man's immense scientific and academic rigor, mankind's scientific grit and determination and the futuristic vision of engineering science are all the forerunners toward a greater visionary future in environmental engineering science today. In South Asia, particularly Indian state of West Bengal and Bangladesh, heavy metal and arsenic contamination of groundwater are creating immense global water challenges for the future scientific generations.[3] Water supplies are today in a state of immense scientific fortitude and deep scientific vision. Human scientific generations are challenged today as water crisis and global water shortage deeply derail the progress of a nation whether it is developed or developing.[1]

10.4.2   WATER STORAGE, WATER BANKING, AND THE ENVIRONMENTAL ENGINEERING SCENARIO

Water storage and water banking are of immense importance in the academic and scientific rigor of environmental engineering science today. The challenge and vision of science and engineering are widening the visionary scientific candor and the vast and versatile domain of scientific forbearance.

Environmental engineering science today stands in the midst of immense scientific profundity and scientific sagacity. Technology and engineering science are immensely challenged today. Lack of clean drinking water supplies are a bane to human civilization and scientific rigor today. Scientific challenges, scientific endurance, and scientific profundity are today at a stage of immense catastrophe as provision of pure drinking water stands in the midst of deep crisis. The march of science and engineering is a stunning bane to human advancement as provision of clean drinking water derails the scientific fabric and scientific landscape (www.google.com; www.wikipedia.com).[1,3]

10.4.3 WATER REUSE AND THE SUCCESS OF GLOBAL WATER CHALLENGES

Water reuse today stands in the critical juncture of scientific vision, scientific success, and at the same time scientific derailment. Human mankind's immense scientific prowess and wide scientific candor are changing the scientific scenario as provision of clean drinking water is a bane toward scientific progress. Global water challenges and global water research and development initiatives are facing immense scientific and technological challenges in this century. Human scientific and academic rigor are today at a state of immense distress as global water issues derail the progress of a nation and destroys the scientific fabric. In this treatise, the author repeatedly focuses on the immense scientific rigor behind water purification and provision of clean drinking water. Human mankind's wide vision, man's immense scientific girth, and the technological advancements are the visionary parameters toward a newer dimension of scientific thoughts and scientific vision.[1,3]

10.4.4 NONPOINT SOURCE POLLUTION OF GROUNDWATER AND THE VISION OF FUTURE

Nonpoint source pollution of groundwater is of immense concern and scientific relevance as human civilization ponders toward a greater scientific era. In this treatise, the author rigorously points out to the immense scientific potential and the scientific applications of environmental engineering science as human civilization moves toward a newer scientific life and scientific regeneration.[1] Science and engineering are gearing forward toward newer scientific rejuvenation. Nonpoint source pollution of groundwater is of major environmental concern. Technology and engineering science are immensely challenged as scientific endeavor ushers in a newer era in environmental engineering. Success of science, the futuristic vision of engineering, and the immense technological challenges are all leading a long and visionary way in the true realization of environmental sustainability. Sustainable development whether it is energy or environmental are the true forerunners of a nation's progress in our planet.[1]

10.4.5 VISIONARY WATER PURIFICATION AND ENVIRONMENTAL ENGINEERING TECHNIQUES

Water purification and its scientific endeavor today stands in the midst of immense scientific vision and technological forbearance. Global water issues and heavy metal groundwater remediation are the crucial and vexing problems of human civilization today. Scientific girth and scientific determination in solving water shortage problems are challenging the scientific and engineering community throughout the world. Technology should be indigenous and grassroots entrenched with the progress of human mankind. The world of science and technology is facing immense challenges and wide scientific and engineering hurdles. Environmental engineering frontiers need to be surpassed at this crucial juncture of human history and time.[1] The overarching goal toward scientific emancipation and human civilization is the provision of basic human needs such as water, energy, and sustainable development. Science and engineering of water technology need to be restructured and revamped at the utmost and immediate effect (www.google.com; www.wikipedia.com).[1]

Nonconventional environmental engineering techniques are revolutionizing and reframing the environmental protection scenario. Advanced oxidation processes (AOPs) encompass the domain of nonconventional environmental engineering paradigm. Nowadays, there is a continuously increasing worldwide concern for research and development of alternative water reuse innovations, mainly focused on agriculture, industry, and human society. Technological motivation and validation are of utmost importance to the progress of academic and scientific rigor in global water issues. In this section, the author repeatedly urges upon the wide success of nonconventional environmental engineering tools particularly AOPs and integrated AOPs. The tools of environmental engineering science are slowly evolving toward a newer paradigm of scientific vision and scientific rejuvenation. In the relevant concept of water reuse, AOPs are considered a highly competitive water treatment technology for the removal of those organic pollutants not treatable by conventional techniques due to their high chemical stability and low biodegradability. Science, mankind, and vision are ushering in a new era in environmental engineering science. The author in this treatise pointedly focuses on the immense scientific potential behind AOPs primarily ozone oxidation or ozonation. Although chemical oxidation for complete mineralization is usually expensive, its combination with a biological treatment is vastly reported due to its reduced operating cost. Oller et al.[4] delineated with deep and cogent details combination of AOPs and biological treatments for wastewater decontamination in a widely researched review. Integrated AOPs are the utmost background of this review work. Technology needs to be re-envisioned and restructured as scientific endeavor and scientific research forays recalls immense scientific and technological validation.[4] This chapter reviews recent research combining AOPs (as a pretreatment or post-treatment stage) and bioremediation/biochemical technologies for the decontamination of a wide range of synthetic and real industrial wastewater. Oller et al.[4] also emphasized on recent studies and large-scale combination schemes developed in Mediterranean countries for nonbiodegradable wastewater treatment and reuse. Water reuse, water desalination, and water disinfection stands as a major component in a nation's scientific rigor particularly the Middle-Eastern countries and water-stressed nations. Technology and science are retrogressive today as regards immense global water challenges. Scientific validation, technological adroitness, and scientific motivation are of prime importance in the application of integrated AOPs. Oller et al.[4] deeply discussed on the work that needs to be done on degradation kinetics and reactor modeling of the combined process and also dynamics of the initial attack on primary components and intermediate species generation. This chapter rigorously presents the immense scientific success behind integrated AOPs with the sole aim of progress of science and engineering. Other areas of scientific endeavor in this chapter are better economic models which need to be developed to estimate how the cost of this combined process varies with specific industrial wastewater characteristics, the overall decontamination efficiency, and the relative cost of the AOP versus biological treatment.[4]

10.5 GROUNDWATER HEAVY METAL AND ARSENIC CONTAMINATION AND THE IMMENSE SCIENTIFIC CHALLENGES

Groundwater heavy metal and arsenic contamination are immense challenges toward the progress of human civilization today. Water purification today stands in the crucial juncture of scientific vision and scientific forbearance. Today, environmental engineering science is a huge colossus with a definite vision of its own. Technology and engineering science are today moving at a rapid pace crossing visionary frontiers. Water science and water technology are the frontiers of science today. The immense scientific vision and the scientific challenges will lead a long and visionary way in the true emancipation of environmental engineering science today.

10.5.1 SCIENTIFIC ENDEAVOR IN THE FIELD OF HEAVY METAL GROUNDWATER REMEDIATION

Technology needs to be revamped at this state of human civilization today. Groundwater heavy metal contamination stands as a major technological defeat in the progress of human scientific and academic rigor. Science has few answers to this marauding crisis in South Asia particularly Bangladesh and the state of West Bengal in India. Scientific vision, scientific forbearance, and deep scientific introspection are of utmost need at this stage of scientific rigor in groundwater quality research and development initiatives. The concern for this large-scale environmental disaster needs to be re-envisioned and re-envisaged at each step of human life today.[3]

Hashim et al.[3] delineated with deep and cogent insight in a review paper remediation technologies for heavy metal contaminated groundwater. The contamination of groundwater by heavy metal, originating either from natural soil sources or from anthropogenic sources is a matter of immense concern to the public health. Remediation of poisoned groundwater is of highest priority since billions of people around the world use it for drinking water purpose. Science and engineering needs to be revamped and re-organized as human civilization passes through a difficult phase in human history and time. In this treatise, the author repeatedly stresses upon the immense importance and the utmost need of remediation technologies with the sole and immediate need of the progression of science and engineering. In the paper by Hashim et al.,[3] 35 approaches for groundwater treatment have been reviewed and classified under three large categories, namely, chemical, biochemical/biological/biosorption, and physicochemical treatment processes.[3] Selection of a suitable technology for contamination remediation at a particular site is of utmost importance as science and engineering moves from one environmental engineering paradigm over another. The groundwater contamination issue in Bangladesh and the state of West Bengal in India is extremely grave.[3] Technology is deeply challenged and scientific introspection needs to be re-envisaged at each step of human life. Selection of a suitable technology for groundwater contamination remediation at a particular site is one of the most challenging job due to extremely intricate soil chemistry and aquifer characteristics and no thumb rule can be suggested regarding this issue. In the recent past, iron-based technologies, microbial remediation, biological sulfate reduction, and various adsorbents played a pivotal role in efficient groundwater remediation. Technology needs to be restructured and re-envisaged as science of groundwater remediation gains new heights and visionary realms. The authors reiterated the cause of effective groundwater remediation with the primary and pivotal aim toward furtherance of science and engineering. The technologies encompassing natural chemistry, bioremediation, and biosorption are highly recommended to be implemented keeping the sustainability issues and wide ethics of environmental protection in mind. Heavy metal is a widely collective term, which applies to the group of metals and metalloids with atomic density greater than 4000 kg m^{-3} or 5 times more than water and they are natural components of the earth's crust.[3] Although some of them act as micronutrients for living beings, at higher concentrations, they can lead to severe poisoning. In the environment, the heavy metals are generally more persistent than organic contaminants such as pesticides or petroleum byproducts. Technology revamping and scientific motivation are the pillars toward success of science today. Water issues and global water shortages are the immense struggles of human civilization today. Technology stands baffled and science strained as water technology achieves new dimensions in its applications and scientific vision. The heavy metals in the environment can become mobile in soils depending on soil pH and their speciation. The struggles of science and the challenges behind it are of utmost importance as science delves into the murky depths. The heavy metals can leach to aquifer or can become available to living organisms. Heavy metal poisoning can eventually result from drinking-water contamination (e.g., Pb pipes, industrial and consumer wastes), intake via the food chain, or high ambient air concentrations near emission sources.[3] Over the past few years, many remediation technologies have been re-envisioned and re-envisaged with the sole and pivotal aim for the alleviation of global water crisis and enhance global water research and development initiatives. Groundwater contamination has veritably challenged the vast and versatile scientific panorama. The main target areas of remediation technologies are contaminated soil and aquifers. In this review, Hashim et al.[3] rigorously point out the technologies for the removal of only heavy metals from groundwater, that is, the water which is located in soil pore spaces and in the fractures of rock units. Human scientific endeavor, the vision of engineering science and the fruits of environmental engineering science will all lead a long and visionary way in the true emancipation of remediation science and true realization of science of sustainability in decades to come. This treatise widely proves and subsequently delineates the success, the scientific aura, and the immense scientific profundity behind global remediation technologies and its effective applications. In the near past, some technologies were applied for removing only petroleum products, some for inorganic solvent removal, while some were earmarked for heavy metal removal. Of late, these hurdles have been slowly diminishing as researchers around the world are combining various technologies to achieve the desired result. Science and engineering of water technologies are today surpassing vast and versatile visionary frontiers. Technology and science today stands devastated as human mankind trudges ahead in the sole quest for scientific forbearance and scientific sagacity.[3]

Heavy metals occur in the earth's crust and may get solubilized in groundwater through natural processes or by change in soil pH in an effective manner.[3] Engineering science stands immensely strained as human civilization moves toward a newer visionary eon with every step of human life. Groundwater can get highly contaminated with heavy metals from landfill leachate, sewage, leachate from mine tailings, deep-well disposal of liquid wastes, seepage from industrial waste lagoons, or from industrial spills and leaks. Technology is highly advanced today yet highly retrogressive and strained. Water technology needs to be restructured at such point of juncture.[3] A variety of reactions in soil environment, for example, acid/base, precipitation/dissolution, oxidation/reduction, sorption, or ion exchange processes can influence the speciation and high mobility of metal contaminants. The rate and extent of these reactions depend on factors such as pH, Eh, complexation with other dissolved constituents, sorption and ion exchange capacity of the geological materials, and organic matter content.[3]

10.5.2 TECHNOLOGIES FOR THE TREATMENT OF HEAVY METAL CONTAMINATED GROUNDWATER AND THE VISIONARY FUTURE

Science and engineering of treatment of heavy-metal-contaminated water are witnessing immense drastic challenges with every step of human life and every step of scientific endeavor. Several technologies are prevalent for the remediation of heavy metals contaminated groundwater and soil and they have definite outcomes such as (1) complete or substantial degradation of the pollutants, (2) extraction of pollutants for further treatment and disposal, (3) stabilization of pollutants in forms less toxic, (4) separation of noncontaminated materials,[3] and (5) containment of the polluted materials that require further treatment (www.google.com; www.wikipedia.com).[3]

The overall classification involves[3]:

- chemical treatment technologies;
- in-situ treatments by using reductants;
- reduction by dithionite;
- reduction by gaseous hydrogen sulfide;
- reduction by using iron-based technologies;
- soil washing;
- in-situ soil flushing;
- in-situ chelate flushing;
- biological, biochemical, and biosorptive treatment technologies;
- biorestoration; and
- bioprecipitation.

10.6 SCIENTIFIC RESEARCH PURSUIT IN THE FIELD OF WATER PURIFICATION AND THE VISIONARY FUTURE

Science and engineering of environmental protection are challenged today with the progress of human civilization and scientific achievements. Water is a vital need for the developed and the developing economies in our scientific paradigm. Environmental catastrophes, technological and scientific validation, and deep scientific forays are the forerunners toward a greater visionary emancipation of the science of sustainability today. Water purification is gaining new heights as technology of zero-discharge norms and water reuse assumes immense and vital importance. In this treatise, the author evokes the immense understanding of water science and technology with the sole aim of furtherance of science and technology. Technology of water engineering is today retrogressive as environmental catastrophes and the environmental restrictions redefine the wide scientific fabric. The author repeatedly assures the readers the greatness of the science of environmental and energy sustainability in the progress of a nation.

Shannon et al.[1] deal lucidly with foremost foresight the science and technology for water purification in the coming decades. One of the veritable problems facing human civilization today is inadequate access of clean and pure drinking water to the starving millions in our day-to-day human civilization.[1] Sanitation is the other side of the visionary coin of deep scientific endeavor today. Problems with the water technology are expected to worsen in the coming decades with water shortage occurring globally even in regions which are considered to be water rich. Science and engineering are baffled as human civilization enters a newer eon. Research frontiers are vast and varied today. So addressing these problems needs a concerted scientific effort. Robust new technologies for purifying water at a lower cost and with less energy needs to be identified with the passage of scientific history and time. In this treatise, the authors highlighted some of the science and technology being developed through the safe reuse of wastewater and efficient desalination of sea and brackish water. The widespread worldwide problems associated with the lack of clean, fresh water are well known: 1.2 billion people lack access to safe drinking water, 2.6 billion have little or no sanitation, millions of people die annually—3900 children a day—from diseases transmitted through unsafe water or human excreta.[1] Shannon et al.[1] deeply comprehended the success of human civilization and the success of scientific rigor in solving global water problems. The authors with deep and cogent insight delineated the immense scientific girth and scientific profundity in solving today's global water quality issues. Water quality is of vital importance in the progress of human civilization and the progress of mankind's academic rigor. This treatise deeply ponders on the course of action in the implementation of global environmental sustainability. Sustainable development whether it is energy, environmental, or social is the bedrock of human progress today. Global water status today stands in the midst of immense social catastrophe and unimaginable scientific hiatus. Technology needs to be reframed as academic and scientific rigor marches forward. The water crisis stands immensely devastated. Countless people are sickened from disease and contamination of drinking water. Intestinal parasitic infections and diarrheal diseases caused by waterborne bacteria and enteric viruses have become a primordial cause of malnutrition owing to poor digestion of the food eaten by people sickened by water. Technological profundity is veritably baffled with the passage of scientific history, scientific vision, and time. In both developing and industrialized nations, a growing number of contaminants are entering water supplies from human activity: from traditional compounds such as heavy metals and distillates to emerging micropollutants such as endocrine disrupters and nitrosoamines.[1] Human mankind's immense scientific determination, technological forbearance, and the wide academic rigor will go a long and visionary way in the true emancipation and true realization of sustainable development today. Energy and environmental sustainability are the technological directions of human mankind and human scientific endeavor today. The challenge and vision are immense and groundbreaking. More effective, lower cost, robust methods to disinfect and decontaminate waters from source to point of use are veritably needed, without further stressing the environment or endangering human health by the treatment itself. Human mankind is at a state of immense distress and veritable crisis today.[1] The environmental concern, the concerns for ecological biodiversity, and the wide vision of sustainability science will all lead a long and visionary way in the true advocacy of environmental sustainability today. The authors in this treatise detailed the importance of water purification and the wide scientific endeavor in the present century. The overarching goal of this century as regards water purification and its scientific vision are the domains of desalination, disinfection, and effective water reuse (www.google.com; www.wikipedia.com).[1]

10.7   GLOBAL WATER RESEARCH AND DEVELOPMENT INITIATIVES

Global water research and development initiatives today are standing between immense scientific vision and wide avenues of scientific endeavor. Novel separation techniques and nonconventional environmental engineering procedures are today the main pillars behind environmental protection. The success of human scientific endeavor depends on the vast domain of scientific validation and technological motivation. Scientific barriers and scientific challenges are immense as environmental engineering science moves from one paradigmatic shift toward another. Environmental restrictions, stringent environmental regulations and the futuristic vision of environmental engineering separation processes are the forerunners toward a newer visionary future of environmental protection. Global drinking water shortage and heavy metal contamination of groundwater are the torchbearers toward all the governmental policies of developed and developing nations of the world. Global water challenges are moving toward a newer dimension of scientific thought and scientific vision. Technology needs to be restructured as human civilization stands in the midst of environmental devastations.[1]

10.8   CONVENTIONAL ENVIRONMENTAL ENGINEERING TECHNIQUES

Conventional environmental engineering techniques are re-envisioned as human scientific endeavor faces tremendous challenges and scientific barriers. Technology is so much baffled as there are few solutions to the marauding global water crisis. Engineering science today stands in the midst of scientific hope and scientific determination. The success of scientific endeavor in heavy metal and arsenic groundwater remediation are limited as the crisis of groundwater contamination evolves new scientific dimension in South Asia and developing countries. The crisis in Bangladesh and West Bengal state in India is evergrowing. Science and engineering have few answers to this widely researched crisis. Water purification and drinking water treatment today stands in the midst of immense scientific candor and deep scientific understanding. Membrane science is a widely researched arena which comes under conventional environmental engineering techniques. Membrane-separation phenomenon is veritably challenging the future of environmental engineering science. Desalination science and drinking water treatment in Middle East countries in our planet are changing the scientific landscape.

Novel separation processes such as membrane science are the fountainhead of the domain of conventional environmental engineering technique.[1] Membrane science encompasses reverse osmosis (RO), nanofiltration, ultrafiltration, microfiltration, and electrodialysis. This technology today is surpassing wide and vast visionary boundaries. Global water challenges today stands in the midst of wide scientific introspection and deep scientific understanding. Industrial water pollution and drinking water crisis are today challenging the scientific fabric, the deep scientific profundity, and the scientific sagacity. Science today has few answers to groundwater heavy metal contamination and drinking water issues. In this treatise, the author pointedly focuses on the immense scientific potential of conventional and nonconventional environmental engineering tools with the sole aim of emancipation of global water challenges. Membrane science needs to be reorganized and research trends needs to be re-envisioned as human civilization ushers in a newer eon in scientific research pursuit. Technological and scientific validation assumes equal importance today with the passage of scientific history and time.[1,2]

10.8.1 NOVEL SEPARATION PROCESSES, MEMBRANE SCIENCE, AND THE VISIONARY FUTURE

Membrane science encompasses novel separation processes. The visionary future of membrane science is far-reaching and surpassing frontiers of scientific validation and scientific vision. Human mankind's immense scientific girth, scientific determination, and the progress of science are all leading a long way in the true emancipation of environmental sustainability today. Technology needs to be revalidated as environmental engineering science faces scientific challenges. Water is an essential component of a nation's growth. Provision of clean drinking water needs to be re-envisioned as science and engineering ushers in a newer eon of scientific validation and scientific rejuvenation.[1,2]

Novel separation processes are the next generation science and engineering endeavor. Scientific discernment, scientific understanding, and scientific vision are the forerunners toward a greater visionary future in the field of environmental engineering science. Today membrane science is linked by an unsevered umbilical cord with global water research and development initiatives. Zero-discharge norms and concerns for environmental sustainability have urged the global scientific domain to garner resources toward newer innovations and newer visionary avenues. Today the human civilization is moving toward newer scientific regeneration. The challenge and the vision of science is reaching a state of immense scientific imagination and scientific understanding. Global water shortages stand as a major hiatus to the progress of scientific and academic rigor. Thus, in such a crucial juncture of human history and time, environmental and energy sustainability assumes immense importance.

10.9 TECHNOLOGICAL DEVELOPMENT OF MEMBRANE BIOREACTORS

Membrane bioreactors (MBRs) are next generation scientific innovation and a veritable environmental engineering tool. Environmental protection needs to be redefined as scientific and academic rigor opens up a new era of environmental engineering and chemical process engineering. Chemical process engineering and the application areas of membrane science and technology are today linked by an unsevered umbilical cord. In a similar vein, water science and technology are connected to membrane-separation processes by a veritable umbilical cord. MBRs are ushering in a new era in the field of environmental engineering science today. This type of bioreactors works with the principle of membrane science. Technological validation and deep scientific motivation has urged the scientific domain toward greater validation of the science of biotechnology. MBRs works under the principle of biotechnology science (www.google.com; www.wikipedia.com).[2]

MBR is the combination of a membrane process like microfiltration or ultrafiltration with a suspended growth bioreactor and is widely implemented in municipal and industrial wastewater treatment. The plant sizes go up to 80,000 population equivalent (48 million liters per day) (www.google.com; www.wikipedia.com). When used with domestic wastewater, MBR processes can produce effluent of highest quality enough to be discharged to coastal, surface, or brackish waterways or to be reclaimed for urban irrigation and other urban environmental science areas. Science of environmental protection today is highly challenged as human mankind moves toward a world of scientific regeneration and scientific vision. This technology evolves into a newer future dimension in the field of membrane science and environmental engineering science (www.google.com; www.wikipedia.com).[2]

10.10 NONCONVENTIONAL ENVIRONMENTAL ENGINEERING TECHNIQUES

Nonconventional environmental engineering techniques encompass chemical oxidation and AOPs. Science of nonconventional environmental engineering tools is facing immense challenges and a wider vision. Ozone oxidation technique falls under advanced oxidation techniques. Today is the scientific world of integrated advanced oxidation techniques. AOPs are more cost-effective and sludge-free processes. This technique has immense scientific potential. The challenge and vision of science are today evolving into a new era of chemical oxidation science. Technology of AOPs is highly advanced today. Provision of clean drinking water is a global issue and an immense challenge. Science and technology has few answers to this ever-growing crisis.

Global water crisis can be solved only with the veritable advancements in conventional and nonconventional environmental engineering techniques. Technology needs to be transformed and re-envisioned. Global water crisis and heavy metal contamination of groundwater are serious environmental engineering issues. The concept of sustainable development defined by Dr. Gro Harlem Brundtland, former Prime Minister of Norway, needs to be reshaped and readjudicated with the passage of scientific history and time. Environmental sustainability and global sustainable development goals are the major parameters of a nation's growth today. Progress of technology and environmental sustainability are in today's human civilization linked by an unsevered umbilical cord (www.google.com; www.wikipedia.com). Sustainability in today's world also encompasses provision of basic human needs such as water and also veritably involves industrial wastewater treatment. The scientific revelation today is slowly evolving into new dimensions of research and development initiatives in global water initiatives.[2]

In such a crucial juncture of human history and time, environmental engineering techniques, whether it is conventional or nonconventional, are of utmost importance. Science and engineering are immensely challenged and are replete with immense scientific struggles and deep scientific vision.[2]

10.10.1   ADVANCED OXIDATION PROCESSES AND RECENT ADVANCES

AOPs are challenging the scientific fabric today. Engineering science needs to be redefined as human civilization and human scientific endeavor approach a visionary era. Nontraditional environmental engineering techniques are veritably changing the face of human scientific endeavor. Scientific cognizance, scientific vision, and scientific sagacity are the backbones and veritable pillars toward greater emancipation of environmental engineering today. Recent advances in the field of AOPs are vast and varied. Validation of science is of utmost importance as engineering science passes from one vast paradigm toward another.[2]

In this section, the author deeply comprehends the wide domain of AOPs and immense scientific success and potential. This innovation needs to be restructured and re-envisioned with every step of scientific life and regeneration. AOPs today stand in the midst of immense scientific comprehension and deep scientific vision. Human civilization's immense scientific prowess, the deep vision of science and the technological frontiers all will lead a long and visionary way in the true realization of environmental engineering science today. Some uncommon nonconventional environmental engineering techniques and its wide scientific potential and application are dealt with in the following deliberations.

Peralta-Hernandez et al.[5] discussed with deep and cogent insight recent advances in the application of electro-Fenton and photoelectron-Fenton process for removal of synthetic dyes in wastewater treatment.[5] In recent years, increasing need to preserve environment has arisen, mainly by eliminating chemical compounds produced by many industrial activities such as textile industry discharges.[5] In this context, electrochemical AOPs capable of destroying several organic compounds in industrial wastewater are being developed. The foundations for these methods are the production of free hydroxyl radicals (OH') as a primary oxidant.[5] Technology needs to be revamped today with the evergrowing concern for environmental protection. The challenge, the vision, and the immense scientific sagacity are the forerunners toward a greater technological challenge of environmental engineering tools. In these AOPs, the OH' is generated by water discharge over the anode of a high oxygen overvoltage material and by the electro-Fenton type reaction.[5]

Barrera-Diaz et al.[6] discussed lucidly electrochemical AOPs and gave an overview of the current applications to actual industrial effluents. The world of electrochemical AOPs and other integrated AOPs are witnessing new scientific rejuvenation and scientific regeneration. Technology needs to be restructured and scientific vision needs to be readdressed as environmental engineering science passes through hurdles and scientific barriers.[6] Many human activities are involved in the production of wastewater. Usually, physical, chemical, and biological processes are combined in an integrated manner for the treatment of municipal wastewater, attaining good removal efficiencies. However, some wastewater compounds are recalcitrant and are nondegradable. In such a critical situation, the AOPs assume pivotal importance.[6] These processes rely on generating hydroxyl radicals, which is a powerful oxidant that mineralizes effectively pollutants contained in wastewater. Technology of wastewater treatment is highly advanced and is widely promising. In this review, the authors deeply focuses on the use of electrochemical methods to produce hydroxyl radicals, using directly or indirectly electrochemical technology, within the so-called advanced electrochemical oxidation processes.[6] These processes encompass electrochemical, sonoelectrochemical, and photoelectrochemical technologies and this review work delineates the fundamentals and main case studies in the literature related to actual industrial wastewater treatment.[6] Scientific progress in the field of AOPs has been lucid, promising, and far-reaching. Electro-Fenton was the first technology which could be considered as an electrochemical AOP because of the production and active role of hydroxyl radical on the oxidation of organics. This technology is widely dependent on the promotion of one or several of these processes:

- The electrochemical regeneration of iron(II) from iron(III) species on the cathodic surface.[6]
- The cathodic formation of hydrogen peroxide from the reduction of oxygen and these processes lead to the catalytic decomposition of hydrogen peroxide into hydroxyl radical.[6]

10.11　MEMBRANE-SEPARATION PHENOMENON AND THE SCIENTIFIC VISION

Membrane-separation phenomenon is a wide and revolutionary avenue of scientific endeavor today.[2] This process which includes RO, ultrafiltration, nanofiltration, microfiltration, and dialysis are opening up new avenues in scientific research pursuit today. Scientific vision with respect to global water crisis and global water challenges needs to be reframed as science and engineering moves toward newer innovations in this decade. Loeb–Sourirajan model[2] redefined and revolutionized the diffusion process of membrane-separation processes.[2] Technology and science of membrane-separation phenomenon need to be revamped with the renewed concern for successful environmental sustainability. Sustainable development whether it is social, energy, or environmental is the utmost need of the hour. The vision to move forward in a nation's growth depends on the successful realization of sustainability and the successful implementation of global sustainable development goals. Environmental engineering science in a similar vein needs to be revamped and overhauled as growing concerns for environmental catastrophes gains newer heights. The author in this treatise pointedly focuses on the immense scientific potential and scientific vision behind environmental engineering techniques whether it is conventional or nonconventional.[2]

10.11.1　RECENT SCIENTIFIC ENDEAVOR IN THE FIELD OF WATER PURIFICATION

Science is moving from one visionary paradigm toward another. Environmental protection and environmental engineering science today stands in the midst of immense scientific introspection and deep scientific understanding. In this treatise, the author repeatedly points out toward the scientific vision and the success of scientific research pursuit in the field of water purification.[2]

Shannon et al.[1,2] discussed with deep and cogent insight science and technology for water purification in the coming decades. One of the pervasive and disastrous problems afflicting people throughout the world is inadequate access to clean drinking water and proper sanitation. The technological vision and wide objectives of science and engineering are slowly unfolding as environmental engineering science witnesses the vicious scientific challenges of our times. In the paper of Shannon et al.,[1] the authors rigorously point out toward the domain of desalination, disinfection, and the wide world of water reuse.[2] Technological and scientific validation are of utmost importance as science and engineering moves forward. Water purification veritably stands in the midst of deep scientific understanding and discernment. Today, a rigorous approach is of utmost importance with the progress of scientific and academic rigor. Water reuse is the necessity of the hour as drinking water challenges derails the progress of human civilization. Arsenic groundwater poisoning has been the greatest challenge of the human planet. The challenge, the vision, and the success need to be readdressed as mankind enters into a newer eon.[1]

10.11.2   RECENT SCIENTIFIC ENDEAVOR IN MEMBRANE-SEPARATION PROCESSES

Recent scientific endeavor in membrane-separation processes are far-reaching and crossing visionary boundaries. The author in this treatise rigorously points out the immense scientific vision and scientific fortitude behind today's research pursuit in the field of membrane-separation processes. Technology and engineering science are challenging the path of a nation's development and the success of realization of environmental sustainability. Today, a nation's growth depends on the sustainable development whether it is energy, environment, or social. The vision of human mankind needs to be reshaped as environmental engineering science trudges toward the visionary scientific avenues of this century.[1]

Zhu et al.[7] delineated in deep details colloidal fouling of RO membranes and its measurements and fouling mechanisms. The effect of chemical and physical interactions on the fouling rate of cellulose acetate and the aromatic polyamide composite RO membranes by silica colloids are deliberated in details. Results of fouling experiments using laboratory scale unit demonstrate that colloidal fouling rate increases with increasing solution ionic strength, feed colloid concentration, and permeate flux through the membrane. Technological rejuvenation and scientific regeneration are of utmost importance in today's scientific research pursuit.[7] Fouling stands as a major component in successful membrane operation. Science has few answers to it. The entire chemical engineering world and environmental engineering science are involved in widespread research in membrane science and the intricate issue of membrane fouling. Rate of colloidal fouling is controlled by a unique interplay between permeation drag and electrical double-layer repulsion, that is, colloidal fouling of RO membranes involves interrelationship (coupling) between physical and chemical interactions.[7]

The science of fouling and its intense intricacies needs to be readdressed and re-envisaged with each step of scientific progress.[7]

Ang et al.[8] widely researched on chemical and physical aspects of cleaning of organic-fouled RO membranes. The role of physical and chemical interactions is systematically investigated in details. Fouling and cleaning experiments were performed with organic foulants that simulate effluent organic matter and selected cleaning agents using a laboratory scale crossflow unit.[8] The authors in this paper deeply deliberated on the chemical reaction engineering and mass transfer aspects during the cleaning of the membranes. Science and technology needs to be sharpened and enshrined with the progress of scientific and academic rigor in membrane science.[8] The authors touched upon chemical aspects of cleaning, effect of cleaning solution type, effect of cleaning solution pH, effect of cleaning chemical dose, effect of organic foulant composition, physical aspects of cleaning, and effect of cleaning time.[8] Engineering science is moving from one paradigmatic shift in scientific cognizance over another. The evergrowing concerns of global water issues has plunged the wide scientific world into scientific comprehension and deep introspection. The vast challenges and the scientific motivation need to be reorganized as research and development efforts in water science gains momentum. The technological vision thus is ushering in a new era in membrane technology.[8]

Lee et al.[9] dealt lucidly on the area of fouling of RO membranes by hydrophilic organic matter and its vast implications for water reuse. Validation of engineering science needs to be re-envisaged and re-addressed as membrane science and water technology moves to a new era.[9] Effluent organic matter is suspected as a major cause of fouling of RO membranes in advanced wastewater reclamation. Environmental protection and the wide world of chemical process engineering need to be re-envisioned at each stride of scientific research pursuit. Technology revamping and scientific vision are the order of the day. The authors in this treatise targeted on the wide domain of fouling with the sole aim of unraveling the murky depths of science and engineering.[9]

10.12   TECHNOLOGICAL VISION OF ADVANCED OXIDATION PROCESSES AND OZONE OXIDATION

Technological vision of AOPs and particularly ozone oxidation are in the path of immense and deep scientific introspection. Deliberations on water purification will be veritably incomplete without the mentioning the wide scientific forays in nonconventional and chemical oxidation processes. In this section, the author rigorously points out the immense scientific comprehension and deep introspection into the field of AOPs particularly ozone oxidation. Technology and engineering science are highly advanced today. Environmental engineering science in the similar vein is a wide branch of scientific endeavor which is going beyond scientific imagination and scientific cognizance.[10–12]

10.13   ENVIRONMENTAL SUSTAINABILITY AND GROUNDWATER REMEDIATION

Environmental sustainability today is linked to a nation's economic and social growth. Sustainability issues whether it is environmental or energy is linked by an unsevered umbilical cord with global water issues and challenges. Science and technology are highly advanced today. Groundwater heavy metal contamination is urging the scientific domain to gear forward toward newer challenges and newer innovations. The present day success of human civilization depends on the basic mankind's issues such as provision of clean drinking water. Groundwater remediation and de-poisoning of drinking water should be scientifically re-envisioned as science and engineering of water purification evolves into a newer visionary era and a newer world order. Sustainable development is a vexing issue in the progress of human civilization. Today's science is the wide visionary science of environmental and energy sustainability. Today sustainability issues encompass global water issues also.[10–12]

Sustainable development is defined as a process of meeting human development goals while sustaining the ability of natural systems to continue to provide the natural resources and ecosystem services upon which the economy and society depends. While the present modern concept of sustainable development is derived most strongly from 1987 Brundtland Report, it is rooted in earlier ideas about sustainable forest management and the 20th century environmental concerns.[10–12]

10.13.1   ENERGY SUSTAINABILITY, THE SOCIAL SUSTAINABILITY, AND THE WIDE VISION

Energy sustainability and the provision of basic human needs are of immense importance in the path toward scientific regeneration today. Technological challenges, scientific motivation, and deep scientific understanding are the torchbearers toward a greater visionary future in the domain of sustainable development. Social sustainability is a concept on the other side of the visionary coin. Water and electricity are two pivotal components of a nation's progress. The scientific endeavor needs to be re-envisioned and re-emphasized as regards successful realization of both energy and environmental sustainability. Many developing nations are in the throes of lack of basic needs such as water and power. In such a context, sustainable development needs to be re-envisaged and re-enshrined with the passage of scientific history, scientific validation, and the world of scientific rigor.[10–12]

10.14   FUTURE RESEARCH TRENDS AND FUTURE FLOW OF SCIENTIFIC THOUGHTS

Research and development initiatives in global water issues today stand in the midst of deep scientific understanding and introspection. Future research trends in water science and technology should be toward more inclined toward zero-discharge norms and successful realization of environmental and energy sustainability. The immense challenge of the human civilization and human scientific endeavor lies in the hands of technocrats and environmental scientists. Today human civilization stands in the midst of deep crisis as regards environmental protection. Science has few answers to the vexing water issue. Zero-discharge norms and successful sustainable development are the ever-growing concerns of human mankind today. The future of human scientific endeavor needs to be re-envisaged toward newer innovations and newer discoveries in conventional and nonconventional environmental engineering techniques. In this treatise, the author pointedly focuses on the immense vision behind nonconventional environmental engineering tools such as AOPs. The immediate and the utmost need of the hour are the targets toward environmental sustainability and the emancipation toward industrial wastewater treatment. Mankind's scientific prowess, scientific discernment and the futuristic vision will lead a long and visionary way in the true progress of environmental engineering science (www.google.com; www.wikipedia.com).[10–12]

10.15   CONCLUSION AND ENVIRONMENTAL PERSPECTIVES

Zero-discharge norms and water reuse are the primary motives of today's scientific research pursuit in water purification. Scientific endeavor and scientific forbearance are the pillars of today's science and engineering.

In this treatise, the author repeatedly stresses upon the success of environmental engineering tools in industrial water pollution control and drinking water treatment (www.google.com; www.wikipedia.com).[10–12]

The world of environmental protection is dramatically moving from one definite vision toward another. Water is an important component to the human civilization's advancement. Technological and scientific validation needs to be re-addressed and re-adjudicated as science and engineering of water purification widens toward a visionary eon. Global water research and development initiatives are witnessing the challenge of our times. In this treatise, the author pointedly focuses on the present day environmental engineering perspectives with the ultimate and sole aim toward the furtherance of science and engineering. Technological motivation needs to be readdressed as water issues such as provision of pure drinking water and industrial wastewater treatment expands the whole domain of environmental engineering science. Science of water purification today is redefined and environmental perspectives reshaped as human civilization ushers in newer challenges and innovations. Technology needs to be revamped as environmental engineering science ushers in a new era of water purification and zero-discharge norms.

ACKNOWLEDGMENTS

The author with great respect wishes to acknowledge the contribution of Shri Subimal Palit, the author's late father, and an eminent textile engineer from India from whom the author learnt the rudiments of chemical engineering science.

KEYWORDS

- vision
- water
- purification
- membranes
- oxidation
- advanced

REFERENCES

1. Shannon, M. A.; Bohn, P. W.; Elimelech, M.; Georgiadis, J. A.; Marinas, B. J. *Science and Technology for Water Purification in the Coming Decades*; Nature Publishing Group: London, 2008; pp 301–310.
2. Cheryan, M. *Ultrafiltration and Microfiltration Handbook*; Technomic Publishing Company Inc.: Lancaster, PA, 1998.
3. Hashim, M. A.; Mukhopadhayay, S.; Sahu, J. N.; Sengupta, B. Remediation Technologies for Heavy Metal Contaminated Groundwater. *J. Environ. Manage.* **2011**, *92*, 2355–2388.
4. Oller, I.; Malato, S.; Sanchez-Perez, J. A. Combination of Advanced Oxidation Processes and Biological Treatments for Wastewater Decontamination—A Review. *Sci. Total Environ.* **2011**, *409*, 4141–4166.
5. Peralta-Hernandez, J. M.; Martinez-Huitle, C. A.; Guzman-Mar, J. L.; Hernandez-Ramirez, A. Recent Advances in the Application of Electro-Fenton and Photoelectron–Fenton Process for Removal of Synthetic Dyes in Wastewater Treatment. *J. Environ. Eng. Manage.* **2009**, *19* (5), 257–265.
6. Barrera-Diaz, C.; Canizares, P.; Fernandez, F. J.; Natividad, R.; Rodrigo, M. A.; Electrochemical Advanced Oxidation Processes: An Overview of the Current Applications to Actual Industrial Effluents. *J. Mex. Chem. Soc.* **2014**, *58* (3), 256–275.
7. Zhu, X.; Elimelech, M. Colloidal Fouling of Reverse Osmosis Membranes: Measurements and Fouling Mechanisms. *Environ. Sci. Technol.* **1997**, *31*, 3654–3662.
8. Ang, W. S.; Lee, S.; Elimelech, M. Chemical and Physical Aspects of Cleaning of Organic-Fouled Reverse Osmosis Membranes. *J. Membr. Sci.* **2006**, *272*, 198–210.
9. Lee, S.; Ang, W. S.; Elimelech, M. Fouling of Reverse Osmosis Membranes by Hydrophilic Organic Matter: Implications for Water Reuse. *Desalination* **2006**, *187*, 313–321.
10. Palit, S. Nanofiltration and Ultrafiltration—The Next Generation Environmental Engineering Tool and a Vision for the Future. *Int. J. Chem. Technol. Res.* **2016**, 9 (5), 848–856.
11. Palit, S. Filtration: Frontiers of the Engineering and Science of Nanofiltration—A Far-Reaching Review. In *CRC Concise Encyclopedia of Nanotechnology*; Ortiz-Mendez, U., Kharissova, O. V.; Kharisov, B. I., Eds.; Taylor and Francis: Milton Park, 2016, pp 205–214.
12. Palit, S. Advanced Oxidation Processes, Nanofiltration, and Application of Bubble Column Reactor. In *Nanomaterials for Environmental Protection*; Kharisov, B. I., Kharissova, O. V., Dias, R., Eds.; Wiley: Hoboken, NJ, 2015; pp 207–215.

CHAPTER 11

FOOD-BORNE VIRUSES: ROLE OF ONE HEALTH IN FOOD SAFETY: "FOOD SAFETY IS EVERYBODY'S BUSINESS"

PORTEEN KANNAN[1*,] S. WILFRED RUBAN[2], and M. NITHYA QUINTOIL[3]

[1]*Department of Veterinary Public Health and Epidemiology, Madras Veterinary College, Chennai 600007, India*

[2]*Department of Livestock Products Technology, Veterinary College, Hebbal, Bangalore 560024, India*

[3]*Department of Veterinary Public Health, Rajiv Gandhi Institute of Veterinary Education and Research, Puducherry, India*

*Corresponding author. E-mail: Rajavet2002@gmail.com

CONTENTS

ABSTRACT

Food as a potential vehicle for disease transmission is well documented and as a result food safety is a prime public health imperative. Food-borne illness caused by new pathogens and expansion of types of food pose an expanded burden of illness and growing spectrum of threats. In this context, food-borne viruses are potential pathogens capable for causing food-borne illness. Therefore, ensuring a safe food supply warrants new levels of collaboration, understanding, and thinking. To counteract the present challenge, there is a need for development of a One Health strategy where potential solutions are viewed and delivered more holistically with an emphasis on prevention in a timely manner.

Food-borne diseases are among the most widespread public health problems but only a limited proportion of these illnesses come to the notice of health services, and of which only few are investigated. In many parts of the world specially in developing nations, limited outbreaks are documented and it is attributed due to various reasons like poverty and lack of facilities for food safety management and control services. In spite of underreporting, increases in food-borne diseases in many parts of the world and the emergence of new or newly recognized food-borne problems have been identified.[22]

11.1   ONE HEALTH AND FOOD SAFETY

One Health is the collaborative effort of multiple disciplines—working locally, nationally, and globally to attain optimal health for people, animals, and our environment.[9] The degree and complexity of food safety issues claim that scientists, researchers, and should think beyond their boundaries, disciplines, professions, and mindsets to explore novel modes of team science, and the One Health concept represent this affirmation. The scope of One Health is impressive, broad, and growing. In recent times, primary concern of One Health is limited to various zoonotic and infectious diseases, but there is an urge to include and interact with different fields, namely, environmental studies, social sciences, ecology, metabolic and nutritional deficiency diseases, biodiversity, land use, antimicrobial resistance, and much more. The competition of survival at human–animal interface is expanding, and becoming increasingly more consequential.

Issues and problems connected with food safety, food security and sustainable production systems that ensure environmental protections, and the capacity to help feed the increasing population is an important and unanswerable. In the present day context satisfying the needs of consumers with respect to safety, accessibility, affordability, and nutritional value of food is becoming increasingly difficult. One should be aware that contamination of food by bacterial pathogens is a prime issue; however, various etiological agents namely viruses, parasites, prions, chemicals, toxins, metals, and allergens may also be transmitted by food and water resulting in an expanded burden of illness and increased spectrum of threats. The FoodNet system analysis has shown that adenoviruses, sapoviruses, scaffold viruses, and picobirna viruses as potential pathogens. To further understand and analyze the safety of our food, transmission patterns and adaptation of pathogen tends to be critical. To overcome various hurdles in relation to food safety, a unique One Health measure is required to employ intervention strategies which will help in devising various regulatory measures. One Health, although not new, is certainly a renewed field of inquiry and transdisciplinary thinking. Our own prejudice and false separation between veterinary and animal health and public health is a critical barrier to the acceptance of One Health.

Dr. Gro Brundtland, former Director of the World Health Organization, stated, "In the modern world, pathogens travel almost as fast as money." With globalization, a single microbial sea washes all of humankind. In this context, the dynamic approach of One Health will lead to interaction of human, animal, and environmental domains in issues never formerly experienced. Hence, actions in one domain may significantly impact other domain resulting in multiplier effects in the others. Hence, One Health is the need of the hour to bring about food safety and bring about paradigm shift in the mindset, but however changing the habitual thinking guided by diverse cultures and interests still remains as a challenge.

Dr. Lederberg, a Nobel Laureate remarked that the future of humanity and microbes will be based on "our understanding in opposition to their genes." This is a visionary statement underpinning One Health's application to food safety.[14]

11.2 FOOD-BORNE VIRUSES

Microbiological quality control criteria for food globally still rely on standard counts of coliform bacteria that were developed as indicators for fecal contamination. In recent times, it has been proved that these criteria are inadequate to protect against viral food-borne infections. For example, foods at retail that passed all microbiological control criteria contains high loads of infectious human pathogenic viruses which warrants the need revisiting the existing protocols. It is, therefore, important to understand the fundamental properties of food-borne viruses.

The Centers for Disease Control and Prevention had listed viruses as one of the emerging causes of food-borne disease. Viruses cause a wide range of infection in plants, animals, and humans and each group of viruses has its own typical host range and cell preference. The role of viruses in causing food-borne illnesses had been happening from decades but due to lack of detection systems, these illnesses have been undocumented/unreported or reported to have been caused by an unknown causative agent. However, recent developments in detection systems have led to the detection and confirmation for the presence of viruses.

11.3   FEATURES OF VIRUSES AS FOOD-BORNE DISEASE AGENTS

- Viruses which contaminate food do not exhibit deteriorative and organoleptic changes.
- Low infectious dose is sufficient to cause infection and produce illness.[11]
- During clinical illness, the animal infected sheds 10^7–10^{11} viral particles per gram of stool.[19]
- Majority of the food-borne viruses are quite stable and resistant to extreme physical and chemical treatments (pH, drying, radiation).
- Transmission of viruses through food is uncommon, except Hepatitis E virus (HEV).
- Person-to-person transmission is the most common route for food-borne viruses.
- Viruses require living cells to replicate, and almost all food-borne viruses are strictly human pathogens. Hence, transmission via food reflects fecal contamination, with the persistence of viruses in the product, but without replication.

Sources of contamination: Humans get the infection by eating food products that have been contaminated during processing. As per Koopmans and Duizer,[11] contamination of foods by virus occurs due to contact with:

- water contaminated with human feces;
- materials soiled with feces (including hands);
- vomit or water contaminated with vomit;

- environments in which infected people were present, even if the surface was not directly contaminated with stool or vomit; and
- aerosols generated by infected people.

However, there is no proof that animal contact, directly or indirectly as a source of food-borne infection (pigs, calves, surface-contaminated meat, meat products, or other products derived from those animals). The major source of contamination of food is through human food handlers. They include:

- infected food handlers showing symptoms and shedding of virus occurs during the period of illness;
- infected food handlers who have recovered from illness—Carriers;
  - o *Norovirus* (NoV) may persist for at least 3 weeks after recovery;
  - o carriers of Hepatitis A typically shed high quantities of the virus 10–14 days after infection; in the weeks following this period, carriers may or may not develop symptoms;
- asymptomatic food handlers;
- food handlers with contacts with sick people (e.g., people with sick children or relatives).

Food-borne viruses can be divided into three main categories[8]:

1. Viruses that cause gastroenteritis—Astrovirus, rotavirus, adenovirus, NoV (formally Norwalk-like viruses), and SLV (Sapporo-like viruses).
2. Viruses that are transmitted by fecal–oral route—hepatitis A and E.
3. Viruses that cause other illnesses—Enteroviruses.

Even though food-borne outbreaks are rarely encountered with rotavirus and astrovirus, these viruses typically affect children as opposed to adults. The viruses at the highest risk for food-borne transmission are NoV and hepatitis A virus (HAV). The reasons may be due to their extreme stability in the environment and their highly infectious nature.[12]

11.3.1 NoV: PREVIOUSLY NORWALK-LIKE VIRUSES

NoVs (genus *Norovirus*, family Calciviridae) are a group of related, single-stranded plus sense RNA, nonenveloped viruses that cause acute gastroenteritis in humans. Currently, human NoVs belong to one of the three NoV genogroups (GI, GII, or GIV), which is further divided into >25 genetic clusters. Majority (75 %) of the confirmed human NoV infections are associated with genotype GII.[5]

NoVs are emerging as one of the foremost enteric pathogens of food-borne disease worldwide since its first recognition by Albert Kapikian and colleagues in 1972;[7] however, the prevalence of NoV infections has been underestimated due to the limited availability of detection methods.

General features

- Causes acute but self-limited gastroenteritis.
- There is no long-term immunity to the virus and reinfection by the same strain can occur several months after the initial population.
- Can affect nearly everyone in the population (especially, the elderly and children under 5 years old).
- Genetic susceptibility to NoV infections is associated to the expression of histoblood group antigens on the mucosal surface of intestinal epithelial cells.[13,16]

Infective dose: As low as 1–10 viral particles.[24] In addition, symptomatic and asymptomatic people excrete 10^{10}–10^{12} million viral particles/g feces.

Route and source of infection: Fecal–oral and through consumption of contaminated ready-to-eat foods, produce or molluskan shellfish. Airborne transmission via aerosolization with vomiting has been implicated as source of infection.[17]

Onset: A mild illness develops between 12 and 48 h after ingestion of contaminated food or water (median in outbreaks: 33–36 h). 30% NoV infection are asymptomatic.

Illness/complications: Self-limited but very debilitating disease. Recovery is usually complete within 2–5 days. Dehydration is the most common complication.

Symptoms: Commonly known as "winter-vomiting" disease, NoV gastroenteritis often accompanied with nausea, vomiting, and watery diarrhea, following an incubation period of 1–2 days. Other associated symptoms include abdominal pain or cramps, anorexia, malaise, and low-grade fever. NoVs are not invasive pathogens, and dysenteric symptoms including bloody or mucoid diarrhea or high fever are uncommon.[10] The clinical manifestation of *Norovirus* infection, however, is relatively mild. Asymptomatic infections are common and may contribute to the spread of the infection.[25] These symptoms are more severe in hospitalized and immunocompromised patients, and elderly people.

Mortality: Accounts for 11% deaths associated with food consumption.

Therapy: No specific therapy for NoV infection. Treatment is supportive (oral rehydration and intravenous replacement of electrolytes). No vaccines available.

Detection of Norovirus

Nomenclature and genetic classification of the viruses have been a major issue due to the vast genetic diversity of the group.[4] Till date, four main "genogroups" of NoV have been described and humans are mainly infected by either genogroup I or genogroup II. Within genogroup I, there are at least seven distinct gene clusters and within genogroup II, there are at least eight distinct gene clusters.[3]

Due to the great genetic diversity of the virus, detection is difficult. NoV was first detected using electron microscopy; however, this technology can be quite insensitive. In recent years, detection has primarily been done using reverse transcriptase polymerase chain reaction (RT-PCR). The problem with RT-PCR is due to the genetic diverse population of NoVs, there is not one set of primers that can routinely detect the virus. To further complicate detection, cultivation of NoV is not possible under lab conditions (Table 11.1).

11.3.2   HEPATOVIRUS

Hepatitis A virus: HAV is the only member of the genus *Hepatovirus* of the family Picornaviridae. HAV is nonenveloped and approximately 27–32 nm in diameter was first described by Feinstone. HAV occurs as a single antigenic type; nonetheless, four human genotypes and three genotypes naturally affecting other primates (chimpanzee and nonhuman primates) can be discriminated. HAV differs from enteroviruses by certain biological characteristics such as marked tropism for liver cells, exceptional thermostability (it survives heating for 30 min to 56°C), acid-resistance (it tolerates pH 1) or slow replication without cytopathic effect on the host cell.[28] HAV has been found to persist days and even months in contaminated freshwater, seawater, live oysters, and even crème filled cookies.[21] HAV is inactivated by UV radiation (1.1 W for 1 min), formalin (8% at 25°C for 1 min), iodine (3 mg/L), and by free chlorine, 2.5 mL/L for 15 min.[19]

TABLE 11.1 Important Characteristics of *Norovirus* Facilitating its Transmission.

Characteristics	Observations	Consequence
Stability	Highly resistant to freezing, heating (up to 60°C) and disinfections such as chlorine	Difficult to eliminate in water, leading to infections from oysters, bathing water, and food irrigated with sewage. Increased risk of infections in closed settings such as hospitals
Asymptomatic shedding	Patients can shed NoV up to 3 weeks after resolution of symptoms	Increased risk of secondary spread is especially a problem concerning food handlers
Diversity	Multiple genetic, antigenic, and receptor specific strains exist	Developed detection methods may not be sensitive for all strains. Reinfections can occur more easily
Low infectious dose	Less than 10 virus particles are needed for symptomatic infection	Increases risk of infection from person-to-person spread, droplets, secondary spread, food contamination
Lack of long-term immunity	Symptomatic reinfection with the same strain can occur	Adults are not protected although infected as children. Hinders development of effective vaccines

Sources:

Human: Human feces are the major reservoir.

Food: Contaminated shellfish,[4] salads, fresh fruits and vegetables,[3] water, and any manually prepared food products. Poor hygiene practices and poor sanitation are major risk factors. Presymptomatic food handlers excreting HAV pose a risk. Food is rarely available for analysis because of the long incubation period.

Environment: Human fecal pollution from sewage discharges, septic tank, and boat discharges has caused contamination of shellfish beds, recreational water, irrigation water, and drinking water.

Transmission routes: The fecal/oral route is the established route of transmission and infection occurs following ingestion of fecally contaminated food and water. Viral contamination of fresh fruits and salad vegetables through the global marketplace is becoming a significant route of exposure, especially in countries with low endemicity of hepatitis A. Person-to-person transmission is also important especially among young children in overcrowded living conditions, daycare centers or institutions. Parenteral transmission occurs in the drug-using population and via contaminated blood products.

Incubation: 2–6 weeks (28 days).

Symptoms: In initial stages of infection, the symptoms are nonspecific, namely, fever, headache, fatigue, anorexia, nausea, and vomiting; however, during the 2nd week viremia, jaundice, and hepatitis symptoms appear. Virus circulates in the blood during first 2–4 weeks and is shed in feces ($>10^6$ particles/g) from 2 to 5 weeks of the incubation period. Jaundice is usually evident from 4 to 7 weeks, and virus shedding generally continues throughout this period. Acute hepatitis is usually self-limiting but can occasionally cause death. Estimated hospitalization rate is 13%.

Risk groups: All age groups are susceptible. The disease is milder in young children under 6 years than older children and adults. Mortality risk increases with age and unexposed older people are at higher risk.

Prevention: Based on hygiene (e.g., hand washing), sanitation (e.g., clean water sources), hepatitis A vaccine (pre-exposure), and immune globulin (pre- and postexposure).

11.3.3 HEPATITIS E VIRUS

It is a small nonenveloped virus, which was previously classified as a member of the Caliciviridae family, but recently classified in a separate genus "HEV-like viruses."[2] HEV is a major cause of outbreaks and sporadic cases of viral hepatitis in tropical and subtropical countries, but its occurrence is not frequent in industrialized countries. The virus is transmitted by the fecal–oral route via drinking water contaminated with fecal material being the usual vehicle. Direct contamination is rare. Young adults between 15 and 30 years of age are the main targets of infection, and the overall death rate ranges from 0.5% to 3.0%. The disease is usually mild, except in pregnant women, wherein high fatality rate have been recorded due to hepatic failure.[20]

11.3.4 OTHER VIRUSES OCCASIONALLY TRANSMITTED VIA FOODS

In addition to the above group of viruses discussed, there are several other groups of human enteric viruses but their transmission via to be transmitted via foods only infrequently or not at all. Factors such as duration and level of fecal shedding of the virus, efficiency of peroral infection, or stability of the virus in the food vehicle may play a role. Given that some of these viruses are able to replicate in laboratory cell cultures and cause cytopathic effects, they may be better characterized than the more important food-borne viruses discussed above.

a) **Astroviruses:** Astroviruses are small, 28-nm diameter nonenveloped, single-stranded RNA viruses comprising the only members of the family Astroviridae. Usual features of the disease vary slightly from those of the Norwalk-like viruses; the incubation period is somewhat longer, vomiting is less common, and very young (<1 year) children are more often affected. Epidemiologic evidence of transmission via foods is limited. However, the likely route of transmission is fecal–oral via food or water.

b) **Rotaviruses:** Double-stranded RNA virus of the family Reoviridae. The viruses are not enveloped, and thus have a degree of robustness in the environment outside of a host. Rotaviral infection may develop directly after consumption of meat from an infected animal, or indirectly by consumption of contaminated food usually eaten raw (fruit and vegetables). Transmission is Fecal–oral spread, that is, close person-to-person contact, fomites (environmental surfaces contaminated by stool), contaminated food and water, and possibly by respiratory droplets.

c) **Enterovirus:** The genus *Enterovirus* is a member of the broad Picornaviridae family of RNA viruses. The genus *Enterovirus* is divided into five major groups: polioviruses, group A coxsackieviruses, group B coxsackieviruses, echoviruses, and newer identified enteroviruses. The human enteroviruses are ubiquitous, enterically transmitted viruses that cause a wide spectrum of illnesses among infants and children.[1] They are quite resistant to the impact of the environment where they can survive for several weeks; they are stable in acid conditions (pH 3–5), and consequently also in gastric juices. Viral particles are shed with feces and symptoms of the diseases caused by them are different from typical gastroenteritis. Viruses enter the host with contaminated water or food and multiply in the digestive tract. Symptoms of infection are often slight, moderate but almost the enterovirus infections are asymptomatic. However, viruses may spread into other organs and cause diseases that are serious or even fatal such as aseptic meningitis, and occasionally paralysis.[15]

11.4   DETECTING FOOD-BORNE VIRUSES

Detection of viruses in food is most likely to be undertaken when an outbreak has occurred. It would be helpful, however, if some routine testing method were available to apply to foods, such as shellfish, that often serve as vehicles for viruses. However, detection of viruses in foods is challenging due to the following issues (Table 11.2):

- Low virus copy number in foods
- Complex food matrices are difficult to analyze and may be inhibitory to polymerase chain reaction (PCR)
- Efficiency of virus recovery is generally poor and can be as low as 1–10%
- Many controls are required to monitor each stage of the detection process
- Infectivity data are not provided by current molecular methods and culture is not an option for NoV and wild-type HAV

A fundamental problem is that the viruses of greatest concern, HAVs and the Norwalk like gastroenteritis viruses replicate slowly in cell cultures under laboratory conditions. The major methods of identification includes their morphology (as seen by electron microscopy), their antigenic specificity (as demonstrated by reactions with homologous antibody), their genetic specificity (as demonstrated with complementary probes or PCR

TABLE 11.2 Overview of Food-borne Viruses.

Etiology	Incubation period	Signs and symptoms	Duration of illness	Associated foods	Lab testing	Treatment
Hepatitis A	30 days	Diarrhea; dark urine; jaundice; and flu-like symptoms (i.e., fever, headache, nausea, and abdominal pain	Variable, 2 weeks–3 months	Shellfish harvested from contaminated waters, raw produce, uncooked foods and cooked foods that are not reheated after contact with infected food handler	Increase ALT, bilirubin, Positive IgM, and antihepatitis A antibodies	Supportive care. Prevention with immunization
Norwalk-like viruses	24–48 h	Nausea, vomiting, watery, large-volume diarrhea, fever rare	24–60 h	Poorly cooked shellfish; ready-to-eat foods touched by infected workers; salads, sandwiches, ice, cookies, fruit	Clinical diagnosis, negative bacterial cultures, >four-fold increase in antibody titers of Norwalk antibodies, acute, and convalescent, special viral assays in reference lab. Stool is negative for WBC's	Supportive care. Bismuth sulfate
Rotavirus	1–3 days	Vomiting, watery diarrhea, low-grade fever, Temporary lactose intolerance may occur. Infants and children, elderly, and immunocompromised are especially vulnerable	4–8 days	Fecally contaminated foods. Ready-to-eat foods touched by infected food workers	Identification of virus in stool via immunoassay	Supportive care. Severe diarrhea may require fluid and electrolyte replacement
Other viral agents (astroviruses, caliciviruses, adenoviruses, parvoviruses)	10–70 h	Nausea, vomiting, diarrhea, malaise, abdominal pain, headache, fever	2–9 days	Fecally contaminated foods. Ready-to-eat foods touched by infected food workers. Some shellfish	Identification of the virus in early acute stool samples. Serology	Supportive care, usually mild, self-limiting

primers), or combinations of these.[11] These methods may be less sensitive than tests based on infectivity, and all these tests can provide false-positive results for virus that has been inactivated (no longer infectious).

11.5   FOOD-BORNE VIRUSES ARE STILL A THREAT

- Microbiological quality control for food mostly relies on coliform bacteria as an indication of food contamination
- Inapt strategies to monitor food contamination (*zoonotic viruses may be also present without fecal contamination*)
- Infections are highly communicable and spread rapidly
- Lack of systematic surveillance
- Few countries have reporting systems of food-borne viral illnesses
- Recognition of outbreaks is complicated by the variable onset of clinical syndromes

11.6   FUTURE CHALLENGES

The social and economic cost of food-borne viral infections is largely unknown and the effort directed at the major food-borne bacteria had little impact on viruses. In addition, susceptible population is increasing because of declining acquired immunity and increasing proportions of elderly and immunocompromised individuals. Hence, the concept of One Health needs to be put in place and a constant transdisciplinary thinking and dialog between public health, veterinary and food safety experts is essential to

- identify new threats
- monitor changing trends in well-recognized diseases
- detect emerging pathogens
- understand transmission routes
- develop control effective strategies
- ensure food hygiene conditions during production and processing

Conclusion: Despite the fact that the surveillance of food-borne viral disease is minimal, significant progress has been made in our understanding of the epidemiology of food-borne viral infections, stressing that their relevance is not likely to diminish in the near future. Current regulations need to be reconsidered, as they are based on bacterial pathogens and do not suffice for control of viruses. The driver for the One Health initiative comes from our knowledge that viruses and other agents of disease can move between species. Most emerging infectious diseases are of animal origin; therefore, surveillance response and preparedness efforts need to incorporate One Health.[23] Clear data gaps have been identified that should be the focus of future research and eventually provide the evidence for prevention and intervention strategies.

> "At either end of any food chain you find a biological system—a patch of soil, a human body—and the health of one is connected—literally—to the health of the other."
>
> *Michael Pollan*
> *The Omnivore's Dilemma, 2006*

Overview
• Food-borne viral disease is a major public health problem.
• Food which conforms to bacterial standards may still be contaminated with viruses.
• Viruses are responsible for the majority of food-borne infections.
• Clinical and economic impact of food-borne disease is underestimated.
• Food-borne viruses are nonenveloped and are resistant to heat, disinfection, and pH changes.
• HAV and NoVs are the leading viral causes of water- and food-borne disease.
• The infectious dose for HAV and NoVs is estimated to be around 10–100 infectious viral particles.
• An effective vaccine is available for the prevention of hepatitis A.
• The analysis of water and food samples for viruses is a complex process and is often not performed.

KEYWORDS

- **One Health**
- **food safety**
- **food-borne virus**
- **food-borne diseases**
- **transdisciplinary**

REFERENCES

1. Acha, P. N.; Szyfres, B. *Zoonoses and Communicable Diseases Common to Man and Animals*, 3rd ed.; Pan American Health Organization: Washington, 2003.
2. Berke, T.; Matson, D. O. Reclassification of the Caliciviridae into Distinct Genera and Exclusion of Hepatitis E Virus from the Family on the Basis of Comparative Phylogenetic Analysis. *Arch. Virol.* **2000,** *145*, 1421–1436.
3. Dentinger C.M., Bower W.A., Nainan O. V. An outbreak of hepatitis A associated with green onions. *J. Infect. Dis.* **2001,** *183*, 1273–1276.
4. Di Pinto A., Conversano M. C., Forte V.T., La Salandra G., Montervino C., Tantillo G.M. A Comparison of RT-PCR-based Assays for the Detection of HAV from Shellfish. *New Microbiol.* **2004,** *27*, 119–124.
5. Fankhauser, R. L.; Monroe, S. S.; Noel, J. S.; Humphrey, C. D.; Bresee, J. S.; Parashar, U. D.; Ando, T.; Glass, R. I. Epidemiologic and Molecular Trends of "Norwalk-Like Viruses" Associated with Outbreaks of Gastroenteritis in the United States. *J. Infect. Dis.* **2002,** *186*, 1–7.
6. Gallimore, C. I.; Green, J.; Lewis, D.; Richards, A. F.; Lopman, B. A.; Hale, A. D.; Eglin, R.; Gray, J. J.; Brown, D. W. G. Diversity of Noroviruses Cocirculating in the North of England from 1998 to 2001. *J. Clin. Microbiol.* **2004,** *42*, 1396–1401.
7. Hutson, A. M.; Atmar, R. L.; Estes, M. K. Norovirus Disease: Changing Epidemiology and Host Susceptibility Factor. *Trends Microbiol.* **2004,** *12*, 279–287.
8. Kapikian, A. Z. The Discovery of the 27-nm Norwalk Virus: An Historic Perspective. *J. Infect. Dis.* **2000,** *181* (Suppl. 2), S295–S302.
9. Kapikian, A. Z.; Wyatt, R. G.; Dolin, R.; Thornhill, T. S.; Kalica, A. R.; Chanock, R. M. Visualization by Immune Electron Microscopy of a 27-nm Particle Associated with Acute Infectious Nonbacterial Gastroenteritis. *J Virol.* **1972,** *10*, 1075–1081.
10. Koo, H. L.; Ajami, N. J.; Jiang, Z. D.; Atmar, R. L.; DuPont, H. L. Norovirus Infection as a Cause of Sporadic Healthcare-Associated Diarrhoea. *J. Hosp. Infect.* **2009,** *72*, 185–187.
11. Koopmans, M.; C. H. von Bonsdorff, J. Vinje, D. de Medici, S. Monroe. Foodborne Viruses. *FEMS Microbiol. Rev.* **2002,** *26*, 187–205.
12. Koopmans, M.; Duizer, E. Foodborne Viruses: An Emerging Problem. *Int. J. Food Microbiol.* **2004,** *90*, 23–41.
13. Le Pendu, J.; Ruvoen-Clouet, N.; Kindberg, E.; Svensson, L. Mendelian Resistance to Human Norovirus Infections. *Semin. Immunol.* **2006,** *18*, 375–386.
14. Lederberg, J. Infections History. *Science* **2000,** *288* (5464), 287–293.
15. Lees, D. Viruses and Bivalve Shellfish. *Int. J. Food Microbiol.* **2000,** *59*, 81–116.
16. Lindesmith, L.; Moe, C.; Marionneau, S.; Ruvoen, N.; Jiang, X.; Lindblad, L.; Stewart, P.; LePendu, J.; Baric, R. Human Susceptibility and Resistance to Norwalk Virus Infection. *Nat. Med.* **2003,** *9*, 548–553.
17. Marks, P. J.; Vipond, I. B.; Carlisle, D.; Deakin, D.; Fey, R. E.; Caul, E. O. Evidence for Airborne Transmission of Norwalk-Like Virus (NLV) in a Hotel Restaurant. *Epidemiol. Infect.* **2000,** *124*, 481–487.
18. Nordgren, J. Prevalence, Transmission, and Determinants of Disease Susceptibility. Linköping Medical Dissertations No. 1111, 2009.
19. Vasickova, P.; Dvorska, L.; Lorencova, A.; Pavlik, I. Viruses as a Cause of Foodborne Diseases: A Review of the Literature. *Vet. Med.—Czech* **2005,** *50* (3), 89–104.

20. Patel, M. M.; Hall, A. J.; Vinje, J.; Parashar, U. D. Noroviruses: A Comprehensive Review. *J. Clin. Virol.* **2009**, *44*, 1–8.
21. Siegl, G.; Weitz, M.; Kronaeur, G. Stability of Hepatitis A Virus. *Intervirology* **1984**, *22*, 218–226.
22. Smith, J. L. A Review of Hepatitis E Virus. *J. Food Protect.* **2001**, *64*, 572–586.
23. Sobsey, M. D.; Shields, P. A.; Hauchman, F. S., et al. Survival and Persistence of Hepatitis A Virus in Environmental Samples. In *Viral Hepatitis and Liver Disease*; Zuckerman, A. J., Ed.; Alan R. Liss: New York, 1988; pp 121–124.
24. Tauxe, R. V. Emerging Foodborne Pathogens. *Int. J. Food Microbiol.* **2002**, *78*, 31–41.
25. Taylor, L. H., et al. Risk Factors for Human Disease Emergence. *Philos. Trans. Royal Soc. Lond. B: Biol. Sci.* **2001**, *356*, 983–989 [cited from: The Importance of a One Health Approach to Public Health and Food Security in Australia by *Chris Baggoley. Microbiol. Austr.* **2012**, *34* (4)].
26. Teunis, P. F.; Moe, C. L.; Liu, P.; Miller, S. E.; Lindesmith, L.; Baric, R. S.; Le Pendu, J.; Calderon, R. L. Norwalk Virus: How Infectious Is It? *J. Med. Virol.* **2008**, *80*, 1468–1476.
27. Ushijima, H. Molecular Epidemiology of Norwalk Virus. *Nippon Rinsho* **2002**, *60*, 1143–1147
28. WHO (World Health Organization) (1993): Prevention of Foodborne Hepatitis A. *Weekly Epidemiological Record* **1993**, *68* (22), 157–158.

PART IV
Nanoscience and Nanotechnology

CHAPTER 12

NANOMATERIALS: NEW INSIGHTS IN CANCER TREATMENT

PRIYA PATEL[1*], MIHIR RAVAL[2], and PARESH PATEL[1]

[1]*Department of Pharmaceutical Sciences, Saurashtra University, Rajkot, Gujarat, India*

[2]*Shivam Institute and Research Centre of Pharmacy, Valasan, Anand, Gujarat, India*

Corresponding author. E-mail: patelpriyav@gmail.com

CONTENTS

ABSTRACT

Nowadays, nanotechnology takes an innovative way at the prevention, diagnosis, and treatment of cancer. The application of nanomaterials (NMs) in biomedicine is increasing rapidly day by day, can enhance the delivery and treatment efficiency, and offers excellent prospects for the development of new noninvasive strategies for the diagnosis and treatment of anticancer drugs. As compared to the conventional dosage form, nanoparticulate system is more effective in treatment of cancer due to the prevention of multidrug resistance followed by improved the therapeutic effects of the anticancer drugs. In this chapter, we provide an information related to pathology of cancer along with the characterization of nanoparticulate drug-delivery system for tumor targeting followed by synthesis of nanoparticles, fabrication of bulk NMs, use of various tools in cancer research, elucidation of nanoscale mechanisms of cellular functions relevant to cancer, and nanoscale computational modeling. Direct interactions between the particles and cellular molecules to cause adverse biological responses are also discussed. Many nanoparticulate dosage forms are already available in markets and some dosage forms are under research but still much more research is required in the field of nanotechnology for the treatment of cancer.

12.1 INTRODUCTION

Cancer is not just one disease but many diseases. Cancer is a very critical disease which refers to a condition where the body's cells start to grow and damage the healthy tissue of the body by reproducing in an uncontrollable way and causing death. Cancer occurs at a molecular level when multiple subsets of genes undergo genetic alterations, either activation of oncogenes or inactivation of tumor suppressor genes. Then, malignant proliferation of cancer cells, tissue infiltration, and dysfunction of organs will appear.[1] It is estimated that due to the cancer, 7 million deaths occur per year. 16 million new cancer cases are expected to occur every year. Due to the beginning of uncontrolled growth of cell primary tumor mass and metastasis of cancer cells can easily spread to healthy tissues of the body. Recently, the field of cancer therapies has been improved by many diverse scientific disciplines to better fight cancer diseases. A new very interesting area of research is nanotechnology that involves the creation, manipulation, and application of structures in the nanometer size range that is revolutionizing cancer diagnosis and therapy.[2] Tumor tissues are characterized with active angiogenesis and high vascular density which keep blood supply for their growth with a defective vascular architecture. When combined with poor lymphatic drainage, it is known as the enhanced permeation and retention (EPR) effect.[3,4] There are more than 100 different types of cancers, and most cancers are named for the organ or type of cell in which they start. For instance, cancer that begins in the lung is called lung cancer; cancer that begins in basal cells of the skin is called basal cell carcinoma; cancer that begins in colon is called colon cancer.

Cancer types can be grouped into broader categories. The main categories of cancer include: carcinoma cancer that begins in the skin or in tissues and cover internal organs; sarcoma cancer that begins in bone, cartilage, fat, muscle, blood, vessels, or other connective or supportive tissue; Leukemia means cancer that starts in blood-forming tissue such as the bone marrow and causes large numbers of abnormal blood cells to be produced and enter the blood; and central nervous system cancers means cancers that begin in the tissues of the brain and spinal cord.

Cancer is a complex group of diseases with many possible causes. First, genetic (biological) factors, age, obesity, declining immune system, exposure to chemicals and radiation, pesticides, and virus are risk factors. Besides these, environment and lifestyle factors have a major impact to constitute the cancer. Relevant research demonstrates that of all cancer-related deaths, nearly 5–10% is due to genetic factors, whereas the rest of it, nearly 90–95% is because of lifestyle and environment factors.[5,6]

Cancer cells differ from the normal cells of the body by their ability to divide indefinitely and evade programmed cell death.[7] In normal cells, the cell cycle is controlled by a complex series of signaling pathways by which a cell grows, replicates its DNA, and divides. This process also includes mechanisms which ensure that any errors therein are corrected, and if not, the cells commit suicide in a systematic cellular process known as programmed cell death or apoptosis. In cancer due to the genetic mutations, malfunctions occur, resulting in uncontrolled cell proliferation. Alteration in these mechanisms commonly affects the expression of cell cycle regulatory proteins, causing overexpression of cyclins and loss of expression of cyclin-dependent kinase inhibitors (CDKIs). Therefore CDKIs and related transcription factors have long been viewed as potential targets for anticancer therapeutics. Cells normally depend upon two important pathways for degradation of cellular proteins, that is, (1) the lysosome pathway for extracellular proteins and (2) the ubiquitin proteasome pathway for intracellular proteins. Proteasomes, key complexes of the pathway, are multicatalytic protease complexes engaged in nonlysosomal recycling of intracellular proteins of short life span.[8,9] Nowadays, most of research groups are studying about cancer formation, progression, and therapeutic approaches. However, cancer still remains one of the leading causes of death after the cardiovascular diseases in the world.[10] Approximately, 12 million people worldwide caught this disease each year and unfortunately 7 million of these patients are dying in the same year. The latest world cancer statistics report that the number of new cancer cases will increase to more than 15 million in 2020.[11] In addition, the efficiency of anticancer drugs is limited by their unsatisfactory properties, such as poor solubility, narrow therapeutic window, and intensive cytotoxicity to normal tissues, which may be the causes of treatment failure in cancer.[12] Several different techniques are used nowadays for the treatment of cancer like surgery, radiation, chemotherapy, targeted therapy, and immunotherapy; among these many techniques are still in their developmental phases.[13] Treatment is selected depending on the general state of the patient as well as the location and growth of the tumor. Main challenging tasks for cancer therapy are targeted and localized drug deliveries. For war against cancer, treatment should only be focused on cancer cells so that they produce lesser toxic effects on healthy or normal cells which is a challenging task to develop such new therapies for cancer. Main limitation of conventional therapy of anticancer drug is produce adverse effect to healthy tissue.[14]

Accordingly, there is a great need for new therapeutic strategies capable of delivering chemical agents and other therapeutic materials specifically to tumor locations.[15] With the development of nanotechnology, the integration of nanomaterials (NMs) into cancer therapeutics is one of the rapidly advancing fields. It can revolutionize the treatment of cancer. Nanomedicine represents an innovative and multidisciplinary field that exploits nanotechnology in disease detection, diagnosis and treatment. NMs are defined as materials ranging from 1 to 100 nm in diameter in two or three dimensions.[16]

There are many reasons for which nanoscale size drug-delivery systems are attractive to formulation scientists. The most important reason is that number of surface atoms or molecules to the total number of atoms or molecules increases in drug-delivery systems. Thus, the surface area increases. This helps to bind adsorb and carry with other compounds such as drug, probes, and proteins. The drug particles itself can be engineered to form nanoscale-size materials.[17] Nanotechnology refers to the interactions of cellular and molecular components. Nanotechnology can provide rapid and sensitive detection of cancer-related molecules, enabling scientists to detect molecular changes even when they occur only in a small percentage of cells. Nanotechnology also has the potential to generate unique and highly effective therapeutic agents. Nanoparticles (NPs) as drug-delivery systems enable unique approaches for cancer treatment. Over the last two decades, a large number of NP delivery systems have been developed for cancer therapy, including organic and inorganic materials. Many liposomal, polymer–drug conjugates and micellar formulations are part of the state of the art in the clinics. More recently developed NPs are demonstrating the potential sophistication of these delivery systems by incorporating multifunctional capabilities and targeting strategies in an effort to increase the efficacy of these systems against the most difficult cancer challenges, including drug resistance and metastatic disease. Nanomedicines are more specific than traditional cancer medicines. Various NP technologies have been used for cancer therapy like polymeric NPs[18,19] polymer conjugates,[19-21] protein carriers,[22,23] micelles,[24-26] vesicle-based carriers such as liposomes,[27,28] dendrimers,[29-32] inorganic NPs,[33] and bacterial nanocarriers. NP has been developed with unique characteristics, that is, their shapes and sizes due to which formulations are able to target specific area.

12.2 POTENTIAL ADVANTAGES OF THERAPEUTIC NANOPARTICLE

Potential advantages of therapeutic NP include the following:

1. The ability to improve the pharmaceutical as well as pharmacological properties of drugs, potentially without the need to alter the drug molecules.
2. Enhancement of therapeutic efficacy by targeted delivery of drugs in a targeted cell in specific manner.
3. Delivery of drugs across a range of various biological barriers.
4. Delivery of drugs to intracellular site of action.
5. The ability to deliver multiple types of therapeutics with potentially different physicochemical properties.
6. The ability to deliver real time monitoring treatments including combination of imaging and therapeutic agents to improve therapeutic efficacy.
7. Possibilities to develop highly differentiated therapeutics protected by intellectual properties.[34,35]

Using targeted NPs to deliver chemotherapeutic agents in cancer therapy offers many advantages to improve drug delivery and to overcome many problems associated with conventional chemotherapy.[35-38] For example, NPs via either passive targeting or active targeting have been shown to enhance the intracellular concentration of drugs in cancer cells while avoiding toxicity in normal cells. In addition, the targeted NPs can also be designed as either pH-sensitive or temperature-sensitive carriers.

The pH-sensitive drug-delivery system can deliver and release drugs within the more acidic microenvironment of the cancer cells or components within cancer cells. The temperature-sensitive system can carry and release drugs with changes in temperature locally in the tumor region provided by sources such as magnetic fields, ultrasound waves, etc., so that combined therapy such as chemotherapy and hyperthermia can be applied ideally for anticancer drugs to be effective in cancer treatment if they reach to the targeted site in the body with minimal loss of their volume or activity in the blood circulation. Second, after reaching the tumor tissue, drugs should have the ability to selectively kill tumor cells without affecting normal cells with a controlled release mechanism of the active form. These two basic strategies are also associated with improvements in patient survival and quality of life by increasing the intracellular concentration of drugs and reducing dose-limiting toxicities simultaneously. Increasingly, NPs seem to have the potential to satisfy both of these requirements for effective drug carrier systems.[37]

Despite extensive research on NP systems for cancer therapeutics, there are only a few NP drug-delivery systems approved by the US Federal Drug Administration and European Medicines Agency to treat cancer. Specifically, the systems that have been approved include liposomal doxorubicin (Myocet™, Elan Pharmaceuticals), PEGylated liposomal doxorubicin (Doxil®, Ortho Biotech, and Caelyx®, Schering Plough), PEGylated liposomal daunorubicin (DaunoXome®, Diatos), and the recently approved albumin-bound paclitaxel-loaded NPs (Abraxane®, Abraxis Bioscience).[39]

12.3 LIMITATIONS OF CONVENTIONAL CHEMOTHERAPY

Conventional chemotherapeutic agents work by destroying rapidly dividing cells, which is the main property of neoplastic cells. Due to this reason, chemotherapy also damages normal healthy cells that divide rapidly such as cells in the bone marrow, macrophages, digestive tract, and hair follicles.[40] Main disadvantages of conventional therapy are that they affect the healthy tissue and produce side effects like myelosuppression mucositis, alopecia (hair loss), organ dysfunction, and thrombocytopenia and mainly do not give targeting action to the cancerous cell only.[41,42]

Traditional chemotherapeutic agents often get washed out from the circulation being engulfed by macrophages. Thus, they remain in the circulation for a very short time and cannot interact with the cancerous cells making the chemotherapy completely ineffective. The poor solubility of the drugs is also a major problem in conventional chemotherapy making them unable to penetrate the biological membranes.[43] Another problem is associated with P-glycoprotein, a multidrug resistance protein that is overexpressed on the surface of the cancerous cells, which prevents drug accumulation inside the tumor, acting as the efflux pump, and often mediates the development of resistance to anticancer drugs. Thus, the administered drugs remain unsuccessful or cannot bring the desired output.[43–47]

12.4 TARGETED NANOPARTICLES

Effective cancer therapy remains one of the most challenging tasks to the scientific community, with little advancement on overall cancer survival landscape during the last two decades. A major limitation inherent to most conventional anticancer chemotherapeutic agents is their lack of tumor selectivity. One way to achieve selective drug targeting to solid tumors is to exploit abnormalities of tumor vasculature, namely hypervascularization, aberrant vascular architecture, extensive production of vascular permeability factors stimulating extravasation within tumor tissues, and lack of lymphatic drainage. One of the major challenges in the development of targeted NPs for cancer therapy is to discover targeting ligands that allow for differential binding and uptake by the target cancer cells. Drug-targeting strategies have frequently been divided into categories of "passive" and "active" targeting.[48–52] Targeting nanoparticulate delivery is shown in Figure 12.1.

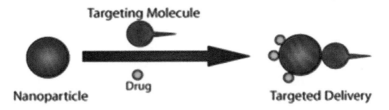

FIGURE 12.1 Targeting nanoparticulate delivery.

12.4.1 PASSIVE TARGETING OF TUMORS USING NANOPARTICLES

Passive targeting systems that target the systemic circulation are generally characterized as "passive" delivery systems (targeting occurs because of the body's natural response to physiochemical characteristics of the drug or drug-carrier system). The ability of some colloids to be taken by the RES vectors for passive hepatic targeting of drugs to these compartments.

NPs that exhibit localization to specific organs or to sites of disease via biological mechanisms such as the RES or the EPR effect are known as passive targeting agents.

This category of targetable devices includes drug-bearing bilayer vesicular systems as well as cellular carriers of micron or submicron size range. The passive targetability of nanoparticulate drug carriers is due to the recognition of these particulates either in the intact or in the opsonized form, by the phagocytic cells of the RES and this behavior is exploited to target cell lines.

NP systems exploit characteristics of tumor growth for the use of a passive form of targeting. The tumor becomes diffusion-limited at a volume of 2 mm³ or above. This diffusion limitation impacts nutrition intake, waste excretion, and oxygen delivery. The tumor is able to overcome the diffusion limitation by increasing the surrounding vasculature in an event called angiogenesis.[53] Passive targeting takes advantage of the unique pathophysiological characteristics of tumor vessels, enabling nanodrugs to accumulate in tumor tissues. Typically, tumor vessels are highly disorganized and dilated with a high number of pores, resulting in enlarged gap junctions between endothelial cells and compromised lymphatic drainage. The "leaky" vascularization, which refers to the EPR effect, allows migration of macromolecules up to 400 nm in diameter into the surrounding tumor region. Passive targeting also involves the use of other innate characteristics of the NP which can induce targeting to the tumor. Likewise, cationic liposomes are found to bind by electrostatic interactions to negatively charged phospholipid head groups preferentially expressed on tumor endothelial cells.[54,55,56,57] NPs can easily accumulate selectively by enhanced permeability and retention (EPR) effect and then diffuse into the cells.[58-63]

12.4.2 ACTIVE TARGETING OF TUMORS USING NANOPARTICLES

Active targeting is based on molecular recognition. Hence, the surface of the NPs is modified to target the cancerous cells. Usually, targeting agents are attached with the surface of NPs for molecular recognition. Active targeting exploits modification or manipulation of drug carriers. The natural distribution pattern of the drug carrier composites is enhanced using chemical, biological and physical means, so that it approaches and identified by particular biosites. Nanotechnology-based targeted delivery system has three main components: (1) an anticancer drug, (2) a targeting moiety- (3) a carrier. The facilitation of the binding of the drug-carrier to target cells through the use of ligands or engineered device to increase receptor mediated localization of the drug and target specific delivery of drug is referred to as active targeting.[64–68]

Attachment of the ligands on the surface of NPs makes them able to target only the cancerous cells. Once NPs bind with the receptors, they rapidly undergo receptor-mediated endocytosis or phagocytosis by cells, resulting in cell internalization of the encapsulated drug.

This targeting approach can further be classified into three different levels of targeting: first-order targeting; second-order targeting; and third-order targeting.

First-order targeting: It refers to restricted distribution of the drug carrier system to the capillary bed of a predetermined target-site organ or tissue. Compartmental targeting in lymphatics, peritoneal cavity, plural cavity, cerebral ventricles, lungs, joints, eyes, etc., represent first-order targeting. The ability of liposomes to extravasate and penetrate into diseased states other than MPS is directly related to their size. Large liposomes (10 μm or above) are rapidly removed via mechanical filtration of lungs and from this size range down up to 150 nm are removed by tissue macrophages originated in the liver and spleen, which are the natural target for these vesicles.

To achieve significant levels in other tissues, liposomes of smaller size with a homogeneous distribution have been developed. They could penetrate into either normal tissue having sinusoidal or fenestrated epithelium or otherwise into diseased tissue having altered capillary permeability. This has in turn increased the chances of achieving targeting in vivo to non-MPS cell linings.[66]

Second-order targeting: The selective delivery of drugs to a specific cell type such as tumor cells as referred to as second-order targeting.

Third-order targeting: Defined as drug delivery specifically to the intracellular site of target cells. An example of third-order targeting is the receptor-based ligand-mediated entry of a drug complex into a cell by endocytosis, lysosomal degradation of carrier followed by release of drug intracellularly or gene delivery to nucleolus.[69] "Active targeting" is used to describe specific interactions between drug carrier and the target cells, usually through specific ligand–receptor interactions.[70–74] Ligand–receptor interactions are possible only when the two components are in close proximity (0.5 nm) One way to overcome the limitations of passive targeting is to attach affinity ligands (antibodies,[75] peptides,[76] aptamers,[77] or small molecules[78] that only bind to specific receptors on the cell surface) to the surface of the nanocarriers by a variety of conjugation. Long circulation times will allow for effective transport of the NPs to the tumor site through the EPR effect, and the targeting molecule can increase endocytosis of the NPs. The internalization of NP drug-delivery systems has shown an increased therapeutic effect.[79–84] If the NP attaches to vascular endothelial cells via a noninternalizing epitope, high local concentrations of the drug will be available on the outer surface of the target cell. Nanodrugs currently approved for clinical use are relatively simple and generally lack active targeting or triggered drug-release components. Active and passive targeting is shown in Figure 12.2.

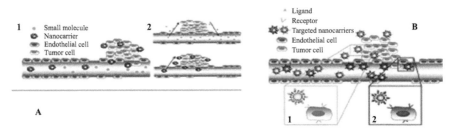

FIGURE 12.2   Passive (A) and active (B) targeting.

12.4.3 SPECIFIC RECEPTOR TARGETING

Folate Receptor: Folate receptors (FRs) are overexpressed in many neoplastic cells providing a target for certain anticancer therapies. Utilizing the concept, researchers are designing the surface of NPs with folic acid (FA). Latalloto et al. test the folate-linked methotrexate dendrimers in immune-deficient athymic nude female mice. The mice were injected with the nanoconjugates twice a week via a lateral tail vein. The results showed that conjugated methotrexate in dendrimers significantly lowered toxicity and resulted in a 10-fold higher efficacy compared to free methotrexate at an equal cumulative dose. As a result, mice survived longer period of time.[85]

Transferrin Receptor: NPs are widely being investigated to target the transferrin receptors for binding and cell entry, as these are overexpressed by certain tumor cells to increase their iron uptake. Transferrin (Tf) can be conjugated to a variety of materials for cancer targeting which include Tf-chemotherapeutic agent, Tf-toxic protein, Tf-RNases, Tf-antibody, and Tf-peptide.[86,87] Bellocq et al. found that at low transferrin modification, the NPs remain stable in physiologic salt concentrations and transfect leukemia cells with increased efficiency. The transferrin-modified NPs are effective for systemic delivery of nucleic acid therapeutics for metastatic cancer.[88,89]

12.4.4   ASIALOGLYCOPROTEIN

Asialoglycoprotein, which is overexpressed in hepatoma, is utilized in cancer targeting by NPs for anticancer drug delivery. Sung and coworkers developed a new strategy by which biodegradable NPs with a mean size of 140 nm can be prepared to target the hepatoma cells. They prepared poly—glutamic acid–poly(lactide) block copolymers loaded with paclitaxel using emulsion solvent evaporation technique.[90]

12.4.5   ANTIBODY-MEDIATED TARGETING

Many tumor cells show unusual antigens due to their genetic defects that are either inappropriate for the cell type, environment, or temporal placement in the organisms' development. Antibody engineering has recently flourished with the outcome of antibody production that contains animal and human origins such as chimeric mAbs, humanized mAbs, and antibody fragments. Antibodies can be used in their original form or as fragments for cancer targeting. However, the presence of two binding sites in single antibody gives higher binding opportunity and makes it advantageous to use the intact mAbs. Moreover, a signaling cascade is initiated to kill the cancer cells when macrophages bind to the Fc segment of the antibody. The Fc portion of an intact mAb can also bind to the Fc receptors on normal cells resulting in increased ability to evoke an immune response and liver and spleen uptake of the nanocarrier.[91,92]

12.5   TOOLS OF NANOTECHNOLOGY FOR CANCER THERAPY

The tools of nanotechnology with applications in early cancer detection and treatment include the following:

- Liposomes
- Polymeric NPs
- Polymeric drug conjugates
- Polymeric micelles
- Dendrimers
- Carbon nanotubes (CNTs)
- Quantum dots (QDs)

12.5.1 LIPOSOMES

Liposomes have become very versatile tools in biology, biochemistry and medicine because of their enormous diversity of structure and compositions.[93-98] Liposomes have been in use for the past several decades and are established as drug and imaging agent carriers with proven clinical efficacy.[99] They are artificial phospholipid vesicles 50 nm to ≥1 μm in size, either unilamellar or multilamellar[100,101] (Fig. 12.3). The cargo can be held in the aqueous compartment or lipid layer.[102] The concept of targeting liposomes to tumor sites was derived from Ehrlich's concept of a "magic bullet" coined in 1906.[103]

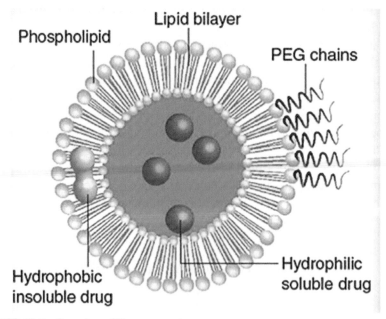

FIGURE 12.3 Overview of liposome system.

Liposomes have gained attention as a carrier system for therapeutically active agents, due to their unique characteristics, including biocompatibility, biodegradability, low toxicity, lack of immune system activation, and capability to incorporate both hydrophilic and hydrophobic drugs also it improves the pharmacokinetic and pharmacodynamic profiles of the therapeutic payload, promote controlled and sustained release of drugs, and exhibit lower systemic toxicity compared with the free drug.

Examples of liposome-mediated drug delivery are doxorubicin (Doxil) and daunorubicin (Daunoxome), which are currently being marketed as liposome delivery systems. Polyethylene glycol (PEG) PEGylated liposomal doxorubicin (Doxil1, Caelyx1; Alza Pharmaceuticals, San Bruno, CA, USA) has achieved the most prolonged circulation, with a terminal half-life of 55 h in humans.[104–106] Leamon et al.[107] have recently evaluated the in vitro and in vivo status of the delivery of oligonucleotides encapsulated in folate-coated liposomes. Moreover, folate-receptor-targeted liposomes have proven effective in delivering doxorubicin in vivo and have been found to bypass multidrug resistance in cultured tumor cells.[108]

In recent years, research has significantly developed in terms of liposomal systems with an improved drug-delivery potential for cancer therapy.[109–112] Subsequent work on improving the therapeutic potential of liposomes has focused mainly on developing strategies for actively targeting the liposomes to a tumor site, intracellular delivery followed by organelle-specific targeting, and triggered release of therapeutic payloads utilizing pathological differences in the tumor's microenvironment. Various strategies have been adopted for targeting liposomes to the tumor sites. Utilization of the EPR effect is an effective strategy for targeting nanopreparations such as liposomes to the site of a tumor. Unlike liposomes and other NPs, low-molecular-weight drugs are not retained in the tumor site for a longer period of time since they re-enter the circulation primarily via diffusion. Targeting of these drugs relies solely on the pathophysiological properties of the tumor tissues, and is referred to as "passive drug targeting." To utilize the EPR effect, the liposomes should usually be smaller than 400 nm in size.[113] In general, actively targeted liposomes are designed to minimize off-target effects. Actively targeted liposomal systems are prepared by conjugating targeting moieties, including small-molecule ligands, peptides, and monoclonal antibodies on the liposomal surface.[114,115] For example, certain receptors, such as folate and transferrin (Tf) receptors (TfR), are overexpressed on many cancer cells and have been used to make liposomes tumor-cell specific.[116,117] *Targeting FRs:* FA has recently been used as a targeting ligand for specialized drug delivery owing to its ease of conjugation to nanocarriers, its high affinity for FRs, and the relatively low frequency of FRs in normal tissues as compared with their overexpression in activated macrophages and cancer cells.[118] FRs are overexpressed in certain ovarian, breast, lung, colon, kidney, and brain tumors.[119]

Chang et al. designed a method to select peptides that bind specifically to the tumor vasculature of human cancer xenografts.[120] Coupling these peptides to a liposome loaded with doxorubicin improved its efficacy against several types of human cancers xenografted on SCID mice. The neovasculature-specific phages, IVO-8 and IVO-24, specifically bound both tumor vessels of xenografts in animal models and the blood vessels of six types of human solid tumors. Coupling of the phage IVO peptides to the PEG terminus of stealth liposomes demonstrated that the phage IVO peptide-anchored liposomal doxorubicin improved therapeutic efficacy, increased cancer cell apoptosis, and decreased tumor angiogenesis in mice, resulting in a decline in tumor growth.[121]

These liposomal systems are able to target various areas of the tumor using specific targeting ligands and circumvent some of the problems associated with multidrug resistance. An additional plus for the use of antibodies is the ability to achieve synergy between the signaling antibodies and chemotherapeutics, since two distinct techniques are being used to target the cells. Many of the ligands employed demonstrate a large overlap in their targets and, thus, could help provide synergistic antitumor effects.

Over all liposomes, optimization of the pharmacokinetics of the encapsulated drug can improve drug accumulation in the tumor and reduce the adverse effects of bolus administration.[122]

12.5.2 POLYMERIC NANOPARTICLES

In polymeric NPs, drug is entrapped, dissolved or encapsulated on the basis of their method of preparation. In polymeric nanoparticulate drug-delivery system, drug is placed in the cavity and core surrounded by a polymer membrane.[123] Polymeric carriers do not interact with the drug containing macromolecules such as proteins, which could sequester the active ingredient by preventing its arrival at the action site. In recent years, polymeric NPs have appeared as a most viable and versatile delivery system for targeted cancer therapy. Various in vivo studies have demonstrated that virus-sized stealth particles are able to circulate for a prolonged time and preferentially accumulate in the tumor site via the EPR effect. The surface decoration of stealth NPs by a specific tumor-homing ligand such as antibody, antibody fragment, peptide, aptamer, polysaccharide, saccharide, FA, etc., might further lead to increased retention and accumulation of NPs in the tumor vasculature as well as selective and efficient internalization by target tumor cells. Polymeric NPs may represent the most effective nanocarriers for prolonged drug delivery. The early in vitro and in vivo development of polymeric NPs loaded with drugs in the 1980s using polyalkylcyanoacrylate-based NPs releasing doxorubicin led to multiple reports using polymer-based materials for drug delivery.[124]

The versatility in physiochemical modification of polymer properties enables it to be tuned to the requirements for drug encapsulation. Apart from being biodegradable and biocompatible, these polymeric systems are capable of giving rise to sustained-release profile of the drugs encapsulated.

A polymer which is used in controlled drug-delivery system must be inert, nontoxic, and free from impurities along with having appropriate physical structure with readily processable and minimal aging and should not interact with macromolecule containing the drug which could sequester the active ingredients preventing its arrival at action site.[125]

Recently, there has been significant interest in employing synthetic polymers like PEG,[126] polylactide (PLA),[127] and poly(D,L-lactide-co-glycolide) (PLGA).[128] Dhar et al.[129] have employed a platimum (pt (IV)-based PLGA–PEG NP to deliver cisplatin in the form of a prodrug showing significantly improved efficacy in vivo. While these polyesters offer excellent biocompatibility and biodegradability, they have limitations with respect to drug release and stability owing to slow degradation of the polymers.[130] Polymer-based NPs had greater advantages than other NMs that is increased in stability of any volatile agents, easily fabricated into larger quantities, oral and intravenous methods of administration with significant efficiency and effectiveness, drug delivery with higher concentration with greater significant target, and the choice of polymer and ability to alter drug release from polymeric NPs. It's an ideal candidate for delivery of targeted drugs, delivery of vaccines, and for cancer therapy.

Polymeric NPs have shown a great deal of promise to provide solutions to such problems in cancer treatment.[128] Among the most commonly used polymers are PLA and PLGA, which have been approved by the Food and Drug Administration (FDA) for the development of drug-delivery systems and other biomedical applications. Degradation by nonenzymatic hydrolysis of PLA and PLGA leads to an accumulation of acidic monomers, which causes a decrease in pH. When the encapsulated therapeutic agent is a protein, it may denaturalize.

The most important challenge in the successful formulation of polymeric drug-delivery systems involves preparing carrier systems that could be capable of encapsulating the preferred drug within its structure and then deliver the drug to the target (cancerous tissues).

Tang reported a novel delivery systems of cholic acid (CA) (polyhydroxy initiator), a star-shaped block copolymers based on PLA and TPGS known as (CA–PLA–TPGS NPs) with unique architectures, revealed sustained and controlled delivery of paclitaxel for breast cancer treatment and achieved higher drug loading content and entrapment efficiency, resulting in faster drug release as well as higher cellular uptake and cytotoxicity. The in vivo cell studies indicated that the PTX-loaded star-shaped CA–PLA–TPGS NPs reported to have significantly superior antitumor activity.[131]

Gonzalo et al. demonstrated a novel nanocapsules prepared using a modified solvent displacement technique where the polyamino acid (PGA) was electrostatically deposited on to the lipid core. The author reveals that the in vivo studies performed in experimental mice model indicated that encapsulation provides the drug with a prolonged blood circulation and significantly reduced the toxicity. Therefore, this study provides a valid claim that PGA could be used as potential nanocapsule-based drug delivery for any anticancer therapy.[132–134]

12.5.3 POLYMER DRUG CONJUGATES

The concept of polymeric macromolecule–drug conjugates was first proposed by Ringsdorf for delivery of hydrophobic small drug molecules to their sites of action.[135,136] Polymer–drug conjugates are a novel class of nanocairrers for anticancer drug, which can protect the drug from premature degradation, prevent drug from prematurely interaction with the biological environment and enhance the absorption of the drugs into tumor (by EPR effect or active targeting).[137,138]

PDCs offer advantages of reduced deleterious effects of anticancer drugs and augmentation of its formulation capability (e.g., solubility). PDCs once entered into the tumor tissue, taking advantage of EPR, are endocytosed into the cell either by simple or receptor mediated endocytosis.

In a polymer–drug conjugate, there are at least three major components: a soluble polymer backbone, a biodegradable linker, and a covalently linked anticancer drug which is deactivated as a conjugate (Fig. 12.4).

Polymer–drug conjugates became necessary because most conventional low molecular weight anticancer drugs have an inherent character to transverse in and out of blood vessels freely and this causes a nonselective distribution in both normal and tumor cells causing undesirable side effects to patients. A polymer–drug conjugate comprises a variety of complex macro-molecular systems, their common feature being the presence of a rationally designed covalent chemical bond between a water-soluble polymeric carrier and the bioactive molecule(s) linking these drugs to a macromolecule such as a polymer (polysaccharides, proteins, polyamino acids) makes it possible for tumor accumulation and confers on them some selectivity as well as enhancing the drugs therapeutic effects.[139]

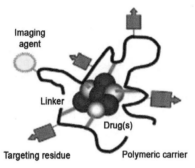

FIGURE 12.4   Major components of a polymer–drug conjugate.

N-(2-hydroxypropyl)methacrylamide (HPMA)-doxorubicin (*N*-(2-hy-droxypropyl) methacrylamide) copolymer (PK1) was the first synthetic polymer–anticancer drug conjugate to enter clinical trials more than a decade ago and the clinical phase II trial for women with advanced breast cancer.[140]

Oxaliplatin drug loading was ~10% (w/w) using a polymer chain of 25 kDa and the drug release was slow. Formulations were injected once a week for 3 weeks and the polymer–drug conjugates significantly retarded tumor growth over 1 month due to higher intracellular concentration of Pt. In the clinical phase I trial, [141] systemic injections of 640 mg Pt injected weekly for 3 weeks resulted in a response by platinum-resistant ovarian cancer. Overall, polymer–drug conjugates are considered simple nanocarrier systems, but tuning the optimal formulation might require extensive devel-opment. For example, small changes in the polymer–drug conjugation effi-ciency may significantly modify the pharmacokinetic parameters and tissue biodistribution.

Polymer–drug conjugates stay longer in circulation than low molecular drugs used in chemotherapy since the endothelium of normal blood cells is typically impermeable to macromolecules.[142] It is necessary that the polymer–drug conjugate is stable during circulation in the bloodstream and the cytotoxic drug should only be released from the conjugate intracellularly or intratumorally.

Normally polymer–drug conjugates achieve tumor-specific targeting by the EPR effect.[143] Hyperpermeable angiogenic tumor vessels allow preferential extravasation of circulating macromolecules and liposomes. This leads to significant tumor targeting (>10–100-fold compared to free drug) and up to 20% dose/g have been reported for HPMA copolymer–doxorubicin conjugates, depending on tumor size. Both polymer- and tumor-related characteristics govern the extent of EPR-mediated targeting.

HPMA copolymer conjugates are one of the nanocarrier families that are extensively evaluated in clinic. HPMA is water soluble, biocompatible, and nonbiodegradable. HPMA has a large number of pendent functional groups that allow the conjugation of hydrophobic anticancer drug to the polymer backbone via an enzymatically degradable linker (usually Gly–Phe–Leu–Gly). Since HPMA is not biodegradable, molecular weight of the conjugate has to be controlled below 45–50 kDa. Branched and graft copolymers with biodegradable spacers are synthesized to achieve EPR effect and at the same time ensure elimination of polymer backbone after drug release.

Polyglutamic acid conjugates are another class of polymeric nanocarriers which contain a high drug loading and are biodegradable in nature. A PGA–paclitaxel conjugate has entered phase III clinical trial and it's by far the most promising conjugate to hit the market in the near future. Preclinical studies in animal tumor models demonstrate enhanced safety and efficacy relative to paclitaxel when administered as a single agent or in conjunction with radiation. Clinical pilot studies with PGA paclitaxel showed improved outcomes compared to standard taxanes and allowed a more convenient administration schedule. Human pharmacokinetic data are consistent with prolonged tumor exposure to active drug and a limited systemic exposure. Recent studies have already reported the use of polymer conjugation techniques with drugs acting on new and promising molecular targets such as the inhibition of specific kinases, apoptotic pathways, heat-shock proteins, and angiogenesis.

12.5.4   POLYMERIC MICELLES

Polymeric micelles are nanoscopic (>100 nm) amphiphilic block copolymers with a core–shell structure. Polymeric micelles, self-assemblies of block copolymers, are one of the most refined and promising modalities of drug delivery systems, since the critical parameters such as size, drug loading, and release can be controlled by engineering the constituent block copolymers. It has been demonstrated that polymeric micelles show unique disposition characteristics in the body suitable for drug targeting (e.g., prolonged blood circulation and significant tumor accumulation).

Polymeric NPs designate cores of biodegradable hydrophobic polymers protected by an amphiphilic block copolymer that stabilizes their dispersion in aqueous media. Liposomes are vesicles consisting of one or more phospholipidic bilayer(s), with an aqueous core.[144]

In addition, polymeric micelles display larger cores than surfactant micelles, leading to higher solubilization capacity than the regular micelles.[145] Polymeric micelles and NPs have been investigated extensively for drug delivery. polymeric NPs core–shell structure with micelles, are matrix-type, solid-colloidal particles and, exhibit greater stability than micelles. They are typically larger (100–500 nm) than polymeric micelles (10–100 nm) and may display somewhat more polydisperse size distributions. Polymeric micelles have several advantages over other nanosized drug-delivery systems, such as a smaller size which is important to reach up to targeted site, for example, percutaneous lymphatic delivery or extravasation from blood vessels into the tumor tissue.[146]

Recently, the use of micelles prepared from amphiphilic copolymers (Fig. 12.5) for solubilization of poorly soluble drugs has attracted much attention.[147–149] Amphiphilic block copolymers with having a large solubility difference between hydrophilic and hydrophobic segments, have a tendency to self-assemble into micelles in a selective solvent. In an aqueous solution, micelles with core–shell structures are formed through the segregation of insoluble hydrophobic blocks into the core, which is surrounded by a shell composed of hydrophilic blocks. This core–shell structure facilitates their utilization, where depending upon the polarity the drug molecule can be entrapped in the (1) core (nonpolar molecule), (2) shell (polar molecule), and (3) in-between the core and shell (intermediate polarity).[150–152]

The polymeric micelles are categorized into two groups depending on drug-loading methods including "physical drug entrapment type micelles" and "covalent drug conjugation type micelle." For the physical drug entrapment-type micelles, they incorporate drug payloads through the hydrophobic interaction in the micelle core.[153] Drugs can be entrapped also in gel-like amorphous core. In either case, the equilibrium rates determine the physicochemical stability and drug-release patterns of the polymeric micelles, which are controlled time-dependently. Covalent drug conjugation type micelles have drug-binding linkers and drugs are stable in the micelle core until the polymeric micelles accumulate in the site of action and are exposed to the in vivo stimuli such as ions, endogenous signal peptides, enzymes, and pH that trigger drug release.[154] Covalent drug-conjugation-type micelles appear to be more stable than physical drug-entrapment-type micelles as long as the linkage remains intact.

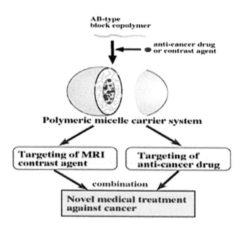

FIGURE 12.5　Polymeric micelle system for cancer treatment.

Several factors are effective in the loading of drugs in polymeric micelles, including the physicochemical characteristics of the drug and core-forming polymer and the loading method. Other factors such as chemical composition of the core-forming polymer, polymer–drug compatibility as well as physical state of the micelle core, can substantially alter drug loading and release kinetics.

Spherical polymeric micelles with diameters in the size range of 15–80 nm structures have been suggested as promising long circulating carriers of poorly water soluble and amphiphilic drugs.

The stability of micellar systems in vasculature and their extent of interaction with blood and cellular components and the control of drug liberation became a subject of interest for application on anticancer drug.

NK012 is a polymeric micellar formulation that consists of a block copolymer of PEG and polyglutamate (PGlu) conjugated with 7-ethyl-10-hydroxy-campothecin (SN-38).[155] SN-38 is a campotothecin analog and acts as a DNA topoisomerase I inhibitor but cannot be administered by i.v. due to its water-insolubility and high toxicity. SN-38 is covalently coupled to the PGlu segment by the condensation reaction between the carboxylic acid of PGlu and the phenol of SN-38 using 1,3-diisopropylcarbodiimide and

N,*N*-dimethylaminopyridine as coupling agent and catalyst, respectively. Consequently, the PGlu segment is rendered hydrophobically to induce micelle formation.[156,157] Preclinical in vivo studies with NK012 showed potent antitumor activity in mice. A pharmacokinetic study revealed that the plasma area under the curve of micellar SN-38 after i.v. administration (30 mg/kg) to HT-29 tumor-cell-bearing mice was around 200 times higher as compared to CPT-11 at a dose of 66.7 mg/kg. The IC50 values of NK012 were up to 5.8 times higher than those of free SN-38. In addition, the clearance of NK012 in the HT-29 tumors was significantly slower compared to CPT-11 and SN-38. The highest tumor-to-plasma concentration ratio of micellar SN-38 was up to 10 times higher compared to free SN-38. Moreover, NK012 clearance was significantly lower compared to CPT-11.[158,159] On another hand, a combination of NK012 with 5-fluorouracil (5-FU) showed a significantly higher antitumor effect in human colon cancer xenografts compared to CPT-11/5-FU.[160]

Polymeric micelle carrier systems are electrically neutral and so have the so-called stealth property that evades rapid clearance at the reticuloendothelial systems, which substantially improves targeting of murine solid tumors due to the EPR effect that depends on the hyperpermeable vasculature and absence of effective lymphatic drainage that prevents efficient clearance of micromolecules in the solid tumor tissues.

In addition to passive targeting, micelles can be modified with ligands for active targeting to increase the selectivity for tumor cells and enhance intracellular drug-delivery while reducing systemic toxicity and adverse side effects compared to untargeted micelles and systemic chemotherapy.[161] The concept behind this approach is based on receptor-mediated endocytosis. When the ligands conjugated to the micelles bind to their specific receptors on the cell membrane, the micelles are internalized by endocytosis.[162] In this way, higher intracellular drug concentrations are obtained. Active targeting can be achieved by conjugation of specific ligands, like monoclonal antibodies (mAbs) or their Fab fragments, oligosaccharides or peptides to the shell-forming block.[163] This allows micelles to specifically bind to antigens or receptors that are overexpressed on the tumor cells.

12.5.5   DENDRIMERS

The term dendrimer, first proposed by Tomalia in 1985, was chosen due to its structural shape, with highly branched, three-dimensional features that resemble the architecture of a tree.[164,165] A typical dendrimer (Fig. 12.6) consists of three main structural components: (1) a focal core, (2) building blocks with several interior layers composed of repeating units, and (3) multiple peripheral functional groups. The branched units are organized in layers called "generations" and represent the repeating monomer unit of these macromolecules.

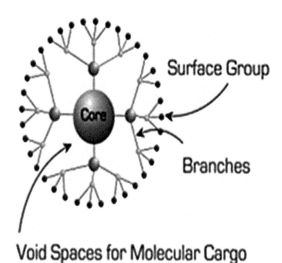

FIGURE 12.6   Schematic representation of dendrimer.

It is a synthetic polymer with a branching, tree-like structure. The name comes from the Greek word "dendron," which translates to "tree." Dendrimers are highly branched, star-shaped macromolecules with nanometer-scale dimensions. Dendrimers are defined by three components: a central core, an interior dendritic structure (the branches), and an exterior surface with functional surface groups.

The varied combination of these components yields products of different shapes and sizes with shielded interior cores that are ideal candidates for applications in both biological and materials sciences.

12.5.5.1   DENDRIMER: SYNTHESIS METHOD

The dendrimer is assembled from a multifunctional core, which is extended outward by a series of reactions, commonly a Michael reaction. Figure 12.7 shows the synthesis step of dendrimers.

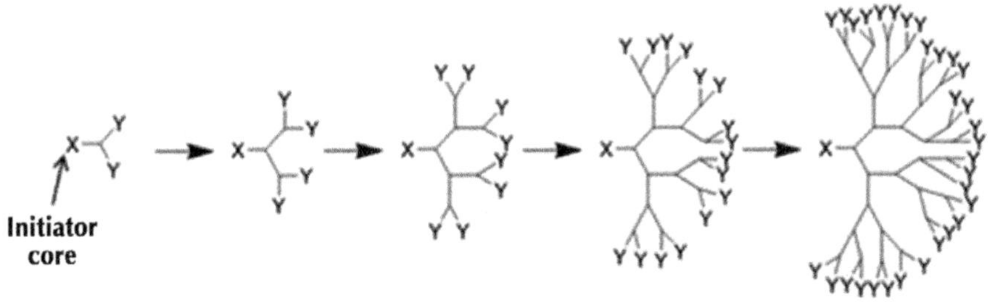

FIGURE 12.7   Schematic of divergent synthesis of dendrimers.

Dendrimers are built from small molecules that end up at the surface of the sphere, and reactions proceed inward building inward and are eventually attached to a core. Figure 12.8 explains the schematic of convergent synthesis of dendrimers.

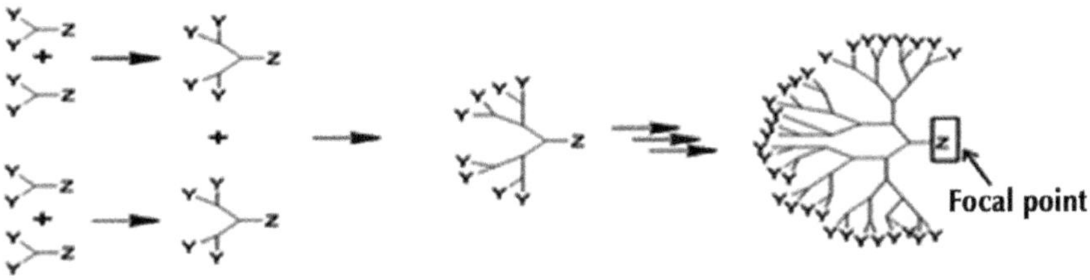

FIGURE 12.8   Schematic of convergent synthesis of dendrimers.

Size and molecular mass of dendrimers can be specifically controlled during synthesis. Dendrimers' solubility is strongly influenced by the nature of surface groups. Dendrimers terminated in hydrophilic groups are soluble in polar solvents, while dendrimers having hydrophobic end groups are soluble in nonpolar solvents. Dendrimers have some unique properties because of their globular shape and the presence of internal cavities.

The most important one is the possibility to encapsulate guest molecules in the macromolecule interior. Two strategies are used for the application of dendrimers to drug delivery: drug encapsulation by dendritic structure and drug conjugation to dendrimers. First, the drug molecules can be physically entrapped inside the dendrimers; second, the drug molecules can be covalently attached onto the surface or other functionalities to afford dendrimer–drug conjugates.[166] Different types of dendrimers, including polyamidoamine (PAMAM), polypropylene imine, polylysine dendrimers, have been used as host for both hydrophilic and hydrophobic drugs. An ideal dendritic drug-carrier must be nontoxic, nonimmunogenic, and preferably biodegradable; present an adequate biodistribution; and allow tissue targeting.[167]

Conjugation of drugs to the dendrimer is an attractive approach for intelligent drug delivery because a single dendrimer molecule can stably carry many drug molecules using many functional groups on the outer shell and reach the target cancer site through EPR effects.

12.6 METHODS FOR TARGETING SPECIFIC BIOMARKERS OF CANCER

As discussed above, dendrimers can achieve passive EPR-mediated targeting to a tumor simply by control of their size and physicochemical properties. Passive targeting, which localizes the nanoparticle in the close vicinity of a cancer cell, can be immediately useful for diagnostic purposes or for the delivery of radioisotopes capable of killing any cell within a defined radius. In general, however, most delivery strategies require that the anticancer agent directly attached to, or be taken up by, the target cell. The ability to append more than one type of functionality to a dendrimer allows the inclusion of ligands intended to bind specifically to cancer cells in the design of a multifunctional drug-delivery nanodevices. Although a wide range of targeting ligands have been considered, including natural biopolymers such as oligopeptides, oligosaccharides, and polysaccharides such as hyaluronic acid, or polyunsaturated fatty acids.

12.6.1 TARGETING BY FOLATE, A SMALL-MOLECULE LIGAND

Folate is an attractive small molecule for use as a tumor-targeting ligand because the membrane-bound FR is overexpressed on a wide range of human cancers, including those originating in ovary, lung, breast, endometrium, kidney, and brain.[168] As a small molecule, it is presumed to be nonimmunogenic, it has good solubility, binds to its receptor with high affinity when conjugated to a wide array of conjugates, including protein toxins, radioactive imaging agents, MRI contrast agents, liposomes, gene transfer vectors, antisense oligonucleotides, ribozymes, antibodies, and even activated T cells.[169,170] Upon binding to the FR, folate-conjugated drug conjugates are shuttled into the cell via an endocytic mechanism, resulting in major enhancements in cancer cell specificity and selectivity over their nontargeted formulation counterparts. Recently, folate has been enlisted in an innovative dendrimer-based targeting schemes.[171]

12.6.2   TARGETING BY MONOCLONAL ANTIBODIES

Of the many strategies devised to selectively direct drugs to cancer cells, perhaps the most commonly is the use of monoclonal antibodies that recognize and selectively bind to tumor-associated antigens (TAAs).[172-175] TAA-targeting monoclonal antibodies have been exploited as delivery agents for conjugated "payloads" such as small molecule drugs and prodrugs, radioisotopes, and cytokines.[176,177] Current prospects remain mixed but hopeful; optimistically, progress marked by commercial interest with companies providing their immunotherapeutic drug candidates with flashy trademarked names such as "Armed Antibodies™."[178 179] Perhaps, more realistically, one recent synopsis holds out "hope" for a major clinical impact for this strategy within the next 10 years. One key issue readily addressed by dendrimers is the requirement that an extremely potent cytotoxic drug be used in targeted antibody therapy. This point is illustrated by the fact that the greatest progress in this field has occurred for immunotoxins, which are antibody–toxin chimeric.

Molecules that kill cancer cells via binding to a surface antigen, internalization, and delivery of the toxin moiety to the cell cytosol, inhibit cell function and cause cell death. The high potency of immunotoxins for killing cancer cells is dramatically illustrated by ricin, where the catalytic activity of this ribosome-inactivating enzyme allows a single immunotoxin conjugate to kill a cell upon successful uptake and trafficking to the site of action.[180,181] A drawback of immunotoxins is their significant immunogenicity; from a broader perspective, their repeated use is made necessary by difficulties in providing a sufficiently high drug load to eradicate all cancer cells despite the high potency of conjugated toxin. An alternative approach of radioimmunotherapy, where high energy radionuclides are conjugated to TAA-targeting antibodies, also shows promise but suffers from indiscriminate toxicity.[182] A third possible approach for immunotherapy, the conjugation of commonly used small molecule drugs to TAAs, is hindered by the relatively low potency of most low molecular weight therapeutics. The widely used anticancer drug cisplatin, requires internalization of at least 50× this level of drug molecules for therapeutic efficacy.

A numerical analysis of the cisplatin presented above indicates that each tumor-targeting antibody would have to be modified with a large number of small molecules to be effective as an anticancer drug. Modification of an antibody with multiple radioisotopes, toxins, or even small molecules to increase the efficacy of cell killing, however, diminishes or eliminates the inherent specific antigen-binding affinity of an antibody. Therefore, to maximize drug loading with minimum side effects on the biological integrity of the host antibody, using linker molecule is nowadays becoming an attractive approach.[183] Methodology to covalently attach antibodies to dendrimers that preserve the activity of the antigen–antibody binding site,[184,185] for example, by chemical modification of their carbohydrates and subsequent linkage to PAMAM,[186] has opened the door for the inclusion of dendrimers in immunotherapy.[187,188]

12.6.3   CARBON NANOTUBES

CNTs are a new form of carbon molecule around in a hexagonal network of carbon atoms, these hollow cylinders can have diameter as a small as 0.7 nm and reach several millimeters in length.[189] Each end can be opened or closed by a fullerene half molecule. The small dimensions of nanotubes, combined with their remarkable physical, mechanical, and electrical properties, make them unique materials. The mechanical strength of CNTs is more than 60 times greater than that of the best steels, even though they weigh six times less. They also represent a very large specific surface area, are excellent heat conductors, and display unique electronic properties, offering three-dimensional configurations.

They have higher capacity for molecular absorption.[190] Typically, CNTs are classified as single-walled (SWCNT) or multiwalled (MWCNT). Figure 12.9 depicted single walled and multiwalled CNT. SWCNTs consist of a single cylindrical carbon layer with a diameter in the range of 0.4–2 nm, depending on the temperature at which they have been synthesized. It has been observed that a higher growth temperature gives a larger diameter. In contrast, MWCNTs are usually made from several cylindrical carbon layers with diameters in the range of 1–3 nm for the inner tubes and 2–100 nm for the outer tubes.[191,192]

CNTs, with their unique physical and chemical properties, hold great hopes for cancer imaging and treatment.[193,194] In pharmacological applications, CNTs have primarily been explored as potential drug carriers and delivery vehicles. In particular, recent data suggest that CNTs can deliver intracellularly apoptotic agents.[195] Nevertheless, besides precise tumor targeting and toxicity concerns, drug resistance remains a major obstacle for the treatment of advanced cancerous tumors.

Filling CNTs with an appropriate anticancer drug is another method of delivering anticancer therapy. According to Arsawang et al., a CNT with a diameter of 80 nm can hold up to 5 million drug molecules.[196] Several strategies have been used to incorporate drugs into CNTs. One of these methods is steered molecular dynamic simulation. The general principle of steered molecular dynamics involves applying an external force to particles in a specific direction by use of harmonic (spring-like) restraint to create greater change of the particle coordinates. In an experiment carried out by researchers in Thailand, gemcitabine, an anticancer drug, was loaded onto SWCNTs using a steered molecular dynamic technique. Following application of force to the gemcitabine, it was shown that the cytosine ring of gemcitabine formed π–π stacking on the internal surface of the CNT with 25 Å far from one end of the SWCNT.[197]

The wet chemical technique is also commonly used. An example of this method comes from a study in which a 1-mg/ suspension of open-ended CNTs was placed in a 10-mg/ carboplatin solution, with sonication of the mixture for 10 min, followed by stirring for 24 h. Optical investigations of the sample obtained was performed using transmission electron microscopy, energy dispersive X-ray analysis, electron energy loss spectroscopy, and X-ray photoelectron spectroscopy, all of which established the presence of carboplatin inside the CNTs.

According to the National Institute of Standards and Technology, all antibodies that have been used for cell targeting have been monoclonal IgG antibodies. However, experiments have recently been carried out using IgY as a substitute for IgG. IgY has shown some biochemical, immunological, and production-related advantages in comparison with IgG.[198,199] Other observations show that the attachment of antibodies to the CNT surface does not lead to alteration of antibody specificity for the target cell. It has been shown that the antibody can successfully deliver anticancer drug-loaded CNTs to the site of action. For example, Ashcroft et al. found that more than 40 CNT–anticancer drug complexes could be targeted as a result of coating the CNT with ZME-108, a specific type of skin cancer antibody.[200] In another experiment, a SWCNT functionalized by PEG and rituxan (the monoclonal antibody against CD20, found primarily on B cells), selectively targeted the CD20 cell surface receptor on B cells with little binding to T cells.[201]

It is well known that cancer cells overexpress FA receptors, and several research groups have designed nanocarriers with engineered surfaces to which FA derivatives can be attached. Moreover, nonspherical nanocarriers (e.g., CNTs) have been reported to be retained in the lymph nodes for longer periods of time compared to spherical nanocarriers[202] (e.g., liposomes).

Thus, CNTs might be used for targeting lymph node cancers as shown by various investigators.[203–205]

Dhar and coworkers[206] have developed longboat delivery system, which is a complex of cisplatin and FA derivative. This complex was attached to a functionalized SWNT via a number of amide bonds to comprise the "longboat" which has been reported to be taken up by cancer cells via endocytosis, followed by the release of the drug and its subsequent interaction with the nuclear DNA. Another platinum anticancer, namely, carboplatin, after being incorporated into CNTs has been shown to inhibit the proliferation of urinary bladder cancer cells in vitro. In another study, anticancer effects have been shown to be dependent on the method used to entrap the drug in the CNTs, which highlighted the possible effects of preparation conditions on the therapeutic activity of therapeutic molecules associated with CNTs.[207] Similar findings of anticancer activity have been recently shown when paclitaxel was loaded into PEGylated SWNTs or MWNTs using HeLa cells and MCF-7 cancer cell lines.[208] Multidrug resistance is a significant obstacle to successful anticancer drug therapy since the P-glycoprotein efflux transporter can interfere with the accumulation of anticancer drugs in the target cells, resulting in reduced effectiveness of therapy.[209,210]

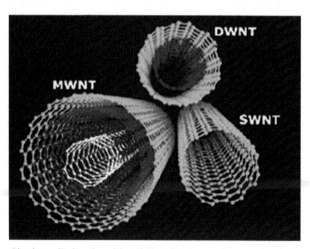

FIGURE 1.9   Single-walled and multiwalled carbon nanotube.

12.6.4   QUANTUM DOTS

An emerging branch of NP technologies is the development of optical-based functional NPs[211,212–220] such as QDs.[221]

The QD is defined as an artificially structured system with the capacity to load electrons.[222] Its special physicochemical properties differentiate it from other naturally occurring biogenic and anthropogenic NPs.[223] QDs are one type of NPs with three characteristic properties: semiconductors, zero-dimension, and strong fluorescence. Colloidal semiconductor QDs are single crystals with size ranging in nanometers, and their sizes and shapes can be precisely controlled by the duration, temperature, and ligand molecules during the synthetic processes.[224]

The ability to internalize QD conjugates, along with the increased multi-plexing capabilities, offers a major advancement in time and cost-effective-ness over single experiments. Ultimately, these advantages will contribute to the detection of various cell proteins or other components of heterogeneous tumor samples and used in cancer diagnostics. Labeling cells with QDs for multiplexed analysis have much potential in the clinical setting. Compared to traditional immunochemistry assays, QD immunostaining has been shown to be more accurate and precise at low protein expression levels.[225,226]

Photostability is one of the greatest advantages of QDs for in vivo appli-cations since it allows images to be recorded over a longer period of time than available with the use of fluorescent dyes or proteins, due to resistance in photobleaching.[227]

QDs plays an increasingly important role in tumor imaging, especially near-infrared (NIR, 700–900 nm) imaging. Figure 12.10 shows the QDs effectivity in cancer cells. NIR-fluorescence imaging of tumor is expected to have a major impact in biomedical imaging because in the NIR region, the absorbance spectra for all the biomolecules in tumor reach their minima, which provides a clear window for in vivo optical imaging of tumor.[228] In 2010, Gao et al. reported that QD800-MPA (a NIR noncadmium QDs coated with mercaptopropionic acid with an emission wavelength of about 800 nm) had high tumor uptake and excellent contrast of tumor to surrounding tissues due to the EPR effect of this kind of ultrasmall NPs.[229]

QDs, tiny light-emitting particles on nanometer scale, are new type of fluo-rescent probes for molecular and cellular imaging. Compared with organic dyes and fluorescent proteins, QDs have unique optical and electronic prop-erties in cellular imaging: wavelength-tunable emission, improved bright-ness of signal, resistance against photobleaching, etc.[230] Such preponderant optical properties were not realized until the QD-based probes are equipped with warheads targeting tumor. Xingyong Wu and coworkers synthesized immunofluorescent probes by conjugating the QDs with streptavidin or IgGs (immunoglobulin Gs).[231] Using the conjugates, they conducted comprehen-sive investigations on cell imaging at the targets of interest including cell surface receptors, cytoskeleton components, and nuclear antigens.[232] Up to date, QDs have been rapidly developed in tumor imaging such as imaging tumor vasculature[233] and sentinel lymph node.[234]

Although relatively fewer researches on QDs for tumor therapy were reported, it is conceivable that QDs have the potentialities for tumor therapy due to their large surface areas available for the modification of functional groups or therapeutic agents such as anticancer drugs[235] and PDT photosensitizers (PS).[236] Moreover, QDs themselves can also functionalize as PDT PSs for tumor therapy.[237]. Barberi-Heyo et al. established that QDs conjugated with FA could be used as PS for PDT of cancer.[238]FA is an optimal targeting ligand which selectively deliver the attached therapeutic agents) by Quantam dot to cancer tissues. MTT assay indicated that the survival rate of cancer cell was decreased and intensity was increased with FA conjugated QD (shown in Fig. 12.10). The results demonstrated that CdTe(S)-type QDs had photosensitizing properties, which could be used to promote PDT effect. In their study, they mentioned that the concentration of QDs should be inferior to 10 nm and the incubation time less than 8 h to avoid the intrinsic cytotoxicity of QDs without light irradiation.

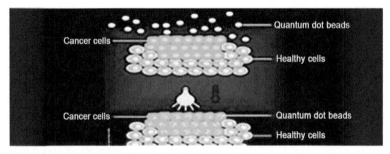

FIGURE 12.10   Quantum dots effective in cancer cells.

12.7   REGULATORY ASPECT

One of the main areas related to the safety aspects of drug–nanocarrier systems is to encourage academic organizations industry, and regulatory governmental agencies to establish convincing testing procedures on the safety aspects of the NMs. The global importance of trade for NMs has established new international organizations, such as the International Council on Nanotechnology (ICON), the International Organization for Standardization (Geneva, Switzerland), etc., for sharing responsibilities in this field. In the year 1996, the NNI was established in the United States of

America to coordinate governmental multi-agencies, such as the FDA, the Department of Labor through the Occupational Safety and Health Administration (OSHA), the National Institute for Occupational Safety and Health (NIOSH), and the Environmental Protection Agency (EPA), for the development of nanoscience and technology.

12.8   A GLIMPSE TO FUTURE OF NANOSIZE DRUG-DELIVERY SYSTEMS

Advancement of nanosize drug-delivery systems establishes a new paradigm in pharmaceutical field. Convergence of science and engineering leads a new era of hope where medicines will act with increase efficacy, high bioavailability, and less toxicity. Several nanoscale drug-delivery systems are currently in clinical trials and few of them are already commercially available. Examples of such products are Abeicet (for fungal infection), Doxil (antineoplastic), Abraxane (metastatic breast cancer), Emend (antiemetic), etc. Despite the impressive progress in the field, very few nanoformulations have been approved by US FDA and even reached market in recent years. Although nanocarriers have lots of advantages because of the unique properties they have, there are many clinical, toxicological, and regulatory aspects which are the matters of concern too. The biocompatibility of NMs is of atmost importance because of the effect of the NMs in the body ranging from cytotoxicity to hypersensitivity.[239] With the advancement of nanotechnology, the biological phenomenon such as host response to a specific NM should also be clinically transparent.[240] Therefore, it is quite essential to introduce cost-effective, better, and safer nanobiomaterials which will provide efficient drug loading and controlled drug release of some challenging drug moieties for which there is no other suitable delivery available yet.

Nanoliposomes are well developed and presently possess the highest amount of clinical trials among other NMs with some formulations currently in the market. This may be due to the fact that other materials have not been investigated for the same duration and are relatively newer in comparison. However, polymer-based NM, CNTs, gold NPs, etc., should not be overlooked because of less number of clinical trials.[241]

Genexol-PM is an example which was undergone recent clinical trial. This is an amphiphilic diblock copolymer (PEG-D,L-lactic acid) that delivers paclitaxel. Clinical trial currently is in phase IV using Genexol-PM for recurrent breast cancer and phase III for breast cancer.[242] Some recent clinical trials are shown in Table 12.1.

TABLE 12.1   Recent Clinical Trials

Product name	Delivery material	Phase	Condition	Therapeutic Delivered	Sponsor	Clinical trials. gov Identifier
Genexol-PM	Amphilic diblock Copolymer forming micelle	I	Non small Cell lung cancer	Paclitaxel	Samyang Biopharmaceutical Corp	NCT01023347
Docetaxel-PNP	Polymeric nanoparticles	I	Advanced solid malignancies	Docetaxel	Samyang Biopharmaceutical Corp	NCT01103791
Kogenate FS	PEG-liposome	I	Hemophilia A	Recombinant factor VIII	Bayer	NCT00629837
Long-circulating liposomal prednisolone disodium phosphate	Liposome	II	Rheumatoid arthritis	Prednisolone	Radboud University	NCT00241982
Cisplatin and Liposomal Doxorubicin	Liposome	I	Advanced cancer	Cisplatin and doxorubicin	M.D. Anderson Cancer Center	NCT00507962
Liposomal doxorubicin and bevacizumab	Liposome	II	Kaposi's sarcoma	Doxorubicin and bevacizumab	NCI	NCT00923936

Nanomachines are also largely in the research-and-development phase, but some primitive molecular machines have been tested. An example is nanorobot which is capable of penetrating the various biological barriers of human body to identify the cancer cells. Thus, nanodrug-delivery systems have a leading role to play in nanomedicine in near future.

12.9   MULTIFUNCTIONAL NANOPARTICLES

The application of NPs for the delivery and targeting of pharmaceutical, therapeutic, and diagnostic agents in cancer therapy has received significant attention in recent years. NPs may be constructed from a wide range of materials and used to encapsulate or solubilize chemotherapeutic agents for improved delivery in vivo or to provide unique optical, magnetic, and electrical properties for imaging and therapy. Several functional NPs have already been demonstrated, including some clinically approved liposome drug formulations and metallic imaging agents.

The next generation of NP-based research is directed at the consolidation of functions into strategically engineered multifunctional systems, which may ultimately facilitate the realization of individual therapy. These multiplexed NPs may be capable of identifying malignant cells by means of molecular detection, visualizing their location in the body by providing enhanced contrast in medical imaging techniques, killing diseased cells with minimal side effects through selective drug targeting, and monitoring treatment in real time.

12.10   CONCLUSION

Cancer is an inherently biological disease, in which cell replication fails to be regulated by the usual mechanisms. Last few years, several new technologies have been developed for the treatment of various diseases. The use of nanotechnology in developing nanocarriers for drug delivery is bringing lots of hope and enthusiasm in the field of drug-delivery research. Nanoscale drug-delivery devices present some advantages which show higher intracellular uptake than the other conventional form of drug-delivery systems. Nanocarriers can be conjugated with a ligand such as antibody to favor a targeted therapeutic approach. Thus, nanoscale size drug-delivery systems may revolutionize the entire drug therapy strategy and bring it to a new height in near future. However, toxicity concerns of the nanosize formulations should not be ignored. Full-proof methods should be established to evaluate both the short-term and long-term toxicity analysis of the nanosize drug-delivery system. So overall, nanocarriers may lead to a solution of major unsolved medical problems which will aggressively enhance quality of life and it may become an attractive approach in future to fight with cancer.

KEYWORDS

- cancer
- nanotechnology
- cancer targeting
- polymeric nanoparticle
- liposomes
- cancer therapy

REFERENCES

1. Sarkar, F. H.; Banerjee, S.; Li, Y. W. Pancreatic Cancer: Pathogenesis, Prevention and Treatment. *Toxicol. Appl. Pharmacol.* **2007,** *224* (3), 326–336.
2. Salamanca-Buentello, F.; Persad, D. L.; Court, E. B.; Martin, D. K.; Daar, A. S.; Singer, P. A. Nanotechnology and the Developing World. *PLoS* **2005,** *2* (5), e97.
3. Byrne, J. D.; Betancourt, T.; Brannon-Peppas, L. Active Targeting Schemes for Nanoparticle Systems in Cancer Therapeutics. *Adv. Drug Deliv. Rev.* **2008,** *60* (15), 1615–1626.
4. Iyer, A. K.; Khaled, G.; Fang, J.; Maeda, H. Exploiting the Enhanced Permeability and Retention Effect for Tumor Targeting. *Drug Discov. Today* **2006,** *11* (17), 812–818.
5. Anand, P.; Kunnumakkara, A. B.; Sundaram, C.; Harikumar, K. B.; Tharakan, S. T.; Lai, O. S.; Sung, B.; Aggarwal, B. B. Cancer is a Preventable Disease that Requires Major Lifestyle Changes. *Pharm Res.* **2008,** *25* (9), 2097–2116.
6. Kolonel, L. N.; Altshuler, D.; Henderson, B. E. The Multiethnic Cohort Study: Exploring Genes, Lifestyle and Cancer Risk. *Nat. Rev. Cancer* **2004,** *4,* 519–527.
7. http://sphweb.bumc.bu.edu/otlt/MPHModules/PH/PH709_Cancer/PH709_Cancer_print.html (last accessed on November 2015).
8. Gronostajski, R. M.; Goldberg, A. L.; Pardee, A. B. The Role of Increased Proteolysis in the Atrophy and Arrest of Cell Division in Fibroblasts Following Serum Deprivation. *J. Cell. Biol.* 1984, *12,* 189–198.
9. Muratani, M.; Tansey, W. P. How the Ubiquitin–Proteasome System Controls Transcription. *Nat. Rev. Mol. Cell Biol.* **2003,** *4,* 192–201.
10. Jemal, A.; Siegel, R.; Ward, E.; Murray, T.; Xu, J.; Thun, M. J. Cancer Statistics, 2007. *Cancer J. Clin.* **2007,** *57* (1), 43–66.

11. http://www.cancer.gov/statistics (last accessed on November 2015).
12. Pulkkinen, M.; Pikkarainen, J.; Wirth, T., et al. Three-Step Tumor Targeting of Pacli-taxel Using Biotinylated PLA–PEG Nanoparticles and Avidin–Biotin Technology: Formulation Development and In Vitro Anticancer Activity. *Eur. J. Pharm. Biopharma-ceut.* **2008,** *70* (1), 66–74.
13. www.Biomedtracker.Com/Cancer_Immunotherapies_Report.pdf (last accessed on December 2015).
14. Patel, P. V.; Soni, T. G.; Thakkar, V. T.; Gandhi, T. R. Nanoparticle as an Emerging Tool in Pulmonary Drug Delivery System. *Micro Nanosyst.* **2013,** *5,* 288–302.
15. Panyala, N. R.; Pena-Mendez, E. M.; Havel, J. Gold and Nano-Gold in Medicine: Over-view, Toxicology and Perspectives. *J. Appl. Biomed.* **2009,** *7* (2), 75–91.
16. Sobha, D. K.; Surendranath, K.; Meena, V.; Jwala, K. T.; Swetha, N.; Latha, K. S. M. Emerging Trends in Nanobiotechnology. *J. Biotechnol. Mol. Biol. Rev.* **2010,** *5* (1), 001–012.
17. Iijima, S. Helical Microtubules of Graphitic Carbon. *Nature* **1991,** *354* (6348), 56–58, 1991.
18. Kreuter, J.; Ramge, P.; Petrov, V.; Hamm, S.; Gelperina, S. E.; Engelhardt, B.; Alyautdin, R.; von Briesen, H.; Begley, D. J. Direct Evidence that Polysorbate-80-Coated Poly(butylcyanoacrylate) Nanoparticles Deliver Drugs to the CNS via Specific Mecha-nisms Requiring Prior Binding of Drug to the Nanoparticles. *Pharm. Res.* **2003,** *20,* 409–416.
19. Zauner, W.; Farrow, N. A.; Haines, A. M. In Vitro Uptake of Polystyrene Microspheres: Effect of Particle Size, Cell Line and Cell Density. *J. Control. Release* **2001,** *71,* 39–51.
20. Redhead, H. M.; Davis, S. S.; Illum, L. Drug Delivery in Poly(lactide-*co*-glycolide) Nanoparticles Surface Modified with Poloxamer 407 and Poloxamine 908: In Vitro Characterisation and In Vivo Evaluation. *J. Control. Release* **2001,** *70,* 353–363.
21. Dunne, M.; Corrigan, O. I.; Ramtoola, Z. Influence of Particle Size and Dissolution Conditions on the Degradation Properties of Polylactide-*co*-glycolide Particles. *Bioma-terials* **2000,** *21,* 1659–1668.
22. Mohanraj, V. J.; Chen, Y. Nanoparticles—A Review. *Trop. J. Pharm. Res.* **2006,** *5* (1), 561–573.
23. Swarbrick, J.; Boylan, J. *Encyclopedia of Pharmaceutical Technology,* 2nd ed.; Marcel Dekker: New York, 2002; pp 84–92.
24. Muller, R. H.; Wallis, K. H. Surface Modification of i.v. Injectable Biodegradable Nanoparticles with Poloxamer Polymers and Poloxamine 908. *Int. J. Pharm.* **1993,** *89,* 25–31.
25. Brigger, I.; Dubernet, C.; Couvreur, P.; Nanoparticles in Cancer Therapy and Diagnosis. *Adv. Drug Deliv. Rev.* **2002,** *54,* 631–651.
26. Grislain, L.; Couvreur, P.; Lenaerts, V.; Roland, M.; Deprez-Decampeneere, D.; Speiser, P.; Pharmacokinetics and Distribution of a Biodegradable Drug-Carrier. *Int. J. Pharm.* **1983,** *15,* 335–345.
27. Olivier, J. C. Drug Transport to Brain with Targeted Nanoparticles. *Neuro Rx* **2005,** *2,* 108–119.
28. Couvreur, P.; Barratt, G.; Fattal, E.; Legrand, P.; Vauthier, C. Nanocapsule Technology: A Review. *Crit. Rev. Ther. Drug Carrier Syst.* **2002,** *19,* 99–134.
29. Govender, T.; Stolnik, S.; Garnett, M. C.; Illum, L.; Davis, S. S. PLGA Nanoparticles Prepared by Nanoprecipitation: Drug Loading and Release Studies of a Water Soluble Drug. *J. Control. Release* **1999,** *57,* 171–185.

30. Govender, T.; Riley, T.; Ehtezazi, T.; Garnett, M. C.; Stolnik, S.; Illum, L.; Davis, S. S. Defining the Drug Incorporation Properties of PLA–PEG Nanoparticles. *Int. J. Pharm.* **2000,** *199,* 95–110.

31. Panyam, J.; Williams, D.; Dash, A.; Leslie-Pelecky, D.; Labhasetwar, V. Solid-State Solubility Influences Encapsulation and Release of Hydrophobic Drugs from PLGA/ PLA Nanoparticles. *J. Pharm. Sci.* **2004,** *93,* 1804–1814.

32. Peracchia, M.; Gref, R.; Minamitake, Y.; Domb, A.; Lotan, N.; Langer, R. PEG-Coated Nanospheres from Amphiphilic Diblock and Multiblock Copolymers: Investigation of their Drug Encapsulation and Release Characteristics. *J. Control. Release* **1997,** *46,* 223–231.

33. Chen, Y.; McCulloch, R. K.; Gray, B. N. Synthesis of Albumindextran Sulfate Micro-spheres Possessing Favourable Loading and Release Characteristics for the Anti-cancer Drug Doxorubicin. *J. Control. Release* **1994,** *31,* 49–54.

34. Chen, Y.; Mohanraj, V. J.; Parkin, J. E. Chitosan-Dextransulfate Nanoparticles for Delivery of an Antiangiogenesis Peptide. *Lett. Pep. Sci.* **2003,** *10,* 621–627.

35. Magenheim, B.; Levy, M. Y.; Benita, S. A New In Vitro Technique for the Evaluation of Drug Release Profile from Colloidal Carriers—Ultrafiltration Technique at Low Pres-sure. *Int. J. Pharm.* **1993,** *94,* 115–123.

36. Fresta, M.; Puglisi, G.; Giammona, G.; Cavallaro, G.; Micali, N.; Furneri, P. M.; Peflox-acinmesilate- and Ofloxacin Loaded Poly Ethylcyanoacrylate Nanoparticles: Character-ization of the Colloidal Drug Carrier Formulation. *J. Pharm. Sci.* **1995,** *84,* 895–902.

37. Terrari, M. Cancer Nanotechnology: Opportunities and Challenges. *Nat. Rev. Cancer* **2005,** *5,* 161–171.

38. Gillies, E. R.; Frechet, J. M. Dendrimers and Dendritic Polymers in Drug Delivery. *Drug Discov. Today* **2005,** *10,* 35–43.

39. Jain, K. K. Applications of Nanobiotechnology in Clinical Diagnostics. *Clin Chem.* **2007,** *53* (11), 2002–2009.

40. Brigger, I.; Dubernet, C.; Couvreur, P. Nanoparticles in Cancer Therapy and Diagnosis. *Adv. Drug Deliv. Rev.* **2002,** *54,* 631–651.

41. Zhao, G.; Rodriguez, B. L. Molecular Targeting of Liposomal Nanoparticles to Tumor Microenvironment. *Int. J. Nanomed.* **2013,** *8,* 61–71.

42. Nguyen, K. T. Targeted Nanoparticles for Cancer Therapy: Promises and Challenges. *J. Nanomed. Nanotechnol.* **2011,** *2* (5), 103–106.

43. Coates, A.; Abraham, S.; Kaye, S. B. On the Receiving End—Patient Perception of the Side-Effects of Cancer Chemotherapy. *Eur. J. Cancer Clin. Oncol.* **1983,** *19* (2), 203–208.

44. Mousa, S. A.; Bharali, D. J. Nanotechnology-Based Detection and Targeted Therapy in Cancer: Nano-bioparadigms and Applications. *Cancers* **2011,** *3* (3), 2888–2903.

45. Krishna, R.; Mayer, L. D. Multidrug Resistance (MDR) in Cancer: Mechanisms, Reversal Using Modulators of MDR and the Role of MDR Modulators in Influencing the Pharmacokinetics of Anticancer Drugs. *Eur. J. Pharm. Sci.* **2000,** *11* (4), 265–283.

46. Links, M.; Brown, R. Clinical Relevance of the Molecular Mechanisms of resistance to Anti-cancer Drugs. *Expert Rev. Mol. Med.* **1999,** *1,* 1–21.

47. Gottesman, M. M.; Hrycyna, C. A.; Schoenlein, P. V.; Germann, U. A.; Pastan, I. Genetic Analysis of the Multidrug Transporter. *Annu. Rev. Genet.* **1995,** *29,* 607–649.

48. Davis, M. E.; Chen, Z.; Shin, D. M. Nanoparticle Therapeutics: An Emerging Treatment Modality for Cancer. *Nat. Rev. Drug Discov.* **2008,** *7* (9), 771–782.

49. Torchilin, V. P. Drug Targeting. *Eur. J. Pharm. Sci.* **2000**, *11*, S81–S91.

50. Gerber, D. E. Targeted Therapies: A New Generation of Cancer Treatments. *Am. Fam. Physician* **2008**, *77*, 311–319.

51. Mimeault, M.; Hauke, R.; Batra, S. K. Recent Advances on the Molecular Mechanisms Involved in the Drug Resistance of Cancer Cells and Novel Targeting Therapies. *Clin. Pharmacol. Ther.* **2008**, *83*, 673–691.

52. Strebhardt, K.; Ullrich, A. Paul Ehrlich's Magic Bullet Concept: 100 Years of Progress. *Nat. Rev. Cancer* **2008**, *8*, 473–480.

53. Moghimi, S. M.; Hunter, A. C.; Murray, J. C. Long-Circulating and Target-Specific Nanoparticles: Theory to Practice. *Pharmacol. Rev.* **2001**, *53*, 283–318.

54. Sajja, H. K.; East, M. P.; Wang, Y. A.; Yang, L. N. Development of Multifunctional Nanoparticles for Targeted Drug Delivery and Noninvasive Imaging of Therapeutic Effect. *Curr. Drug Discov. Technol.* **2009**, *6*, 43–51.

55. Tong, Q.; Li, H.; Li, W.; Chen, H.; Shu, X.; Lu, X.; Wang, G. In Vitro and In Vivo Antitumor Effects of Gemcitabine Loaded with a New Drug Delivery System. *J. Nanosci. Nanotechnol.* **2011**, *11*, 3651–3658.

56. Mu, L.; Feng, S. S. A Novel Controlled Release Formulation for the Anticancer Drug Paclitaxel (Taxol(R)): PLGA Nanoparticles Containing Vitamin E TPGS. *J Control. Release* **2003**, *86*, 33–48.

57. Podila, R.; Brown, J. M. Toxicity of Engineered Nanomaterials: A Physicochemical Perspective. *J. Biochem. Mol. Toxicol.* **2013**, *27* (1), 50–55.

58. Mills, J. K.; Needham, D. Targeted Drug Delivery. *Expert Opin. Ther. Pat.* **1999**, *9*, 1499–1513.

59. Zensi, A.; Begley, D.; Pontikis, C.; Legros, C.; Mihoreanu, L.; Wagner, S.; Büchel, C.; Briesen, H. V.; Kreuter, J. Albumin Nanoparticles Targeted with ApoE Enter the CNS by Transcytosis and Are Delivered to Neurons. *J. Control. Release* **2009**, *137*, 78–86.

60. Lentacker, I.; Vandenbroucke, R. E.; Lucas, B.; Demeester, J.; De Smedt, S. C.; Sanders, N. N. New Strategies for Nucleic Acid Delivery to Conquer Cellular and Nuclear Membranes. *J. Control. Release* **2008**, *132*, 279–288.

61. Lin, C.-Y.; Liu, T.-M.; Chen, C.-Y.; Huang, Y.-L.; Huang, W.-K.; Sun, C.-K.; Chang, F.-H.; Lin, W.-L. Quantitative and Qualitative Investigation into the Impact of Focused Ultrasound with Microbubbles on the Triggered Release of Nanoparticles from Vasculature in Mouse Tumors. *J. Control. Release* **2010**, *146*, 291–298.

62. Nam, H. Y.; Kwon, S. M.; Chung, H.; Lee, S. Y.; Kwon, S. H.; Jeon, H.; Kim, Y.; Park, J. H.; Kim, J.; Her, S.; Oh, Y. K.; Kwon, I. C.; Kim, K.; Jeong, S. Y. Cellular Uptake Mechanism and Intracellular Fate of Hydrophobically Modified Glycol Chitosan Nanoparticles. *J. Control. Release* **2009**, *135*, 259–267.

63. Sauer, A. M.; de Bruin, K. G.; Ruthardt, N.; Mykhaylyk, O.; Plank, C.; Brauchle, C. Dynamics of Magnetic Lipoplexes Studied by Single Particle Tracking in Living Cells. *J. Control. Release* **2009**, *137*, 136–145.

64. Yuan, F. Transvascular Drug Delivery in Solid Tumors. *Semin. Radiat. Oncol.* **1998**, *8* (3), 164–175.

65. Yezhelyev, M. V.; Gao, X.; Xing, Y.; Al-Hajj, A.; Nie, S.; O'Regan, R. M. Emerging Use of Nanoparticles in Diagnosis and Treatment of Breast Cancer. *Lancet Oncol.* **2006**, *7* (8), 657–667.

66. Duncan, R. Polymer Conjugates as Anticancer Nanomedicines. *Nat. Rev. Cancer* **2006**, *6* (9), 688–701.

67. Ferrari, M. Cancer Nanotechnology: Opportunities and Challenges. *Nat. Rev. Cancer* **2005,** *5* (3), 161–171.
68. La Van, D. A.; McGuire, T.; Langer, R. Small-scale Systems for In Vivo Drug Delivery. *Nat. Biotechnol.* **2003,** *21* (10), 1184–1191.
69. Brigger, I.; Dubernet, C.; Couvreur, P. Nanoparticles in Cancer Therapy and Diagnosis. *Adv. Drug Deliv. Rev.* **2002,** *54* (5), 631–651.
70. http://www.authorstream.com/Presentation/nikitaverma1988-715330-drug-targeting/ (last accessed on November 2015).
71. Beduneau, A.; Saulnier, P.; Hindre, F.; Clavreul, A.; Leroux, J. C.; Benoit, J. P. Design of Targeted Lipid Nanocapsules by Conjugation of Whole Antibodies and Antibody Fab Fragments. *Biomaterials* **2007,** *28*, 4978–4990.
72. Deckert, P. M. Current Constructs and Targets in Clinical Development for Antibody Based Cancer Therapy. *Curr. Drug Targets* **2009,** *10*, 158–175.
73. Hong, M.; Zhu, S.; Jiang, Y.; Tang, G.; Pei, Y. Efficient Tumor Targeting of Hydroxyl Camptothecin Loaded PEGylated Niosomes Modified with Transferrin. *J. Control. Release* **2009,** *133*, 96–102.
74. Zensi, A.; Begley, D.; Pontikis, C.; Legros, C.; Mihoreanu, L.; Wagner, S.; Büchel, C.; Briesen, H. V.; Kreuter, J. Albumin Nanoparticles Targeted with ApoE Enter the CNS by Transcytosis and are Delivered to Neurons. *J. Control. Release* **2009,** *137*, 78–86.
75. Canal, F.; Vicent, M. J.; Pasut, G.; Schiavon, O. Relevance of Folic Acid/Polymer Ratio in Targeted PEG-Epirubicin Conjugates. *J. Control. Release* **2010,** *146*, 388–399.
76. Seleverstov, O.; Phang, J. M.; Zabirnyk, O. Semiconductor Nanocrystals in Autophagy Research: Methodology Improvement at Nano Sized Scale. *Methods Enzymol.* **2009,** *452*, 277–296.
77. Joshi, P.; Chakraborti, S.; Ramirez-Vick, J. E.; Ansari, Z. A.; Shanker, V.; Chakrabarti, P.; Singh, S. P. The Anticancer Activity of Chloroquine–Gold Nanoparticles against MCF-7 Breast Cancer Cells. *Colloids Surf. B: Biointerfaces* **2012,** *95*, 195–200.
78. Wei, P.; Zhang, L.; Lu, Y.; Man, N.; Wen, L. Nanoparticles Enhance Chemotherapeutic Susceptibility of Cancer Cells by Modulation of Autophagy. *Nanotechnology* **2010,** *21*, 495–501.
79. Salvador-Morales, C.; Gao, W.; Ghatalia, P.; Murshed, F.; Aizu, W.; Langer, R.; Farokhzad, O. C. Multifunctional Nanoparticles for Prostate Cancer Therapy. *Exp. Rev. Anticancer Ther.* **2009,** *9*, 211–221.
80. Cho, H. J.; Yoon, I. S.; Yoon, H. Y.; Koo, H.; Jin, Y. J., et al. Polyethylene Glycol–Conjugated Hyaluronic Acid–Ceramide Self-assembled Nanoparticles for Targeted Delivery of Doxorubicin. *Biomaterials* **2012,** *33*, 1190–1200.
81. Park, J.-H.; von Maltzahn, G.; Zhang, L.; Derfus, A. M.; Simberg, D.; et al. Systematic Surface Engineering of Magnetic Nano Worms for In vivo Tumor Targeting. *Small* **2009,** *5*, 694–700.
82. Park, J. W.; Hong, K.; Kirpotin, D. B.; Colbern, G.; Shalaby, R.; et al. Anti-HER2 Immunoliposomes: Enhanced Efficacy Attributable to Targeted Delivery. *Clin. Cancer Res.* **2002,** *8*, 1172–1181.
83. Nellis, D. F.; Ekstrom, D. L.; Kirpotin, D. B.; Zhu, J.; Andersson, R.; et al. Preclinical Manufacture of an Anti-HER2 scFv-PEG-DSPE, Liposome-Inserting Conjugate. 1. Gram-Scale Production and Purification. *Biotechnol. Progress* **2005,** *21*, 205–220.
84. Wang, T.; DSouza, G. G. M.; Bedi, D.; Fagbohun, O. A.; Potturi, L. P.; et al. Enhanced Binding and Killing of Target Tumor Cells by Drug-Loaded Liposomes Modified with Tumor-Specific Phage Fusion Coat Protein. *Nanomedicine* **2010,** *5*, 563–574.

85. Shadidi, M.; Sioud, M. Identification of Novel Carrier Peptides for the Specific Delivery of Therapeutics into Cancer Cells. *FASEB J.* **2003**, *17*, 256–258.

86. Kukowska-Latallo, J. F.; Candido, K. A.; Cao, Z.; et al. Nanoparticle Targeting of Anticancer Drug Improves Therapeutic Response in Animal Model of Human Epithelial Cancer. *Cancer Res.* **2005**, *65* (12), 5317–5324.

87. Kawamoto, M.; Horibe, T.; Kohno, M.; Kawakami, K. A Novel Transferrin Receptor-Targeted Hybrid Peptide Disintegrates Cancer Cell Membrane to Induce Rapid Killing of Cancer Cells. *BMC Cancer* **2011**, *11* (359), 98–110.

88. Daniels, T. R.; Bernabeu, B.; Rodríguez, J. A.; et al. Transferrin Receptors and the Targeted Delivery of Therapeutic Agents against Cancer. *Biochim. Biophys. Acta* **2012**, *1820* (3), 291–317.

89. Bellocq, N. C.; Pun, S. H.; Jensen, G. S.; Davis, M. E. Transferrin-Containing, Cyclodextrin Polymer-Based Particles for Tumor-Targeted Gene Delivery. *Bioconj. Chem.* **2003**, *14* (6), 1122–1132.

90. Dass, C. R.; Choong, P. F. M. Targeting of Small Molecule Anticancer Drugs to the Tumour and Its Vasculature Using Cationic Liposomes: Lessons from Gene Therapy. *Cancer Cell Int.* **2006**, *6* (17), 120–135.

91. Wang, X., et al. Application of Nanotechnology in Cancer Therapy and Imaging. *Cancer J. Clin.* **2008**, *58*, 97–110.

92. Vasir, J. K.; Labhasetwar, V. Biodegradable Nanoparticles for Cytosolic Delivery of Therapeutics. *Adv. Drug Deliv. Rev.* **2007**, *59*, 718–728.

93. Gabizon, A., et al. Development of Liposomal Anthracyclines: From Basics to Clinical Applications. *J. Control. Release* **1998**, *53*, 275–279.

94. Sahoo, S. K.; Labhasetwar, V. Nanotech Approaches to Drug Delivery and Imaging. *Drug Discov. Today* **2003**, *8*, 1112–1120.

95. Torchilin, V. Antibody-Modified Liposomes for Cancer Chemotherapy. *Expert Opin. Drug Deliv.* **2008**, *5*, 1003–1025.

96. Talekar, M.; Kendall, J.; Denny, W.; Garg, S. Targeting of Nanoparticles in Cancer: Drug Delivery and Diagnostics. *Anti-Cancer Drugs* **2011**, *22* (10), 949–962.

97. Mattheolabakis, G.; Rigas, B.; Constant Inides PP. Nano delivery Strategies in Cancer Chemotherapy: Biological Rationale and Pharmaceutical Perspectives. *Nanomedicine* **2012**, *7* (10), 1577–1590.

98. Perche, F.; Torchilin, V. P. Recent trends in Multifunctional Liposomal Nanocarriers for Enhanced Tumor Targeting. *J. Drug Deliv.* **2013**, *2013*, 7052–7065.

99. Nazem, A.; Mansoori, G. A. Nanotechnology Solutions for Alzheimer's Disease: Advances in Research Tools, Diagnostic Methods and Therapeutic Agents. *J. Alzheim. Dis.* **2008**, *13* (2), 199–223.

100. Wang, X., et al. Application of Nanotechnology in Cancer Therapy and Imaging. *Cancer J. Clin.* **2008**, *58*, 97–110.

101. Fassas, A.; Anagnostopoulos, A. The Use of Liposomal Daunorubicin (Dauno Xome) in Acute Myeloid Leukemia. *Leuk. Lymphoma* **2005**, *46*, 795–802.

102. Leamon, C. P., et al. Folate-Liposome-Mediated Antisense Oligodeoxynucleotide Targeting to Cancer Cells: Evaluation In Vitro and In Vivo. *Bioconjug. Chem.* **2003**, *14*, 738–747.

103. Immordino, M. L.; et al. Stealth Liposomes: Review of the Basic Science, Rationale, and Clinical Applications, Existing and Potential. *Int. J. Nanomed.* **2006**, *1*, 297–315.

104. Voinea, M.; Simionescu, M. Designing of 'Intelligent' Liposomes for Efficient Delivery of Drugs. *J. Cell. Mol. Med.* **2002**, *6* (4), 465–474.

105. Huwyler, J.; Drewe, J.; Krähenbühl, S. Tumor Targeting Using Liposomal Antineoplastic Drugs. *Int. J. Nanomed.* **2008,** *3* (1), 21–29.
106. Torchilin, V. P. Recent Advances with Liposomes as Pharmaceutical Carriers. *Nat. Rev. Drug. Discov.* **2005,** *4* (2), 145–160.
107. Elbayoumi, T.; Torchilin, V. Current Trends in Liposome Research. In *Liposomes*; Weissig, V., Ed.; Humana Press: Totowa, NJ, 2010; pp 1–27.
108. Sawant, R. R.; Torchilin, V. P. Liposomes as Smart Pharmaceutical Nanocarriers. *Soft Matter* **2010,** *6* (17), 4026–4044.
109. Danhier, F.; Feron, O.; Préat, V. To Exploit the Tumor Microenvironment: Passive and Active Tumor Targeting of Nanocarriers for Anticancer Drug Delivery. *J. Control. Release* **2010,** *148* (2), 135–146.
110. Byrne, J. D.; Betancourt, T.; Brannon-Peppas, L. Active Targeting Schemes for Nanoparticle Systems in Cancer Therapeutics. *Adv. Drug Del. Rev.* **2008,** *60* (15), 1615–1626.
111. Egusquiaguirre, S. P.; Igartua, M.; Hernández, R. M.; Pedraz, J. L. Nanoparticle Delivery Systems for Cancer Therapy: Advances in Clinical and Preclinical Research. *Clin. Transl. Oncol.* **2012,** *14* (2), 83–93.
112. Talekar, M.; Kendall, J.; Denny, W.; Garg, S. Targeting of Nanoparticles in Cancer: Drug Delivery and Diagnostics. *Anticanc. Drugs* **2011,** *22* (10), 949.
113. Low, P. S.; Henne, W. A.; Doorneweerd, D. D. Discovery and Development of Folic-Acid-Based Receptor Targeting for Imaging and Therapy of Cancer and Inflammatory Diseases. *Acc. Chem. Res.* **2007,** *41* (1), 120–129.
114. Strebhardt, K.; Ullrich, A. Paul Ehrlich's Magic Bullet Concept: 100 Years of Progress. *Nat. Rev. Cancer* **2008,** *8* (6), 473–480.
115. Chang, D. K.; Chiu, C. Y.; Kuo, S. Y., et al. Antiangiogenic Targeting Liposomes Increase Therapeutic Efficacy for Solid Tumors. *J. Biol. Chem.* **2009,** *284* (19), 12905–12916.
116. He, X.; Na, M. H.; Kim, J. S., et al. A Novel Peptide Probe for Imaging and Targeted Delivery of Liposomal Doxorubicin to Lung Tumor. *Mol. Pharm.* **2011,** *8* (2), 430–438.
117. Bajpai, A. K.; Shukla, S. K.; Bhanu, S.; Kankane, S. Responsive Polymer in Controlled Drug Delivery. *Prog. Polym. Sci.* **2008,** *33*, 1088–1118.
118. Couvreur, P.; Kante, B.; Lenaerts, V.; Scailteur, V.; Roland, M.; Speiser, P. Tissue Distribution of Antitumor Drugs Associated with Poly-alkylcyano Acrylate Nanoparticles. *J. Pharm. Sci.* **1980,** *69*, 199–202.
119. Deng, X.; Jia, G.; Wang, H.; et al. Translocation and Fate of Multiwalled Carbon Nanotubes In Vivo. *Carbon* **2007,** *45*, 1419–1424.
120. Cheng, L.; He, W.; Gong, H.; Wang, C.; Chen, Q.; et al. PEGylated Micelle Nanoparticles Encapsulating a Non-fluorescent Near-infrared Organic Dye as a Safe and Highly-Effective Photothermal Agent for In Vivo Cancer Therapy. *Adv. Funct. Mater.* **2013,** *21*, 220–235.
121. Jabbari, E.; Yang, X.; Moeinzadeh, S.; He, X. Drug Release Kinetics, Cell Uptake, and Tumor Toxicity of Hybrid Peptide-Assembled Polylactide Nanoparticles. *Eur. J. Pharm. Biopharm.* **2013,** *84*, 49–62.
122. Verderio, P.; Bonetti, P.; Colombo, M.; Pandolfi, L.; Prosperi, D. Intracellular Drug Release from Curcumin-Loaded PLGA Nanoparticles Induces G2/M Block in Breast Cancer Cells. *Biomacromolecules* **2013,** *14*, 672–682.
123. Dhar, S.; Kolishetti, N.; Lippard, S. J.; Farokhzad, O. C. Targeted Delivery of a Cisplatin Prodrug for Safer and more Effective Prostate Cancer Therapy In Vivo. *Proc. Nat. Acad. Sci. U.S.A.* **2011,** *108*, 1850–1855.

124. Deng, C.; Jiang, Y.; Cheng, R.; Meng, F.; Zhong, Z. Biodegradable Polymeric Micelles for Targeted and Controlled Anticancer Drug Delivery: Promises, Progress and Prospects. *Nano Today* **2012**, *7*, 467–480.

125. Tang, X.; Cai, S.; Zhang, R.; Liu, P.; Chen, H.; Zheng, Y.; Sun, L. Paclitaxel-Loaded Nanoparticles of Star-Shaped Cholic Acid-Core PLA–TPGS Copolymer for Breast Cancer Treatment. *Nanoscale Res. Lett.* **2013**, *17*, 8 (1), 420–439.

126. Gonzalo, T.; Lollo, G.; Garcia-Fuentes, M.; Torres, D.; Correa, J.; Riguera, R.; Fernandez-Megia, E.; Calvo, P.; Aviles, P.; Guillen, M. J.; Alonso, M. J. A New Potential Nanooncological Therapy Based on Polyamino Acid Nanocapsules. *J. Control. Release* **2013**, *10* (169), 10–16.

127. Zheng, Y.; Chen, H.; Zeng, X.; Liu, Z.; Xiao, X.; Zhu, Y.; Gu, D.; Mei, L. Surface Modification of TPGS-*b*-(PCL-ran-PGA) Nanoparticles with Polyethyleneimine as a Co-delivery System of TRAIL and Endostatin for Cervical Cancer Gene Therapy. *Nanoscale Res. Lett.* **2013**, *8* (1), 161.

128. Martin, D. T.; Hoimes, C. J.; Kaimakliotis, H. Z.; Cheng, C. J.; Zhang, K.; Liu, J.; Wheeler, M. A.; Kelly, W. K.; Tew, G. N.; Saltzman, W. M.; Weiss, R. M. Nanoparticles for Urothelium Penetration and Delivery of the Histone Deacetylase Inhibitor Beninostat for Treatment of Bladder Cancer. *Nanomedicine* **2013**, *9* (8), 112–134.

129. Hoste, K.; De Winne, K.; Schacht, E. Polymeric Prodrugs. *Int. J. Pharm.* **2004**, *277*, 119–131.

130. Goodarzi, N.; Varshochian, R.; Kamalinia, G.; Atyabi, F.; Dinarvand, R. A Review of Polysaccharide Cytotoxic Drug Conjugates for Cancer Therapy. *Carbohydr. Polym.* **2013**, *92*, 1280–1293.

131. Goodarzi, N.; Varshochian, R.; Kamalinia, G.; Atyabi, F.; Dinarvand, R. A Review of Polysaccharide Cytotoxic Drug Conjugates for Cancer Therapy. *Carbohydr. Polym.* **2013**, *92* (2), 1280–1293.

132. Feng, X. Polymeric Conjugates for Anti-cancer Drug Delivery. *Literature Seminar*, September 16, 2010.

133. Vasey, P. A.; Kaye, S. B.; Morrison, R.; Twelves, C.; Wilson, P.; Duncan, R.; Thomson, A. H.; Murray, L. S.; Hilditch, T. E.; Murray, T. Phase I Clinical and Pharmacokinetic Study of PK1 [*N*-(2-hydroxypropyl)methacrylamide Copolymer Doxorubicin]: First Member of a New Class of Chemotherapeutic Agents–Drug–Polymer Conjugates. *Clin. Cancer Res.* **1999**, *5*, 83–94.

134. Campone, M.; Rademaker-Lakhai, J. M.; Bennouna, J.; Howell, S. B.; Nowotnik, D. P.; Beijnen, J. H.; Schellens, J. H. Phase I and Pharmacokinetic Trial of AP5346, a DACH–Platinum–Polymer Conjugate, Administered Weekly for Three Out of Every 4 Weeks to Advanced Solid Tumor Patients. *Cancer Chemother. Pharmacol.* **2007**, *60*, 523–533.

135. Peer, D.; Karp, J. M.; Langer, R. Nanocarriers as an Emerging Platform for Cancer Therapy. *Nat. Nano* **2007**, *2*, 751–760.

136. Matsumura, Y.; Maeda, H. A New Concept for Macromolecular Therapeutics in Cancer Chemotherapy: Mechanism of Tumoritropic Accumulation of Proteins and the Antitumor Agent Smancs. *Cancer Res.* **1986**, *46*, 6387–6392.

137. Van Butsele, K.; Jerome, R.; Jerome, C. Functional Amphiphilic and Biodegradable Copolymers for Intravenous Vectorisation. *Polymer* **2007**, *48* (26), 7431–7443.

138. He, G.; Ma, L. L.; Pan, J.; Venkatraman, S. ABA and BAB Type Triblock Copolymers of PEG and PLA: A Comparative Study of Drug Release Properties and Stealth Particle Characteristics. *Int. J. Pharm.* **2007**, *334* (1–2), 48–55.

139. Ruddy, K.; Mayer, E.; Partridge, A. Patient Adherence and Persistence with Oral Anti-cancer Treatment. *CA Cancer J. Clin.* **2009**, *59*, 56–66.

140. Kuppens, I.; Bosch, T. M.; van Maanen, M. J.; Rosing, H.; Fitzpatrick, A.; Beijnen, J. H.; Schellens, J. H. M. Oral Bioavailability of Docetaxel in Combination with OC144-093. *Cancer Chemother. Pharmacol.* **2005**, *55*, 72–78.

141. Liu, S. Q.; Wiradharma, N.; Gao, S. J.; Tong, Y. W.; Yang, Y. Y. Bio-functional Micelles Self-assembled from a Folate-Conjugated Block Copolymer for Targeted Intracellular Delivery of Anticancer Drugs. *Biomacromolecules* **2007**, *28*, 1423–1433.

142. Luo, Z.; Jiang, J. pH-Sensitive Drug Loading/Releasing in Amphiphilic Copolymer PAE–PEG: Integrating Molecular Dynamics and Dissipative Particle Dynamics Simu-lations. *J. Control. Release* **2012**, *162*, 185–193.

143. Xiong, X. B.; Xiong, A.; Falamarzian, S. M.; Garg, A.; Lavasanifar. Disulfide Cross-Linked Micelles for the Targeted Delivery of Vincristine to B-Cell Lymphoma. *J. Control. Release* **2011**, *155*, 248–261.

144. Kataoka, K.; Harada, A.; Nagasaki, Y. Block Copolymer Micelles for Drug Delivery: Design, Characterization and Biological Significance. *Adv. Drug Deliv. Rev.* **2001**, *47*, 113–131.

145. Kopecek, J.; Kopeckova, P.; Minko, T.; Lu, Z. R.; Peterson, C. M. Water Soluble Poly-mers in Tumor Targeted Delivery. *J. Control. Release* **2001**, *74*, 147–158.

146. Kong, G.; Braun, R. D.; Dewhirst, M. W. Hyperthermia Enables Tumor-Specific Nanoparticle Delivery: Effect of Particle Size. *Cancer Res.* **2000**, *60*, 4440–4445.

147. Gong, J.; Chen, M.; Zheng, Y.; Wang, S.; Wang, Y. Polymeric Micelles Drug Delivery System in Oncology. *J. Control. Release* **2012**, *159*, 312–323.

148. Nishiyama, N.; Bae, Y.; Miyata, K.; Fukushima, S.; Kataoka, K. Smart Polymeric Micelles for Gene and Drug Delivery. *Drug Discov. Today Technol.* **2005**, *2*, 21–26.

149. Matsumura, Y. Poly(amino Acid) Micelle Nanocarriers in Preclinical and Clinical Studies. *Adv. Drug Deliv. Rev.* **2008**, *60*, 899–914.

150. Matsumura, Y. Polymeric Micellar Delivery Systems in Oncology. *J. Clin. Oncol.* **2008**, *38*, 793–802.

151. Koizumi, F.; Kitagawa, M.; Negishi, T.; Onda, T.; Matsumoto, S.; Hamaguchi, T.; et al. Novel SN-38-Incorporating Polymeric Micelles, NK012, Eradicate Vascular Endo-thelial Growth Factor-Secreting Bulky Tumors. *Cancer Res.* **2006**, *66*, 10048–10056.

152. Saito, Y.; Yasunaga, M.; Kuroda, J.; Koga, Y.; Matsumura, Y. Enhanced Distribution of NK012, a Polymeric Micelle-Encapsulated SN-38, and Sustained Release of SN-38 within Tumors Can Beat a Hypovascular Tumor. *Cancer Sci.* **2008**, *99*, 1258–1264.

153. Nakajima, T. E.; Yasunaga, M.; Kano, Y.; Koizumi, F.; Kato, K.; Hamaguchi, T.; et al. Synergistic Antitumor Activity of the Novel SN-38-Incorporating Polymeric Micelles, NK012, Combined with 5-Fluorouracil in a Mouse Model of Colorectal Cancer, as Compared with That of Irinotecan Plus 5-Fluorouracil. *Int. J. Cancer* **2008**, *122*, 2148–2153.

154. Mahmud, A.; Xiong, X. B.; Aliabadi, H. M.; Lavasanifar, A. Polymeric Micelles for Drug Targeting. *J Drug Target.* **2007**, *15*, 553–584.

155. Torchilin, V. P. Cell Penetrating Peptide-Modified Pharmaceutical Nanocarriers for Intracellular Drug and Gene Delivery. *Biopolymers* **2008**, *90*, 604–610.

156. Tomalia, D. A.; Naylor, A. M.; Goddard, W. A. Starburst Dendrimers, Molecular-Level of Size, Shape, Surface Chemistry, Morphology, and Flexibility from Atoms to Macro-scopic Matter. *Angew. Chem. Int. Ed. Engl.* **1990**, *29*, 138–175.

157. Kuchkina, N. V.; Morgan, D. G.; Stein, B. D.; Puntus, L. N.; Sergeev, A. M.; Peregudov, A. S.; Bronstein, L. M.; Shifrina, Z. B. Polyphenylenepyridyl Dendrimers as Stabilizing and Controlling Agents for CdS Nanoparticle Formation. *Nanoscale* **2012**, *4* (7), 2378–2386.

158. Liu, M.; et al. Water-Soluble Unimolecular Micelles: Their Potential as Drug Delivery Agents. *J. Control. Release* **2000**, *65*, 121–131.

159. Pan, B.; Cui, D.; Sheng, Y.; Ozkan, C.; Gao, F.; He, R.; Li, Q.; Xu, P.; Huang, T. Dendrimer-Modified Magnetic Nanoparticles Enhance Efficiency of Gene Delivery System. *Cancer Res.* **2007**, *67*, 8156–8163.

160. Reddy, J. A.; Allagadda, V. M.; Leamon, C. P. Targeting Therapeutic and Imaging Agents to Folate Receptor Positive Tumors. *Curr. Pharm. Biotechnol.* **2005**, *6*, 131–150.

161. Leamon, C. P.; Reddy, J. A. Folate Targeted Chemotherapy. *Adv. Drug Deliv. Rev.* **2004**, *56*, 1127–1141.

162. Roy, E. J.; Gawlick, U.; Orr, B. A. Folate-Mediated Targeting of T Cells to Tumors. *Adv. Drug Deliv. Rev.* **2004**, *56*, 1219–1231.

163. Choi, Y.; Thomas, T.; Kotlyar, A.; Islam, M. T. Synthesis and Functional Evaluation of DNA Assembled Polyamidoamine (PAMAM) Dendrimer Clusters with Cancer Cell Specific Targeting. *Chem. Biol.* **2005**, *12*, 35–43.

164. Laheru, D.; Jaffee, E. M. Immunotherapy for Pancreatic Cancer—Science Driving Clinical Progress. *Nat. Rev. Cancer* **2005**, *5*, 549–467.

165. Harris, M. Monoclonal Antibodies as Therapeutic Agents for Cancer. *Lancet Oncol.* **2004**, *5*, 292–302.

166. Lin, M. Z.; Teitell, M. A.; Schiller, G. J. The Evolution of Antibodies into Versatile Tumor-Targeting Agents. *Clin. Cancer Res.* **2005**, *11*, 129–138.

167. Zhang, J. Y. Tumor-Associated Antigen Arrays to Enhance Antibody Detection for Cancer Diagnosis. *Cancer Detect. Prev.* **2004**, *28*, 114–118.

168. Kontermann, R. E. Recombinant Bispecific Antibodies for Cancer Therapy. *Acta Pharmacol. Sin.* **2005**, *26*, 1–9.

169. Trail, P. A.; King, D. H.; Dubowchik, G. M. Monoclonal Antibody Drug Immune Conjugates for Targeted Treatment of Cancer. *Cancer Immunol. Immunother.* **2003**, *52*, 328–337.

170. McDonald, G. C.; Glover, N. Effective Tumor Targeting: Strategies for the Delivery of Armed Antibodies. *Curr. Opin. Drug Discov. Dev.* **2005**, *8*, 177–183.

171. Govindan, S. V.; Griffiths, G. L.; Hansen, H. J.; Horak, I. D.; Goldenberg, D. M. Cancer Therapy with Radiolabeled and Drug/Toxin Conjugated Antibodies. *Technol. Cancer Res. Treat.* **2005**, *4*, 375–392.

172. FitzGerald, D. J.; Kreitman, R.; Wilson, W.; Squires, D. Recombinant Immune Toxins for Treating Cancer. *Int. J. Med. Microbiol.* **2004**, *293*, 577–582.

173. Sandvig, K.; Grimmer, S.; Iversen, T. G.; Rodal, K.; Torgersen, M. L. Ricin Transport into Cells: Studies of Endocytosis and Intracellular Transport. *Int. J. Med. Microbiol.* **2000**, *290*, 415–420.

174. Gruaz-Guyon, A.; Raguin, O.; Barbet, J. Recent Advances in Pretargeted Radio Immunotherapy. *Curr. Med. Chem.* **2005**, *12*, 319–338.

175. Barth, R. F.; Adams, D. M.; Soloway, A. H. Boronated Starburst Dendrimer Monoclonal Antibody Immune Conjugates: Evaluation as a Potential Delivery System for Neutron Capture Therapy. *Bioconj. Chem.* **1994**, *5*, 58–66.

176. Roberts, J. C.; Adams, Y. E.; Tomalia, D. A.; Mercer-Smith, J. A. Using Starburst Dendrimers as Linker Molecules to Radiolabel Antibodies. *Bioconj. Chem.* **1990**, *1*, 305–308.

177. Singh, P. Terminal Groups in Starburst Dendrimers: Activation and Reactions with Proteins. *Bioconj. Chem.* **1998**, *9*, 54–63.
178. Kobayashi, H.; Sato, N.; Saga, T.; Nakamoto, Y.; Ishimori, T. Monoclonal Antibody Dendrimer Conjugates Enable Radiolabeling of Antibody with Markedly High Specific Activity with Minimal Loss of Immune Reactivity. *Eur. J. Nucl. Med. Mol. Imaging* **2000**, *27*, 1334–1339.
179. Fischer-Durand, N.; Salmain, M.; Rudolf, B. Synthesis of Metalcarbonyl–Dendrimer–Antibody Immune Conjugates: Towards a New Format for Carbonyl Metallo-immunoassay. *Chem. Biol. Chem.* **2004**, *5*, 519–525.
180. Thomas, T. P.; Patri, A. K.; Myc, A.; Myaing, M. T. In Vitro Targeting of Synthesized Antibody-Conjugated Dendrimer Nanoparticles. *Biomacromolecules* **2004**, *5*, 2269–2274.
181. Patri, A. K.; Myc, A.; Beals, J.; Thomas, T. P.; Bander, N. H. Synthesis and In Vitro Testing of J591 Antibody-Dendrimer Conjugates for Targeted Prostate Cancer Therapy. *Bioconj. Chem.* **2004**, *15*, 1174–1181.
182. Kolhe, P.; Khandare, J.; Pillai, O.; Kannan, S.; Lieh-Lai, M.; Kannan, R. M. Preparation, Cellular Transport, and Activity of Polyamidoamine-Based Dendritic Nano Devices with a High Drug Payload. *Biomaterials* **2006**, *27*, 660–669.
183. Nigavekar, S. S.; Sung, L. Y.; Llanes, M.; El-Jawahri, A.; Lawrence, T. S. 3H Dendrimer Nano Nanoparticles Organ/Tumor Distribution. *Pharm Res.* **2004**, *21*, 476–483.
184. http://www.swissre.com (last accessed on December 2015).
185. Kang, H. C.; Lee, M.; Bae, Y. H. Polymeric Gene Carriers. *Crit. Rev. Eukaryot. Gene Expr.* **2005**, *15*, 317–342.
186. Klumpp, C.; Kostarelos, K.; Prato, M.; Bianco, A. Functionalized Carbon Nanotubes as Emerging Nano Vectors for the Delivery of Therapeutics. *Biochim. Biophys. Acta* **2006**, *1758*, 404–412.
187. Bekyarova, E.; Ni, Y.; Malarkey, E. B.; et al. Applications of Carbon Nanotubes in Biotechnology and Biomedicine. *J. Biomed. Nanotechnol.* **2005**, *1*, 3–17.
188. Hampel, S.; Kunze, D.; Haase, D.; Krämer, K.; Rauschenbach, M.; Ritschel, M.; Leonhardt, A.; Thomas, J.; Oswald, S.; Hoffmann, V.; Büchner, B. Carbon Nanotubes Filled with a Chemotherapeutic Agent: A Nanocarrier Mediates Inhibition of Tumor Cell Growth. *Nanomedicine* **2012**, *3* (2), 175–182.
189. Tan, A.; Yildirimer, L.; Rajadas, J.; De La Peña, H.; Pastorin, G.; Seifalian, A. Quantum Dots and Carbon Nanotubes in Oncology: A Review on Emerging Theranostic Applications in Nanomedicine. *Nanomedicine* 2011, *6* (6), 1101–1114.
190. Fei, B.; Lu, H. F.; Hu, Z. G.; Xin, J. H. Solubilization, Purification and Functionalization of Carbon Nanotubes Using Polyoxometalate. *Nanotechnology* **2006**, *17*, 1589–1593.
191. Arsawang, U.; Saengsawang, O.; Rungrotmongkol, T.; et al. How Do Carbon Nanotubes Serve as Carriers for Gemcitabine Transport in a Drug Delivery System. *J. Mol. Graph. Model.* **2011**, *29*, 591–596.
192. Xiao, Y.; Gao, X. G.; Taratula, O., et al. Anti-HER2 IgY Antibody-Functionalized Single-Walled Carbon Nanotubes for Detection and Selective Destruction of Breast Cancer Cells. *BMC Cancer* **2009**, *95*, 351.
193. Prajapati, V. K.; Awasthi, K.; Gautam, S., et al. Targeted Killing of *Leishmania donovani* In Vivo and In Vitro with Amphotericin B Attached to Functionalized Carbon Nanotubes. *J. Antimicrob. Chemother.* **2011**, *66*, 874–879.
194. Ashcroft, J. M.; Tsyboulski, D. A.; Hartman, K. B.; et al. Fullerene (C60) Immunoconjugates: Interaction of Water-Soluble C60 Derivatives with the Murine Anti-gp240 Melanoma Antibody. *Chem. Commun. (Camb.)* **2006**, *28*, 3004–3006.

195. Beg, S.; Rizwan, M.; Sheikh, A. M.; Hasnain, M. S.; Anwer, K.; Kohli, K. Advancement in Carbon Nanotubes: Basics, Biomedical Applications and Toxicity. *J. Pharm. Pharmacol.* **2011**, *63*, 141–163.

196. Reddy, S. T.; Rehor, A.; Schmoekel, H. G.; Hubbell, J. A.; Swartz, M. A. In Vivo Targeting of Dendritic Cells in Lymph Nodes with Poly(Propylene Sulfide) Nanoparticles. *J. Control. Release* **2006**, *112* (1), 26–34.

197. Yang, F.; Fu, D. L.; Long, J.; Ni, Q. X. Magnetic Lymphatic Targeting Drug Delivery System Using Carbon Nanotubes. *Med. Hypoth.* **2008**, *70* (4), 765–767.

198. Yang, F.; Hu, J.; Yang, D., et al. Pilot Study of Targeting Magnetic Carbon Nanotubes to Lymph Nodes. *Nanomedicine* **2009**, *4* (3), 317–330.

199. Liu, Y.; Ng, K. Y.; Lillehei, K. O. Cell-Mediated Immunotherapy: A New Approach to the Treatment of Malignant Glioma. *Cancer Control* **2003**, *10* (2), 138–147.

200. Dhar, S.; Liu, Z.; Thomale, J.; Dai, H.; Lippard, S. J. Targeted Single-Wall Carbon Nanotube-Mediated Pt(IV) Prodrug Delivery Using Folate as a Homing Device. *J. Am. Chem. Soc.* **2008**, *130* (34), 11467–11476.

201. Hampel, S.; Kunze, D.; Haase, D.; et al. Carbon Nanotubes Filled with a Chemotherapeutic Agent: A Nanocarrier Mediates Inhibition of Tumor Cell Growth. *Nanomedicine* **2008**, *3* (2), 175–182.

202. Lay, C. L.; Liu, H. Q.; Tan, H. R.; Liu, Y. Delivery of Paclitaxel by Physically Loading onto Poly(ethylene glycol) (PEG)–Graft Carbon Nanotubes for Potent Cancer Therapeutics. *Nanotechnology* **2010**, *21* (6), 65–79.

203. Chan, J. Y.-W.; Chu, A. C.-Y.; Fung, K.-P. Inhibition of P-Glycoprotein Expression and Reversal of Drug Resistance of Human Hepatoma HepG2 Cells by Multidrug Resistance Gene (mdr1) Antisense RNA. *Life Sci.* **2000**, *67* (17), 2117–2124.

204. Suri, S. S.; Fenniri, H.; Singh, B. Nanotechnology-Based Drug Delivery Systems. *J. Occup. Med. Toxicol.* **2007**, *2* (1), 114–125.

205. Hanson, R.; Kouwenhoven, L. P.; Petta, J. R.; Tarucha, S.; Vandersypen, L. M. K. Spins in Few-Electron Quantum Dots. *Rev. Mod. Phys.* **2007**, *79*, 1217–1265.

206. Hardman, R. A Toxicologic Review of Quantum Dots: Toxicity Depends on Physicochemical and Environmental Factors. *Environ. Health Perspect.* **2006**, *114*, 165–172.

207. Michalet, X.; Pinaud, F. F.; Bentolila, L. A.; Tsay, J. M.; Doose, S.; Li, J. J.; Sundaresan, G.; Wu, A. M.; Gambhir, S. S.; Weiss, S. Quantum Dots for Live Cells, In Vivo Imaging, and Diagnostics. *Science* **2005**, *307*, 538–544.

208. Bremer, C. Ntziachristos, V.; Weissleder, R. Optical-Based Molecular Imaging: Contrast Agents and Potential Medical Applications. *Eur. Radiol.* **2003**, *13*, 231–243.

209. Medintz, I. L.; Uyeda, H. T.; Goldman, E. R.; Mattoussi, H. Quantum Dot Bioconjugates for Imaging, Labelling and Sensing. *Nat Mater.* **2005**, *4*, 435–446.

210. Ben-Ari, E. T. Nanoscale Quantum Dots Hold Promise for Cancer Applications. *J. Nat. Cancer Inst.* **2003**, *95*, 502–504.

211. Brigger, I.; Morizet, J.; Aubert, G.; Chacun, H.; Terrier-Lacombe, M. J.; Couvreur, P.; Vassal, G. Poly(ethylene glycol)-Coated Hexa-decyl Cyanoacrylate Nanospheres Display a Combined Effect for Brain Tumor Targeting. *J. Pharmacol. Exp. Ther.* **2002**, *303*, 928–936.

212. Licha, K.; Riefke, B.; Ebert, B.; Grötzinger, C. Cyanine Dyes as Contrast Agents in Biomedical Optical Imaging. *Acad. Radiol.* **2002**, *9*, S320–S322.

213. Licha, K.; Olbrich, C. Optical Imaging in Drug Discovery and Diagnostic Applications. *Adv. Drug Deliv. Rev.* **2005**, *57*, 1087–108.

214. Michalet, X.; Pinaud, F. F.; Bentolila, L. A.; Tsay, J. M.; Doose, S.; Li, J. J.; Sundaresan, G.; Wu, A. M.; Gambhir, S. S.; Weiss, S. Quantum Dots for Live Cells, In Vivo Imaging, and Diagnostics. *Science* **2005**, *307*, 538–544.

215. Morgan, N. Y.; English, S.; Chen, W.; Chernomordik, V.; Russo, A.; Smith, P. D.; Gandjbakhche, A. Real Time In Vivo Non-invasive Optical Imaging Using Near-infrared Fluorescent Quantum Dots. *Acad. Radiol.* **2005**, *3*, 313–323.

216. Rosi, N. L.; Mirkin, C. A. Nanostructures in Biodiagnostics. *Chem. Rev.* **2005**, *105*, 1547–1562.

217. Santra, S.; Xu, J.; Wang, K.; Tan, W. J. Luminescent Nanoparticle Probes for Bioimaging. *Nanosci. Nanotechnol.* **2004**, *4*, 590–599.

218. Santra, S.; Dutta, D.; Walter, G. A.; Moudgil, B. M. Fluorescent Nanoparticle Probes for Cancer Imaging. *Technol. Cancer Res. Treat.* **2005**, *4*, 593–602.

219. Yezhelyev, M. V.; Al-Hajj, A.; Morris, C.; Marcus, A. I.; Liu, T.; Lewis, M.; Cohen, C.; Zrazhevsky, P.; Simons, J. W.; Rogatko, A.; Nie, S.; Gao, X.; et al. In Situ Molecular Profiling of Breast Cancer Biomarkers with Multicolor Quantum Dots. *Adv. Mater.* **2007**, *19*, 3146–3151.

220. Hahn, M. A.; Keng, P. C.; Krauss, T. D. Flow Cytometric Analysis to Detect Pathogens in Bacterial Cell Mixtures Using Semiconductor Quantum Dots. *Anal. Chem.* **2008**, *80*, 864–872.

221. Maysinger, D.; Behrendt, M.; Lalancette-Hebert, M.; Kriz, J. Real-Time Imaging of Astrocyte Response to Quantum Dots: In Vivo Screening Model System for Biocompatibility of Nanoparticles. *Nano Lett.* **2007**, *7*, 2513–2520.

222. Cai, W. B.; Shin, D. W.; Chen, K.; Gheysens, O.; Cao, Q. Z.; Wang, S. X.; Gambhir, S. S.; Chen, X. Y. Peptide-Labeled Near-infrared Quantum Dots for Imaging Tumor Vasculature in Living Subjects. *Nano Lett.* **2006**, *6*, 669–676.

223. Gao, J. H.; Chen, K.; Xie, R. G.; Xie, J.; Lee, S.; Cheng, Z.; Peng, X. G.; Chen, X. Y. Ultrasmall Near-infrared Non-cadmium Quantum Dots for In Vivo Tumor Imaging. *Small* **2010**, *6*, 256–261.

224. Gao, X. H.; Yang, L. L.; Petros, J. A.; Marshal, F. F.; Simons, J. W.; Nie, S. M. In Vivo Molecular and Cellular Imaging with Quantum Dots. *Curr. Opin. Biotechnol.* **2005**, *16*, 63–72.

225. Wu, X. Y.; Liu, H. J.; Liu, J. Q.; Haley, K. N.; Treadway, J. A.; Larson, J. P.; Ge, N. F.; Peale, F.; Bruchez, M. P. Immunofluorescent Labeling of Cancer Marker Her2 and other Cellular Targets with Semiconductor Quantum Dots. *Nat. Biotechnol.* **2003**, *21*, 41–46.

226. Ballou, B.; Ernst, L. A.; Andreko, S.; Harper, T.; Fitzpatrick, J. A. J.; Waggoner, A. S.; Bruchez, M. P. Sentinel Lymph Node Imaging Using Quantum Dots in Mouse Tumor Models. *Bioconjug. Chem.* **2007**, *18*, 389–396.

227. Rhyner, M. N.; Smith, A. M.; Gao, X. H.; Mao, H.; Yang, L. L.; Nie, S. M. Quantum Dots and Multifunctional Nanoparticles: New Contrast Agents for Tumor Imaging. *Nanomedicine* **2006**, *1*, 209–217.

228. Juzenas, P.; Chen, W.; Sun, Y. P.; Coelho, M. A. N.; Generalov, R.; Generalova, N.; Christensen, I. L. Quantum Dots and Nanoparticles for Photodynamic and Radiation Therapies of Cancer. *Adv. Drug Deliv. Rev.* **2008**, *60*, 1600–1614.

229. Bakalova, R.; Ohba, H.; Zhelev, Z.; Nagase, T.; Jose, R.; Ishikawa, M.; Baba, Y. Quantum Dot Anti-CD Conjugates: Are They Potential Photosensitizers or Potentiators of Classical Photosensitizing Agents in Photodynamic Therapy of Cancer. *Nano Lett.* **2004**, *4*, 1567–1573.

230. Morosini, V.; Bastogne, T.; Frochot, C.; Schneider, R.; Francois, A.; Guillemin, F.; Barberi-Heyo, M. Quantum Dot–Folic Acid Conjugates as Potential Photosensitizers in Photodynamic Therapy of Cancer. *Photochem. Photobiol. Sci.* **2011,** *10,* 842–851.

231. Chakroborty, G.; Seth, N.; Sharma, V. Nanoparticles and Nanotechnology: Clinical, Toxicological, Social, Regulatory and Other Aspects of Nanotechnology. *J. Drug Deliv. Therap.* **2013,** *3* (4), 138–141.

232. Couvreur, P. Nanoparticles in Drug Delivery: Past, Present and Future. *Adv. Drug Deliv. Rev.* **2013,** *65* (1), 21–23.

233. Webster, D. M.; Sundaram, P.; Byrne, M. E. Injectable Nanomaterials for Drug Delivery: Carriers, Targeting Moieties, and Therapeutics. *Eur. J. Pharm. Biopham.* **2013,** *84* (1), 1–20.

234. Folkman, J. Toward an Understanding of Angiogenesis: Search and Discovery. *Perspect. Biol. Med.* **1985,** *29,* 10–36.

235. Folkman, J.; Gimbrone, M. A. Jr. Perfusion of the Thyroid. In *Karolinska Symposia on Research Methods in Reproduction Endocrinology, 4th Symposium: Perfusion Techniques*; Diczfalusy, E., Ed.; Karolinska Institute: Stockholm, 1971; pp 237–248.

236. Folkman, J. The Intestine as an Organ Culture. In *Carcinoma of the Colon and Antecedent Epithelium*; Burdette, W. J., Ed.; CC Thomas: Spring field, IL, 1970; pp 113–127.

237. Gimbrone, M. A. Jr.; Cotran, R. S.; Folkman, J. Endothelial Regeneration and Turnover. Studies with Human Endothelial Cell Cultures. *Ser. Haematol.* **1973,** *6,* 453–455.

238. Jaffe, E. A.; Nachman, R. L.; Becker, C. G.; Minick, C. R. Culture of Human Endothelial Cells Derived from Umbilical Veins: Identification by Morphologic and Immunologic Criteria. *J. Clin. Invest.* **1972,** *52,* 2745–2756.

239. Sun, H. K.; Ji, H. J.; Soo, H. L.; Sung, W. K.; Tae, G. P. LHRH Receptor-Mediated Delivery of siRNA Using Polyelectrolyte Complex Micelles Self-Assembled from siRNA–PEG–LHRH Conjugate and PEI. *Bioconj. Chem.* **2008,** *19* (11), 2156–2162.

240. Majuru, S.; Oyewumi, O. Nanotechnology in Drug Development and Life Cycle Management. *Nanotechnol. Drug Deliv.* **2009,** *10* (4), 597–619.

241. Birnbaum, D. T.; Brannon-Peppas, L. Microparticle Drug Delivery Systems. In *Drug Delivery Systems in Cancer Therapy*; Brown, D. M., Ed.; Humana Press: Totowa, NJ, 2004; pp 117–135.

242. Mimeault, M.; Hauke, R.; Batra, S. K. Recent Advances on the Molecular Mechanisms Involved in the Drug Resistance of Cancer Cells and Novel Targeting Therapies. *Clin. Pharmacol. Ther.* **2008,** *83,* 673–691.

CHAPTER 13

NANOTECHNOLOGY IN COSMETICS: HOW SAFE IT IS?

PRIYA PATEL[1*], PARESH PATEL[2], NEHA VADGAMA[1], and MIHIR RAVAL[2]

[1]Department of Pharmaceutical Sciences, Saurashtra University, Rajkot, Gujarat, India

[2]Shivam Institute of Pharmacy, Valasan, Gujarat, India

[*]Corresponding author. E-mail: patelpriyav@gmail.com

CONTENTS

ABSTRACT

Nanotechnology represents one of the most capable technologies of the 21st century. Nanomaterials are being increasingly used in commercial products, a clear indication of the unique potential that nanotechnology represents for industry. This has particularly been the case in the field of cosmetics, where products containing nanomaterials have shown enhanced product performance. Like other sectors, the cosmetics industry resorts to developments in the field of nanotechnologies. The applications of nanotechnology and nanomaterials can be found in many cosmetic products. Recently, nanotechnology is an emerging in the field of cosmetics and dermal preparations as it offers a revolutionize treatment of several skin diseases. Use of carrier system in nanotechnology has added advantage of improved skin penetration, depot effect with sustained release drug action. This chapter discusses advantages, disadvantages, method of preparation, and pharmaceutical applications of various nanoparticulate systems followed by toxicological concern about nanoparticles along with facts regarding the use of nanoparticles in cosmetics.

13.1   INTRODUCTION

It's probably on your face or in your body right now, it's definitely somewhere in your home you can't see it, but it's there just the same.[1] It is nanotechnology and it is more prevalent in our day-to-day lives that we might think this quote by Brindy McNair.[2] The US FDA defines cosmetics by their intended use, as articles intended to be rubbed, poured, sprinkled, or sprayed on, introduced into, or otherwise applied to the human body for cleansing, beautifying, promoting attractiveness, or altering the appearance.[3] Cosmetics have become an important part of our daily life: from the use of deodorant and shampoo, to skin-care products and perfumes. Personal hygiene and beauty products are considered to be essentials today as are health-related products like sunscreens and decay-fighting toothpaste.[4] Today, consumers worldwide are looking for personal-care products that supply multiple benefits with minimal efforts. Not only women but there are also increasing number of males who are using cosmetics usually to enhance their own facial features. Cosmetics are products that are created for application on the body for the purpose of cleansing, beautifying or altering appearance, and enhancing attractive features. Cosmetics are substances used to enhance the appearance or odor of the human body.[5] Cosmetic pharmaceuticals, or cosmeceuticals, are cosmetic products that contain biologically active ingredients and claim to have medicinal or drug-like benefits. Like cosmetics, cosmeceuticals are topically applied, but they contain ingredients that influence the biological function of the skin.[6] Raymond Reed, founding member of the US Society of Cosmetic Chemists, coined the term in 1961,[7] as health products nearly 46 years ago with liposome moisturizing creams. Nanotechnology is most often described as the manufacture and manipulation of purpose-made structures having 100 nm diameter. A cosmetic product shall mean any substance or mixture intended to be placed in contact with the various external parts of the human body (epidermis, hair system, nails, lips, and external genital organs) or with the teeth and the mucous membranes of the oral cavity with a view exclusively or mainly to cleaning them, perfuming them, changing their appearance and keeping them in good condition."[8] Almost all the major cosmetic manufacturers use nanomaterial in their products. L'Oréal has a number of nanotechnology related products in the market and ranks sixth in the United States in the number of nanotech-related patents in the United States. The European Commission estimated in 2006, that 5% of cosmetic products contained nanoparticles.[9] The cosmetics industry therefore uses nanodispersion in the form of encapsulation or carrier systems, so that agents penetrate into deeper skin layers.[10]

Like other sectors, the cosmetics industry resorts to developments in the field of nanotechnologies. The applications of nanotechnology and nano-materials can be found in many cosmetic products including moisturizers, hair-care products, make-up, and sunscreen. Nanomaterials are now being used in leading cosmetic products, most commonly as chemicals used to give the protection in sunscreens. Encapsulation and carrier systems like liposomes, nanoemulsions (NEs), microemulsions, or lipid nanoparticles serve to transport agents to deeper skin layers. Nanoparticles of titanium dioxide and zinc oxide are used as UV filters in sunscreens.[11] Liposomes and niosomes are also used in the cosmetic industry as delivery vehicles. Newer structures such as solid lipid nanoparticles (SLNs) and nanostructured lipid carriers (NLC) have been found to be better performers than liposomes. In particular, NLCs have been identified as a potential next-generation cosmetic delivery agent that can provide enhanced skin hydration, bioavailability, stability of the agent, and controlled occlusion. Encapsulation techniques have been proposed for carrying cosmetic actives. Nanocrystals and NEs are also being investigated for cosmetic applications. Other novel materials, such as fullerene, have also appeared in a small number of beauty products. Recently, nanotechnology is emerging in the field of cosmetics and dermal preparations as it offers a revolutionize treatment of several skin diseases.[5]

It is proved effective in attaining safe and targeted delivery of active medicaments as well cosmetic ingredients. Use of carrier system in nanotechnology has added advantage of improved skin penetration, depot effect with sustained release drug action.

Nanotechnology has found wide applications in diverse commercial products (cosmetics, paints, coasting, textiles, etc.) and industrial applications, and this trend is expected to continue into the future. While the benefits of nanotechnology are beyond debate, concurrently to its growth, there are increasing concerns raised regarding safety and environmental impacts of this rapidly emerging technology. Thus, novel cosmaceutical delivery systems reviewed here possess enormous potential as next-generation smarter carrier systems.[12]

- The functions and benefits of these "encapsulation and carrier systems" are Protection of sensitive agents;
- Controlled release effect;
- Reduction in the amount of agents and additives;
- Longer shelf life and hence greater product effectiveness.[13]

In cosmetics, there are currently two main uses for nanotechnology. The first is the use of nanoparticles as UV filters. Titanium dioxide (TiO_2) and zinc oxide (ZnO) are the main compounds used in these applications and organic alternatives to these have also been developed. The second use is nanotechnology for delivery. Liposomes and niosomes are used in the cosmetic industry as delivery vehicles. Newer structures such as SLNs and NLCs have been found to be better performers than liposomes. Nanocrystals, microemulsions, NEs, and dendrimers are also being investigated for cosmetic applications. Below, we discuss different nanoparticulate drug delivery system used in cosmetics.[14]

13.2   EMULSIONS

Emulsions are the most common type of delivery system used in cosmetics. They enable a wide variety of ingredients to be quickly and conveniently delivered to hair and skin. The best known cosmetic products based on emulsions are creams and lotions.. Emulsion technology continues to expand and the introduction of new w/o emulsifiers which give elegant products without the inherent greasy feel, etc., is of interest to the whole industry.[15]

Following are the different emulsion delivery systems used in cosmetics.

13.2.1 MICROEMULSION

Microemulsions are isotropic, thermodynamically stable transparent systems of oil, water, surfactant, and cosurfactant, with a droplet size usually in the range of 10–200 nm.[16]

Microemulsions represent a promising carrier system for cosmetic active ingredients due to their numerous advantages over the existing conventional formulations. Microemulsion is capable for solubilizing both hydrophilic and hydrophobic ingredients with higher encapsulation efficiency along with improved product efficiency, appearance, and stability.

There is growing recognition of their potential benefits in the field of cosmetic science in addition to the drug delivery.[17] Microemulsions are well suited for the preparation of various cosmetic products for use as moisturizing and soothing agents, as sunscreens, as antiperspirants, and as body cleansing agents. They are also valuable for use in hair-care compositions which ensure a good conditioning of the hair as well as good hair feel and hair gloss,[18] and found wide application in aftershave formulations which upon application to the skin provide reduced stinging and irritation and a comforting effect without tackiness. These newer formulations elicit very good cosmetic attributes and high hydration properties with rapid cutaneous penetration which may accentuate their role in topical products. These smart systems are also suitable for perfuming purposes because minimum amount of organic solvents is required for skin and hair. This chapter highlights the recent innovations in the field of microemulsion technology as claimed by different patents which can bring unique products with great commercial prospects in a very competitive and lucrative global cosmetic market.[19]

Products consisting of these systems are valued due to their stability and small particle size, which affords microemulsions special consideration in the market place. They can be used to improve both the performance of presently used cosmetic active ingredients and the commercial "appeal" of the product.

Since microemulsions were discovered approximately six decades ago, their applications in several fields, including cosmetics, have been increased due to their good appearance, thermodynamic stability, high solubilization power, and ease of preparation. In addition, microemulsions can enhance skin permeation of the loaded substances. They are classified into three types: oil-in-water, bicontinuous, and water-in-oil (W/O), and the efficiency of microemulsion in topical application is related to the type of microemulsion. All types of microemulsions can be formed spontaneously when the ratios of oil, water, and surfactant in the systems are appropriate.[20] These proper ratios can be found in a microemulsion region of a phase diagram. Numerous applications of cosmetic microemulsions are available in market, including skin-care, hair-care, and personal-care products with good product efficiency and stability. Moreover, new materials have been developed to be used in cosmetic microemulsion formulations for increasing the product efficiency and reducing the toxicity.[21]

13.2.2 NANOEMULSION

NEs can be defined as "ultrafine emulsions" because of the formation of droplets in the submicron range. The average droplet size of NEs has been ranging from 50 to 1000 nm. NEs have recently become increasingly important as potential vehicles for the controlled delivery of cosmetics and for the optimized dispersion of active ingredients in particular skin layers.

Due to their lipophilic interior, NEs are more suitable for the transport of lipophilic compounds than liposomes. They have attracted considerable attention in recent years for application in personal-care products as potential vehicles for the controlled delivery of cosmetics. Several cosmetic products are available in the form of NEs, including Korres' Red Vine Hair sunscreen. Several companies supply ready to use emulsifiers for creating stable NEs for cosmetic applications, including Nano cream® from Sinerga and Nano Gel from Kemira.[22,23] NEs are transparent due to the tiny droplets size and they also remain stable for a longer period of time. They are mostly used in deodorants, sunscreens, shampoos, skin, and hair-care products. The NEs are easily valued in skin care because of their good sensorial properties, such as rapid penetration, merging textures, and their biophysical properties such as hydrating power. A significant improvement in dry hair aspect (after several shampoos) is obtained with a prolonged effect after a cationic NE use, and hair becomes more fluid and shiny, less brittle, and nongreasy. Yamazaki et al. provided new W/O emulsion-type nail enamel using human sections, a series of model experiments were performed confirming that moisture is essential for flexible and non brittle nails. The researchers developed a new, nitrocellulose containing W/O emulsion nail enamel, which kept the nails in good condition. L'Oreal has several patents on cosmetics by using nanoemulsion.[24]

Due to their lipophilic interior, NEs are more suitable for the transport of lipophilic compounds than liposomes. Similar to liposomes, they support the skin penetration of active ingredients and thus increase their concentration in the skin and due to its high surface area allowing effective transport to the skin. Furthermore, NEs gain increasing interest due to their own bioactive effects. This may reduce the transepidermal water loss, indicating that the barrier function of the skin is strengthened. NEs are acceptable in cosmetics because there is no inherent creaming, sedimentation, flocculation, or coalescence that is observed with macroemulsions. The incorporation of potentially irritating surfactants can often be avoided by using high-energy equipment during manufacturing.[22,25]

13.2.2.1 PEG-FREE NEs FOR COSMETICS

Another example includes manufacturing and processing of low-viscosity oil-in-water NEs that are free from emulsifiers based on polyethylene glycol (PEG). Such blends are highly attractive in the growing market for impregnating emulsions for moisturized tissue. It is relatively recent but fast growing field of application: emulsion-based wet wipes for such applications as baby care and make-up removal. The key components in these products are low viscosity O/W emulsions with good storage stability. Classical emulsions have typical particle of between 0.5 and 10 μm which causes their typical white appearance and usually show viscosities of over 1000 m Pa s. They are kinetically stable, and can be manufactured with help of a homogenizer.[26] Due to their relatively large particles, comparable viscosity systems are unstable and cream up. Alternatively, O/W microemulsions are easy to produce because of their thermodynamic stability. They are translucent, and particle range between 10 and 40 nm. However, microemulsion formation usually requires large quantities of emulsifiers and surfactants. In terms of their properties, "nanoemulsions" are positioned between microemulsion and traditional emulsions. Their typical particle range is between 30 and 100 nm which causes their typical blue-shinning experience. At these small particle sizes, the Brownian motion prevents creaming, and as a result, NEs often have a long-term good stability. They are typically not easy-to-produce as they require either high-pressure homogenizers or very specific manufacturing processes.[27] Various types of emulsion-based drug delivery systems and their uses are described in Table 13.1.

TABLE 13.1   Formulation of Various Emulsion-Based Drug Delivery Systems and Their Use.

Type of drug delivery system	Active ingredients	Use	Reference
Microemulsion	Silicone quaternary polymer	Hair conditioning as well as protection from heat, improved color retention	[30]
Nano emulsion	Nitrocellulose	Nail enamel, kept the nails in good condition	[31]
Multiple emulsion	Vitamin C, wheat proteins	Anti-aging cream, increased the moisture of the skin	[32]

13.2.2.2   PATENTED NEs

- Patent name: NE based on phosphoric acid fatty acid esters and its uses in cosmetics, dermatological, pharmaceutical, and ophthalmological fields. Assignee: L'Oreal (Paris, FR). US Patent Number: 6,274,150.[28]
- Patent name: NE based on ethylene oxide and propylene oxide block copolymers and its uses in the cosmetics, dermatological fields. Assignee: L'Oreal (Paris, FR). US Patent Number: 6,464,990.[29]

13.3   VESICULAR DELIVERY SYSTEMS

13.3.1   LIPOSOMES

Liposomes are vesicular structures with an aqueous core surrounded by a hydrophobic lipid bilayer membrane composed of a phospholipid and cholesterol as shown in Figure 13.1.

Liposomes are hollow spheres that are enclosed by one or more bilayer membranes. These bilayer membranes consist of natural components as for example phospholipids—in particular phosphatidylcholine (PC), which make these carrier systems biocompatible. PC is obtained either from soybeans or eggs, which differs in its composition of fatty acids. Egg-derived lipids have a higher content of saturated fatty acids (40% of 16:0 and 18:0) in comparison to soybean (80% of 18:1 and 18:2). The amphiphilic nature of PC allows them to self-aggregate in an aqueous solution and to form their spherical structures. The lipid bilayer of liposomes promotes release of its contents, making them useful for drug delivery and cosmetic delivery applications. Liposomes can vary in size, from 15 nm up to several micrometers and can have either a single layer (unilamellar) or multilayer (multi-lamellar) structure. The first liposomal cosmetic product to appear on the market was the anti-ageing cream "Capture" launched by Dior in 1986.[33,34] Phosphatidyl choline, one of the main ingredients of liposomes, has been widely used in skin-care products and shampoos due to its softening and conditioning properties. Liposomes have been formed that facilitate the continuous supply of agents into the cells over a sustained period of time, making them an ideal candidate for the delivery of vitamins and other molecules to regenerate the epidermis.. Minoxidil, a vasodilator, is in the active ingredient in products like Regaine that claim to prevent hair loss.[35] The skincare preparations with empty or moisture-loaded liposome reduce the transdermal water loss and are suitable for the treatment of dry skin. They also enhance the supply of lipids and water to stratum corneum.[36] Westerh reviewed the possible use of liposomes as a dynamic dosage form in dermatology and suggested that liposomes preparation could provide a reservoir for the drug and permit its sustained and regular release into skin. Mezei and Gulasekharam reported for the first time the effectiveness of vesicles for skin delivery, suggesting that the lipid formulations can enhance the topical release of drugs. According to the manufacturers, liposomes may deliver moisture and a novel supply of lipid molecules to skin tissue in a superior fashion. In addition, they can entrap a variety of active molecules and can therefore be utilized for skin creams, anti-aging creams, aftershave, lipstick, sun screen, and make-up.[37,38]

They are capable of delivering either hydrophilic (in the aqueous inner core) or lipophilic substances (in the lipid bilayer). Increased rates of skin permeability have been found for various active ingredients, for example, progesterone and hydrocortisone, when they were applied topically in liposomal form, with less frequent side effects.[39,40] In contrast to these observations, other formulations could not find improved skin penetration of substances when they were applied in liposomes.[41] The conflicting results may stem from different lipid compositions of the liposomes employed as in addition to size. The lipid composition determines the physical characteristics of the liposomes and, also the interaction of these carrier systems with the skin.

In the last few years, some papers highlighted important factors that influence the penetration of active ingredients encapsulated in liposomes. Liquid-state, flexible liposomes showed greater skin penetration than those in a gel-state, small-sized, and unilamellar vesicles seem to result in a higher degree of skin penetration.[42,43] The application formed can also influence the penetration kinetics.[41]

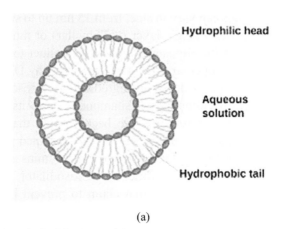

(a)

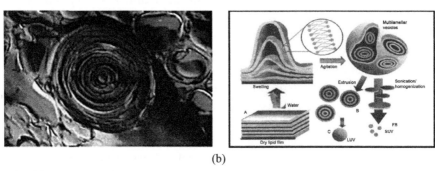

(b)

FIGURE 13.1   (a) Unilamellar liposome and (b) multilamellar liposome.

13.3.1.1   FLEXIBLE LIPOSOMES

Flexible liposomes are small-sized unilamellar vesicles (80–250 nm) prepared of soybean PC (>80%) having a high content of linoleic acid. They provide the skin with essential polyunsaturated fatty acids (vitamin F) which support the formation of ceramide and with choline which is a part of the natural moisturizing factor (NMF).[44] In a clinical study it was proven that these liposomes have cosmetic properties like wrinkle reduction and an increase in skin smoothness and furthermore show pharmaceutical effects like decreasing of efflorescence in the acne treatment.[45] Various types of liposomal drug delivery system and their uses described in Table 13.2.

TABLE 13.2 Formulation of Various Liposomal Dug Delivery Systems and Their Use.

Type of drug delivery system	Active ingredients	Formulation	Use	Reference
Liposome	*Aloe barbadensis* (organic aloe) Juice, *Cocos nucifera*	Deep moisturizing cream	Prevent premature	[46]
Phytosome	*Ginkgo biloba* Dimeric flavonoids	Beauty cream	Not only improves the absorption of the compounds exerting the biological activity but also increases the duration of the activity	[47]
Fullersomes	Vitamins	Skin cream	To refresh dark circles under the eyes	[48]
Ultrasome	coQ10	Stopping wrinkles	To prevent the damage to collagen and elastin production	[49]
Niosomes	Methotrexate	Methotrexate gel	In the treatment of localized psoriasis	[50]

13.3.2 ULTRASOMES

Ultrasomes are specialized liposomes encapsulating an endonuclease enzyme extracted from *Micrococcus luteus*. Endonuclease recognizes UV damage and is reported to accelerate its repair activities four times more as compare with standard conventional formulations.. Ultrasomes also protect the immune system by repairing UV–DNA damage and reducing the expression of TNF-α, IL-1, IL-6, and IL-8.[51]

13.3.3 PHOTOSOMES

Photosomes are incorporated in sun-care product to protect the sun-exposed skin by releasing a photoreactivating enzyme extracted from a marine plant, *Anacystis nidulans*. Photosomes on light activation reverse the cell DNA damage, reducing immune suppression and cancer induction.[33,52]

13.3.4 ETHOSOMES

Ethosomes are noninvasive delivery carriers composed of phospholipids, with 20–50% ethanol and water, which enable drugs to reach the deep skin layers.[25] These are soft, malleable vesicles tailored for enhanced delivery of active agents.

13.3.5 NIOSOMES

Niosomes are microscopic nonionic surfactant vesicles formed by the self-assembly of nonionic surfactant (Fig. 13.2). Niosomes and liposomes have similar physical properties but differ in the chemical nature. Niosomal vesicle is formed by nonionic surfactants, whereas liposomal vesicles by lipids. Niosomes are superior to liposomes because of higher chemical stability of surfactants than lipids.[53] Niosomes can be utilized in the treatment of several diseases like psoriasis, leishmaniasis, cancer, migraine, Parkinson, etc. Niosomes can be used as a diagnostic aid also. Various methods are employed for niosomal administration such as intramuscular, intravenous, peroral, and transdermal. Still researchers have to focus a lot on the commercial utility of niosomes in drug delivery.[54]

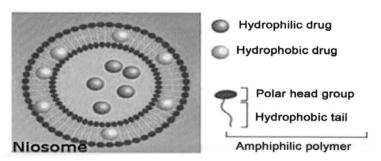

FIGURE 13.2 View of niosomes.

The niosomes have been mainly studied because of their advantages such as higher chemical stability of surfactant than phospholipid, require no special conditions for preparation and storage, they have no purity problems, and the manufacturing costs are low.[55] The advantages of using niosomes in cosmetic and skin-care applications include their ability to increase the stability of entrapped drugs, improved bioavailability of poorly absorbed ingredients and enhanced skin penetration. Niosomes were developed and patented by L'Oréal in the 1970s and 1980s. The first product "Niosome" was introduced in 1987 by Lancôme.[56] Manconi et al. investigated that unilamellar niosomes containing Brij® conferred best protection of tretinoin against photodegradation. Gopinath et al. provided the possibility of converting ascorbyl palmitate into bilayered vesicles with a view to exploit them as carriers for drug delivery. Van Hal et al. reported that niosome encapsulated estradiol can be delivered through the stratum corneum, which is known to be a highly impermeable protective barrier.[55,57]

13.3.6 TRANSFERSOMES

A new type of liposomes called transferosomes, which are more elastic than liposomes.[58] Transferosomes with sizes in the range of 200–300 nm can penetrate the skin with improved efficiency than liposomes.

In the 1990s, transfersomes, that is, lipid vesicles containing large fractions of fatty acids, were introduced by Cevc and coworkers. Transfersomes are vesicles composed of phospholipids as their main ingredient with 10–25% surfactant and 3–10% ethanol. In consequence, their bilayers are much more elastic than those of liposomes and thus well suited for the skin penetration.[59] Transfersomes consist of phospholipids, cholesterol, and additional surfactant molecules such as sodium cholate. The inventors claim that transfersomes are ultra deformable and squeeze through pores less than one-tenth of their diameter. Therefore, 200–300-nm-sized transfersomes are claimed to penetrate intact skin (Fig. 13.3). Higher membrane hydrophilicity and flexibility both help transfersomes to avoid aggregation and fusion, which are observed with liposomes. When applied on occlusively, they significantly improve skin deposition and photostability.[60]

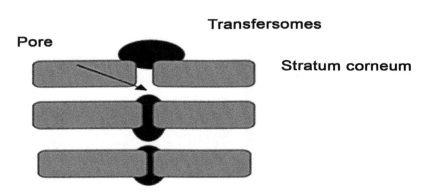

FIGURE 13.3 Transferosome penetration through the pores in stratum corneum.

13.3.7 MARINOSOMES

Marinosomes (Bordeaux, France) are liposomes based on a natural marine lipid extract containing high ratio of polyunsaturated fatty acids like eicosapentaenoic acid and docosahexaenoic acid. They are not present in normal skin epidermis. However, they are metabolized by skin epidermal enzymes into anti-inflammatory and antiproliferative metabolites that are associated with a variety of benefits with respect to inflammatory skin disorders. Marinosomes as potential candidates for cosmeceutical in view of the prevention and treatment of skin diseases.[61]

13.3.8 CUBOSOMES

Cubosomes are discrete, submicron, nanostructured particles of bicontinuous cubic liquid crystalline phase (Fig. 13.4). Bicontinuous cubic liquid crystalline phase is an optically clear, very viscous material that has a unique structure at the nanometer scale.[62] It is formed by the self-assembly of liquid crystalline particles of certain surfactants when mixed with water and a microstructure at a certain ratio. Cubosomes offer a large surface area, low viscosity, and can exist at almost any dilution level. They have high heat stability and are capable of carrying hydrophilic and hydrophobic molecules.[63] Combined with the low cost of the raw materials and the potential for controlled release through functionalization. They are an attractive choice for cosmetic applications as well as for drug delivery. The presence of large amounts of water during cubosome formation makes it

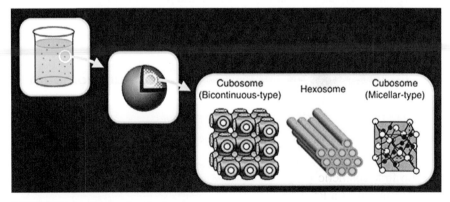

FIGURE 13.4   Schematic of cubosomes.

difficult to load water-soluble actives. However, at present, cubosomes do not offer controlled release on their own.[64] They have also been modified using proteins. A number of companies including L'Oréal,[65] Nivia,[20,66,67] and Procter and Gamble are investigating cubosomes for cosmetic applications. Despite this interest, cubosomes have not yet led to products. The methods of formation must be efficient and cost-effective for scale up before this type of technology can be applied..

13.4 NANOCRYSTALS

Nanocrystals have been used in the pharmaceutical industry for the delivery of poorly soluble actives. They are aggregates comprising several hundred to tens of thousands of atoms that combine into a "cluster" (Fig. 13.5). Typical sizes of these aggregates are between 10 and 400 nm and they exhibit physical and chemical properties somewhere between that of bulk solids and molecules. By controlling the size and surface area, other properties such as band gap, charge conductivity, crystalline structure, and melting temperature can be altered. The crystals must be stabilized to prevent larger aggregates from forming.

FIGURE 13.5   Nanocrystal formation.

The first cosmetic products appeared on the market that is; Juvena in 2007 (rutin) and La Prairie in 2008 (hesperidin).[68] Rutin and hesperidin are two, poorly soluble, plant glycoside antioxidants that could not previously be used dermally. Once formulated in the form of nanocrystals, they became dermally available as measured by antioxidant effect. This dermal use of nanocrystals is protected by patents.[69] Nanocrystals for cosmetic applications are commercially available via contract manufacturer, as incorporation of nanocrystal into cosmetic is a very easy process.

Nanocrystals may be able to reduce the dose to be administered, provide a sustained drug release, and increase patient compliance. De Waard et al. investigated nanocrystals of ibuprofen and fenofibrate. He claimed that shape of the crystals increases the drug absorption to a great extent.[70] Shegokar et al. prepared prolonged release formulation for dermaluse incorporating nanocrystals of poorly water-soluble lidocaine. The pearl milling and high-pressure homogenization can be extended for commercial production of nanocrystals and are accepted by the regulatory authorities.[13] The first four marketed products containing nanocrystals such as Rapamune®, Emend®, Tricor®, Megace ES® were prepared by Pearl mill technology by Elan nano-systems.[71-73] Various types of formulation drug delivery systems with their uses are described in Table 13.3.

TABLE 13.3 Formulation of Various Dug Delivery Systems and Their Use.

Type of drug delivery system	Active ingredients	Formulation	Use	Reference
Nanocrystals	Rutin and hesperidin	Nanosuspensions	Antioxidant effect	[74]
Tranferosome	*Curcuma longa*	Cream	Anti-aging, antiwrinkle, anti-irritant	[75]
Cubosomes	Retinol	Cream	Acne	[76]

13.5 LIPID NANOPARTICLE

Lipid nanoparticles have a similar structure to NEs. Their size ranges typically from 50 to 1000 nm. The difference is that the lipid core is in the solid state (Fig. 13.6). The matrix consists of solid lipids or mixtures of lipids. To stabilize the solid lipid particle against aggregation, surfactants or polymers are added, whereby natural lecithins are preferred as is the case with NEs. If lipid nanoparticles are intended to be used as a carrier, the active ingredients are dissolved or finely dispersed in the lipid matrix. SLNs are nanometer-sized particles with a solid lipid matrix. They are oily droplets of lipids which are solid at body temperature and stabilized by surfactants. Their production is a relatively simple process where the liquid lipid (oil) in a NE is exchanged by solid lipids.[77] This process does not require organic solvents. SLNs offer a number of advantages for cosmetic products such as they can protect the encapsulated ingredients from degradation. Compounds, including coenzyme Q1023 and retinol[23] can remain stable in SLNs over a long time period.[78] They can be used for the controlled delivery of cosmetic agents over a prolonged period of time and have been found to improve the penetration of active compounds into the stratum corneum. SLNs have occlusive properties making them ideal for potential use in day creams. In-vivo studies have shown that an SLN-containing formulation is more efficient in skin hydration than placebo.[79] They have also been found to show UV resistant properties which were enhanced when a molecular sunscreen was incorporated and tested. Enhanced UV blocking by 3,4,5-trimethoxybenzoylchitin (a good UV absorber) was seen when incorporated into SLNs. SLNs have also been tested in perfume formulations. Chanel's Allure perfume was incorporated into SLNs and NEs,[80] it shows that SLN formulations delayed the release of perfume over a longer period of time. This slow release profile is also desirable for insect repellents. Although SLNs are promising for cosmetic purposes, they suffer with some drawbacks such as production process needs improvement to increase loading capability and stop expulsion of the contents during storage.[79,81,82] These problems are caused by the tendency for the particle matrix to form a perfect crystal lattice when solid lipids are used.[80,83]

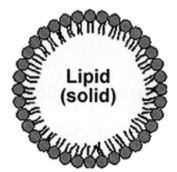

FIGURE 13.6 Lipid nanoparticle: lipid monolayer enclosing a solid lipid core.

SLNs possess some features, which make them promising carriers for cosmetic applications:

1. The protection of labile compounds against chemical degradation (e.g., retinol).

2. Depending on the type of SLNs, controlled release of the active ingredients is possible. SLNs with a drug-enriched shell show burst release characteristics, whereas SLNs with a drug-enriched core lead to sustained release.
3. SLNs act as occlusive, they can be used to increase the water content of the skin.
4. SLNs show a UV-blocking potential. They act as physical sunscreens on their own and can be combined with molecular sunscreens to achieve improved photoprotection.
5. Improved stability of chemically unstable active ingredients.
6. Controlled release of active ingredients.
7. Pigment effect.
8. Improved skin hydration and protection through film formation on the skin.

13.5.1 INCORPORATING LIPID NANOPARTICLES INTO NEW COSMETIC PRODUCTS

In contrast to liposomes and NEs, it is not necessary to develop completely new products if one intends to use lipid nanoparticles.[82] Due to their good physical stability and compatibility with other ingredients they can often be added to existing formulations without any problems. As lipid nanoparticles are patent protected worldwide as Lipopearls® or Nanopearls®, a product exclusivity can be guaranteed when they are used in new products.[84,85]

13.6 NANOSTRUCTURED LIPID CARRIERS

To overcome issues associated with SLNs, a second generation of lipid particles has been developed by mixing solid lipids with liquid lipids. These are known as NLCs. Compared to SLNs, NLCs have a distorted structure which makes the matrix structure imperfect and creates spaces to accommodate active compounds as shown in Figure 13.7. Müller et al.[86] suggest that SLNs are better for applications such as UV protection where a high level of crystallany is required for the carrier. Similar to SLNs, NLCs are also capable of preventing the active compounds from chemical degradation. They also possess a high occlusion factor and high level of skin adherence properties. When the particles adhere to the skin a thin film layer is created which prevents dehydration; as the size of the particles decreases, the occlusion factor increases.[87] Various types of NLC and their uses are described in Table 13.4.

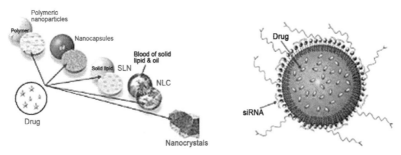

FIGURE 13.7 View of nanostructured lipid carriers.

TABLE 13.4 Formulation of Various Drug Delivery Systems and Their Use.

Type of drug delivery system	Active ingredients	Use	Reference
SLN	3,4,5-tri-methoxy benzoyl chitin	Enhanced UV blocking	[77]
Nanostructured lipid carriers	–	Excellent moisturizer for all types of skin	[82]

13.7 DENDRIMERS AND HYPERBRANCHED POLYMERS

Dendrimers are unimolecular, monodisperse, micellar nanostructures, around 20 nm in size, with a well-defined, regularly branched symmetrical structure, and a high density of functional endgroups at their periphery (Fig. 13.8). A dendrimer is typically symmetric around the core, and often adopts aspherical three-dimensional morphology Dendrimers have a great contribution on cosmetics. Various cosmetics industry used dendrimers in the formulation, one of them is L'Oreal which also has some patents. Unilever also have a patent for dendrimers in the production of formulation for used in spray, gels, and lotions.[88]

Several other patents have been filed for the application of dendrimers in hair-care, skin-care, and nail-care products.[89] Dendrimers have been reported to provide controlled release from the inner core. However, drugs are incorporated both in the interior as well as attached on the surface. Due to their versatility, both hydrophilic and hydrophobic drugs can be incorporated into dendrimers. L'Oréal have a patent for a formulation containing hyperbranched polymers or dendrimers which form a thin film when deposited on a substrate. This formulation could be used for a wide variety of cosmetics, for example, mascara or nail polish. They have also developed a formulation comprising a tanning gent and dendrimers for artificial skin tanning. Unilever have a patent for hydroxyl-functionalized dendrimers from polyester units to create formulations for use in sprays, gels, or lotions.[90,91]

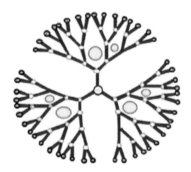

FIGURE 13.8   View of dendrimers.

They are known to be robust, covalently fixed, three-dimensional structures possessing both a solvent-filled interior core (nanoscale container) as well as a homogenous, mathematically defined, exterior surface functionality (nanoscaffold).[92]

They are prepared in a step-wise fashion, with an architecture like a tree branching out from a central point. Hyperbranched polymers are effectively disorganized, unsymmetrical dendrimers that are prepared in a single synthetic polymerization step, making them much more cost-effective than dendrimers. The large number of external groups suitable formulation functionalization which is a requirement for its use as a cosmetic agent carrier.[93]

13.8   CYCLODEXTRIN COMPLEXES

Cyclodextrins (CDs) are cyclic oligosaccharides containing a minimum of six D-(+)-glucopyranose units attached by $\alpha(1{\rightarrow}4)$ glucosidic bonds (Fig. 13.9).

The three natural CDs are α, β, and γ which differ in their ring size and solubility. Most of the molecules fit into the internal CD cavity forming a complex and the resulting structure is called CD clathrates or inclusion complexes. α-CD typically forms inclusion complexes with both aliphatic hydrocarbons and gases. β-CD forms complexes with small aromatic molecules.[94] γ-CD can accept more bulky compounds like vitamin D. Cyclodextrin showed improved light stability compared with uncomplexed form of the compound. Particle containing emulsion produce dull or dry looks on the skin those problems can be overcome by addition of cyclodextrin either β- and α-cyclodextrin.

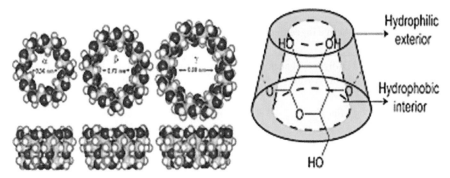

FIGURE 13.9   View of cyclodextrin.

13.9 MICROSPONGE

Microsponge (Fig. 13.10) utilizes microporous beads (10–25 μm in diameter) for the controlled release of topical agents (Table 13.5). These microporous beads are loaded with active agent having properties like inertness with monomer, adequate stability in contact with polymerization catalyst and process, immiscibility or slight solubility in water. Microsponge is being used in cosmetics, over-the-counter skin care, sunscreens, and prescription products. These are stable over pH range of 1–11 and also at the temperature up to 130°C and also having improved thermal, physical, and chemical stability; these are having enhanced material processing and flexibility to develop novel product forms.[95] Microsponge system is having self-sterilizing capacity as particles are very small (0.25 μm) size where bacteria cannot penetrate. So, commonly used in cosmetics formulations.

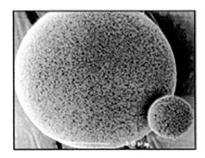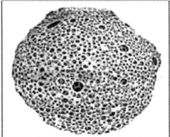

FIGURE 13.10   View of microsponge.

TABLE 13.5   Formulation of Various Dug Delivery Systems and Their Use.

Type of drug delivery system	Active ingredients	Use	Reference
Cyclodextrin	Di-hydroxyl acetone with tyrosine	Increases production of melanin in the skin is an unpleasantly scented substance	[96]
Dendrimers	Minoxidil	Compound stimulating keratinocyte growth and promote hair growth	[97]
Microsponge	Benzoyl peroxide	Antiacne cream	[98]

13.10   PENETRATION ENHANCER

The skin is a complex organ and, as such, must be gently persuaded to allow beneficial substances to pass while still continuing to work as it should. The outer horny layer of the skin, the stratum corneum, has layers of keratin made by keratinocytes (this term refers to cells that make keratin) and is designed to be particularly difficult to penetrate. Vitamin E, however, does penetrate this layer to some degree and enters skin cells at deeper levels to serve as a potent antioxidant, fighting changes of ordinary skin metabolism by products, aging, and sun damage. However, vitamin E acetate (has better antioxidant ability to fight aging and sun damage, and can get into skin cells more easily than plain vitamin E.

Another way is encapsulating or suspending key ingredients in so-called nanospheres or NEs, increases their penetration into the skin. Here, below we mention some examples of renowned cosmetics company.[99]

In 1998, the company formulated Plentitude Revitalift, an antiwrinkle cream using nanoparticles. La Prairie's product, the Dollars 500 Skin Caviar Intensive Ampoule Treatment, claims to minimize the look of uneven skin pigmentation, lines, and wrinkles in 6 weeks using nanotechnology. NEs in the product that "optimize the delivery of functional ingredients into the skin and allow these materials to get to the site of action quicker." Procter & Gamble's Olay brand was designed with NE technology with improved penetration in 2005. Other companies using nanotech in their skin products as of 2005 include Mary Kay and Clinique from Lauder; Neutrogena, from Johnson & Johnson; Avon; and the Estee Lauder brand. Hair products use NEs to encapsulate active ingredients and carry them deeper into hair shafts. Pure Ology began experimenting with NEs in 2000, when the company's founder set out to create a product line especially developed for color-treated hair.

Sunscreens—the zinc and titanium in sunscreens are "micronized", making them transparent, less greasy, less smelly, and more absorbable into the skin.

Colorescience markets a product named Sun forgettable, a powder which contains titanium dioxide nanoparticles that penetrate to the deep layer of the skin.[63,100,101]

13.11 TOXICITY AND SAFETY ASSESSMENT

NP toxicity depends on exposure, the likelihood of sufficient absorption and in sufficient quantities to affect viable cells and the need for an intrinsic toxicity of both the NP ingredient and its coating to be present.[102] This is a very difficult task, however, the research is complicated and incomplete due to its toxicity. Some specific hazards of risk for human health have been identified. These include the possibility of some nanoparticles to induce protein fibrillation, the possible pathological effects caused by specific types of carbon nanotubes, the induction of genotoxicity and size effects in terms of biodistribution. An issue of specific importance is the properties of the nanomaterial as it is actually used in products and to which consumers may be exposed. For the risk assessment, the latter characterization is of highest relevance. Currently, the risk assessment procedure for the evaluation of potential risks of nanomaterials is still under development. To conduct safety assessments for cosmetic products containing nanomaterials, standard safety tests may need to be modified or new methods developed and validated.[103,104]

13.12 REGULATIONS OF NANOCOSMECEUTICALS

Nanocosmetics are more and more popular in skincare products and even cosmetics also. The original Cosmetics Directive was adopted in 1976 with the aim of establishing a single market for cosmetic products, ensuring a high level of protection for consumers. A draft guidance documents from the FDA "Guidance for Industry: Safety of Nanomaterials in Cosmetic Products" discusses the FDA's current thinking on the safety assessment of nanomaterials when used in cosmetic products. Key points include the legal requirements for cosmetics manufactured using nanomaterials are the same as those for any other cosmetics.[105] While cosmetics are not subject to premarket approval, companies and individuals who market cosmetics are legally responsible for the safety of their products and they must be properly labeled.[106]

Regulatory and legislative bodies around the world are beginning to introduce clearer rules about nanotechnology. In the meantime, cosmetics manufacturers are developing more and more nanoformulated products.[107]

The simplification of the Cosmetics Directive had four specific objectives, namely:

1. To improve legal clarity and remove inconsistencies.
2. To remove divergences between national law.
3. To ensure that cosmetic products placed on the market are safe in the light of innovation in this sector.
4. To introduce a possibility in exceptional cases to regulate categories "1" and "2" of carcinogenic, mutagenic, and reprotoxic substances on the basis of their actual risk.[108]

13.13   CONCLUSION

Our society highly values health and beauty. The rationale for the use of nanomaterials in cosmetic products is that they offer added value in terms of product performance (Table 13.6). The unique properties and behavior of nanomaterials mean that nanotechnologies could profoundly transform industry and everyday life. Since about 20 years after the introduction of the liposomes, the cosmetic market is waiting for a similar innovative novel nanocarrier. All systems developed since 1986 were far away from the success of the liposomes. The lipid nanoparticles, both SLNs and NLCs, possess many of the positive advantages of the liposomes and in addition features being clearly superior to liposomes. Novel cosmetic delivery systems reviewed here possess the potential to develop as the "new generation smarter carrier systems." The technical, economic, and sensory aspects should be taken into consideration while selecting an appropriate type of delivery system to enhance the safety, stability, extended efficacy, and to enhance the esthetic appeal of the final product.

TABLE 13.6   Marketed Nanoformulations.

Trade name	Active ingredient	Proposed use	Name of company	Type of nanotechnology used
Nano sun	Zinc oxide	Sunscreen	Microniers Pvt. Ltd.	Nanoscale size
Revitalift	Pro retinol-A	Antiwrinkle	L'oreal	Nanosomes
Multisal salicylic acid 10	Salicylic acid	Acne, psoriasis, dandruff	Salvona	Double-layer encapsulation

13.14   CURRENT AND FUTURE DEVELOPMENTS

Nanotechnology is a rapidly expanding and potentially beneficial field with tremendous implications for medicine and cosmeceuticals. Cosmetic industry is facing new and unfamiliar challenges. These may affect the profitability and, indeed, the survival of some highly successful products. The second straints may be related to health hazards, environmental concerns and product functionality. The commitment to innovation is essential, not only to allow the companies to maintain their global competitiveness, but, more importantly, to improve the performance, safety and environmental impacts of products.

Over all, Cosmetic formulations imparting various nanotechnologies can increase the product efficiency and performance considerably.

KEYWORDS

- **cosmetics**
- **emulsion**
- **vesicular drug delivery system**
- **toxicity**
- **regulatory aspects**

REFERENCES

1. Kaur, I.; Agarwal, R. Nanotechnology: A New Paradigm in Cosmecuticals. *Rec. Pat. Drug Deliv. Formul.* **2007,** *1,* 171–182.
2. Mcnair, B. Nano is Every Where. www.Jour.sc.edu/nanomcnair.html (accessed on 9/12/2016).
3. http://www.fda.gov/cder/drug/infopage/sunscreen/qa.htm (accessed on 10/11/2016).
4. http://www.cosmetics.co.in (accessed on 10/12/2016).
5. Singhal, M.; Khanna, S.; Nasa, A. Cosmeceuticals for the Skin: An Overview. *Asian J. Pharm. Clin. Res.* **2011,** *4* (2), 1–6.
6. Sharma, R. Cosmeceuticals and Herbal Drugs: Practical Uses. *Int. J. Pharm. Sci. Res.* **2012,** *3,*1, 59–65.
7. Srinivas, K. The Current Role of Nanomaterials in Cosmetics. *J. Chem. Pharm. Res.* **2016,** *8* (5), 906–914.
8. Mufti, J.; Cernasov, D.; Macchio, R. New Technologies in Topical Delivery Systems. *Happi* **2002,** *39,* 75–82.
9. Loreal US Patent. Cosmetic Composition Based on Nanoparticles and on Water-Soluble Organic Silicon Compounds. Loreal US patent, US 8377427 b2.
10. Morganti, P. Use and Potential of Nanotechnology in Cosmetic Dermatology. *Clin. Cosmet. Investig. Dermatol.* **2010,** *3,* 5–13.
11. Gautam, A.; Singh, Vijayaraghavan, R. Dermal Exposure of Nanoparticles: An Understanding. *J. Cell Tissue Res.* **2011,** *11* (1), 2703–2708.
12. Dureja, H.; Kaushik, D.; Gupta, M.; Kumar, K.; Lather, V. Cosmeceuticals: An Emerging Concept. *Indian J. Pharmacol.* **2005,** *37* (3), 155–159.
13. Raj, S.; Jose, S.; Sumod, U. S.; Sabitha, M. Nanotechnology in Cosmetics: Opportunities and Challenges. *J. Pharm. Bioallied Sci.* **2012,** *4* (3), 186–193.
14. Institute of Occupational Medicine for the Health and Safety. Available from: http://www.hse.gov.uk (last accessed on December 2015).
15. Royal Society. *Nanoscience and Nanotechnologies: Opportunities and Uncertainties.* Royal Society: London (last accessed on December 2015). Available from http://www.nanotec.org.uk/finalReport.htm.
16. Ghosh, P. K.; Murthy, R. S. R. Microemulsions: A Potential Drug Delivery System. *Curr. Drug Deliv.* **2006,** *3,* 167–180.
17. Narang, A. S.; Delmarre, D.; Gao, D. Stable Drug Encapsulation in Micelles and Microemulsions. *Int. J. Pharm.* **2007,** *345,* 9–25.
18. Kreilgaard, M. Influence of Microemulsion on Cutaneous Drug Delivery. *Adv. Drug Deliv. Rev.* **2002,** 54, S77–S98.
19. Tenjarla, S. Microemulsions: An Overview and Pharmaceutical Applications. *Crit. Rev. Therap. Drug Carrier Syst.* **1999,** *16,* 461–521.
20. Tarl, W. P.; Jeffrey, E. G.; Lynlee, L. L.; Rokhaya, F.; Margaret, B. Nanoparticles and Microparticles for Skin Drug Delivery. *Adv. Drug Deliv. Rev.* **2011,** *63,* 470–491.
21. Maynard, A. *Nanotechnology and Safety.* http://www.cleanroom-technology.co.uk (last accessed on January 2016).
22. Hidaka, H.; Horikoshi, S.; Serpone, N.; Knowland, J. In Vitro Photochemical Damage to DNA, RNA and Their Bases by an Inorganic Sunscreen Agent on Exposure to UVA and UVB Radiation. *J. Photochem. Photobiol. Chem.* **1997,** *111,* 205–213.

23. Jurgen, M. Nanotechnology. Nanoemulsions for PEG-free Cosmetics. *Personal Care* **2008**, 56–57.

24. Mehnert W, Mader K. Solid Lipid Nanoparticles: Production, Characterization and Applications. *Adv Drug Deliv Rev.* **2001**, *47* (2), 165–196.

25. Patravale, V. B.; Mandawgade, S. D. Novel Cosmetic Delivery Systems: An Application Update. *Int. J. Cosmet. Sci.* **2008**, *30*, 19–33.

26. Ohl, L.; Mohaupt, M.; Czeloth, N.; Hintzen, G.; Kiafard, Z.; Zwirner, J.; Blankenstein, T.; Henning, G.; Forster, R. CCR7 Governs Skin Dendritic Cell Migration under Inflammatory and Steady-State Conditions. *Immunity* **2004**, *21*, 279–288.

27. Pernodet, N.; Fang, X. H.; Sun, Y.; Bakhtina, A.; Ramakrishnan, A.; Sokolov, J.; Ulman, A.; Rafailovich, M. Adverse Effects of Citrate/Gold Nanoparticles on Human Dermal Fibroblasts. *Small* **2006**, *2*, 766–773.

28. L'Oreal. US Patent 4830857, 1989.

29. L'Oreal. French Patent 2315991, 1975.

30. Michael, F. Cosmetic Compositions for Hair Treatment Containing Dendrimers or Dendrimer Conjugates. Patent 6068835, 2010.

31. Gers-Barlag, H.; Muller, A.; Beiersdorf, A. G. US Patent Application 20030175221, 2003.

32. Naveed, A.; Yasemin, Y. Formulation and Evaluation of a Cosmetic. Multiple Emulsions. *Pak. J. Pharm. Sci.* **2008**, *21* (1), 45–50.

33. Suzuki, K.; Sakon, K. The Applications of Liposomes to Cosmetics. *Cosmet. Toil.* **1990**, *105*, 65–78.

34. Lauten Schlager, H.; Barel, A. O.; Paye, M.; Maibach, H. I. *Liposomes, Handbook of Cosmetic Science and Technology;* CRC Press: Boca Raton, FL, 2006; pp 155–163.

35. Torchilin, V.; Weissig, V. *Liposomes a Practical Approach*; Oxford University Press: Oxford, UK, 1990, pp 1–67.

36. Ho, N. F. H.; Ganesan, M. G.; Weiner, N. D.; Flynn, G. L. Mechanisms of Topical Delivery of Liposomally Entrapped Drugs. *J. Control. Rel.* **1985**, *11*, 61–65.

37. Lasic, D. D. Novel Applications of Liposomes. *Trends Biotechnol.* **1998**, *16* (7), 307–321.

38. Lasch, J.; Wohlrab, W. Liposome-Bound Cortisol: A New Approach to Cutaneous Therapy. *Biomed. Biochim. Acta* **1986**, *45*, 1295–1299.

39. Knepp, V. M.; Hinz, R.; Szoka, F. C.; Guy, R. Controlled Drug Release from a Novel Liposomal Delivery System. I. Investigation of Transdermal Potential. *J. Control. Rel.* **1988**, *5*, 211–221.

40. Knepp, V. M.; Szoka, F. C.; Guy, R. Controlled Drug Release from a Novel Liposomal Delivery System. II. Transdermal Delivery Characteristics. *J. Control. Release* **1990**, *12*, 25–30.

41. Choi, M. J.; Maibach, H. J. Liposomes and Niosomes as Topical Drud Delivery Systems. *Skin Pharmacol. Physiol.* **2005**, *18*, 209–219.

42. Blume, G.; Sacher, M.; Teichmüller, D.; Schäfer, U. The Role of Liposomes and Their Future Perpective. *SÖFW J.* **2003**, *129*, 10–14.

43. Verma, D. D.; Verma, S.; Blume, G.; Fahr, A. Particle Size of Liposomes Influences Dermal Delivery of Substances into the Skin. *Int. J. Pharm.* **2003**, *258*, 141–151.

44. Fresta, M.; Puglisi, G. Application of Liposomes as Potential Cutaneous Drug Delivery Systems. *J. Drug Target.* **1996**, *4*, 95–101.

45. Cevc, G.; Blume, G. Lipid Vesicles Penetrate into Intact Skin Owing Transdermal Osmotic Gradients and Hydration Force. *Biochim. Biophys. Acta* **1992**, *1104*, 226–232.

46. *PEN Consumer Products Inventory.* http://www.nanotechprojects.org/consumer (last accessed on January 2016).

47. Bombardelli, E.; Cristoni, A.; Morazzoni, P. Phytosome in Functional Cosmetic. *Fitoterapia* **1995**, *5*, 387–401.

48. Thomas, M.; Joshua, C.; Nick, T.; Jason, K.; David, M. B. *The Citizen's Guide to Nanotechnology: Cosmetics*; North Carolina State University: Raleigh, NC, 2008.

49. http://www.nutraingredients.com/Industry/Proprietary-CoQ10-formulation-demonstrates-superior-absorption (last accessed on Jan 2016).

50. Lakshmi, P. K.; Devi, G. S. Niosomal Methotrexate Gel in the Treatment of Localized Psoriasis: Phase I and Phase II Studies. *Indian J. Dermatol.* **2007**, *73* (3), 157–161.

51. http://www.protecingredia.com/products/barnet/ultrasomes.html (last accessed on January 2016).

52. Dayan, N.; Touitou, E. Carriers for Skin Delivery of Trihexyphenidyl HCl: Ethosomes vs Liposomes. *Biomaterials* **2000**, *21*, 1879–1885.

53. Uchegbua, I. F.; Vyas, S. P. Non-ionic Surfactant Based Vesicles (Niosomes) in Drug Delivery. *Int. J. Pharm.* **1998**, *1–2*, 33–70.

54. Mokhtar, M.; Sammour, O. A.; Hammad, M. A.; Megrab, N. A. Effect of Some Formulation Parameters on Flurbiprofen Encapsulation and Release Rates of Niosomes Prepared from Proniosomes. *Int. J. Pharm.* **2008**, *361*, 104–111.

55. Cook, E. J.; Lagace, A. P. *Apparatus for Forming Emulsions*, US Patent 4254553; Mayer, L. D.; Bally, M. B.; Hope, M. J.; Cullis, P. R. Uptake of Antineoplastic Agents into Large Unilamellar Vesicles in Response to a Membrane Potential. *Biochem. Biophys. Acta* **1985**, 816, 294–302.

56. Junyaprasert, V. B.; Teeranachaideekul, V.; Supaperm, T. Effect of Charged and Non-ionic Membrane Additives on Physicochemical Properties and Stability of Niosomes. *AAPS PharmSciTech* **2008**, *9* (3), 851–859.

57. Fang, J. Y.; Hong, C. T.; Chiu, W. T.; Wang, Y. Y. Effect of Liposomes and Niosomes on Skin Permeation of Enoxacin. *Int. J. Pharm.* **2001**, *219*, 61–72.

58. Bauman, L. Botanical Ingredients in Cosmeceuticals. *J. Drugs Dermatol.* **2007**, *6* (11), 1084–8.

59. Dubey, V.; Mishra, D.; Asthana, A.; Jain, N. K. Transdermal Delivery of a Pineal Hormone: Melatonin via Elastic Liposomes. *Biomaterials* **2007**, *27*, 3491–3496.

60. http://www.livestrong.com/article/232965-skin-carebenefits-of-liposomes (last accessed 28/11/2016).

61. Moussaoui, N.; Cansell, M.; Denizot, A. Marinosomes Marine Lipid-Based Liposomes: Physical Characterization and Potential Applications in Cosmetics. *Int. J. Pharm.* **2002**, *242*, 361–365.

62. Spicer, P. T.; Lynch, M. L.; Visscher, M.; Hoath, S. Bicontinuous Cubic Liquid Crystalline Phase and Cubosome Personal Care Delivery Systems. In *Personal Care Delivery Systems and Formulations*; Rosen, M., Ed.; Noyes Publishing: Norwich, 2003.

63. Gopal, G.; Shailendra, S.; Swarnlata, S. Cubosomes: An Overview. *Biol. Pharm. Bull.* **2007**, *30* (2), 350–353.

64. Boyd, B. J. Characterisation of Drug Release from Cubosomes Using the Pressure Ultrafiltration Method. *Int. J. Pharm.* **2003**, *260*, 239–247.

65. Angelova, A.; Angelov, B.; Papahadjopoulos-Sternberg, B.; Ollivon, M.; Bourgaux, C. Proteocubosomes: Nanoporous Vehicles with Tertiary Organized Fluid Interfaces. *Langmuir* 2005, *21*, 4138–4143.

66. Spicer, P. Progress in Liquid Crystalline Dispersions: Cubosomes. *Curr. Opin. Colloid Int. Sci.* **2005,** *10,* 274–279.
67. Afriat, I.; Biatry, B. Use of Cubic Gel Particles as Agents against Pollutants, Especially in a Cosmetic Composition. Eur. Pat. Appl. L'Oréal, 2001.
68. Buzea, C.; Blandino, I. I. P.; Robbie, K. Nanomaterials and Nanoparticles: Sources and Toxicity. *Biointerphases* **2007,** 4MR17–MR172.
69. De Waard, H.; De Beer, T.; Hinrichs, W. L. J.; Vervaet, C.; Remon, J. P.; Frijlink, H. W. Controlled Crystallization of the Lipophilic Drug Fenofibrate during Freeze–Drying: Elucidation of the Mechanism by In-Line Raman Spectroscopy. *AAPS J.* **2010,** 12 (4), 569–575.
70. Shegokar, R.; Keck, C. M.; Müller, R. H.; Gohla, S. *Cosmetic Nanocrystals: Products and Dermal Effects.* In 6th Polish German Symposium on Pharmaceutical Sciences: Perspectives for a New Decade, Düsseldorf, 2011; pp 20–21.
71. Schreiber, A.; Eitrich, Beiersdorf, A. G. Deodorant and Antiperspirant Products with a Content of Disperse Phase Liquid Crystals which Form Cubic Phases. *Ger. Offen., Germany,* 2002.
72. SCENIHR. Risk Assessment of Products of Nanotechnologies. http://www.ec.europa. eu/health/ph_risk/committees/04_scenihr/docs/scenihr-023.pdf (last accessed on Feb 2016).
73. Spicer, P. Cubosome Processing: Industrial Nanoparticle Technology Development. *Chem. Eng. Res. Des.* **2005,** *83,* 1283–1286.
74. Huczko, A.; Lange, H. Fullerenes: Experimental Evidence for a Null Risk of Skin Irritation and Allergy. *Fullerene Sci. Technol.* **1999,** *7,* 935–939.
75. Saraf, S.; Jeswani, G.; Deep Kaur, C.; Saraf, S. Development of Novel Herbal Cosmetic Cream with Curcuma Longa Extract Loaded Transfersomes for Antiwrinkle Effect. *Afr. J. Pharm. Pharacol.* **2011,** *5* (8), 1054–1062.
76. Asensio, J. A.; Grisoni, P.; Gomez, Y.; Martinez, S.; Salvatore Gargano, S.; Andreas, A. *Eco-Friendly Cosmetic Delivery Systems Based on Natural Biopolymers*; Skin Care Forum, BASF Personal Care and Nutrition GmbH, 2011; p 44.
77. Müller, R. H.; Radtke, M.; Wissing, S. A. Solid Lipid Nanoparticles (SLN) and Nano-structured Lipid Carriers (NLC) in Cosmetic and Dermatological Preparations. *Adv. Drug Deliv. Rev. Suppl.* **2002,** *54,* S15–S131.
78. Müller, R. H.; Dingler Feste, A. Lipid-Nanopartikel (Lipopearls) als neuartiger Carrier für kosmetische und, dermatologische Wirkstoffe. *PZ Wiss* **1998,** *49,* 11–15.
79. Jenning, V.; Gohla, S. Encapsulation of Retinoids in Solid Lipid Nanoparticles (SLN). *J. Microencapsul.* **2001,** *18,* 149–158.
80. Wissing, S. A.; Müller, R. H. Cosmetic Applications for Solid Lipid Nanoparticles (SLN). *Int. J. of Pharm.* **2003,** *254,* 65–68.
81. Wissing, S. A.; Müller, R. H. A Novel Sunscreen System Based on Tocopherol Acetate Incorporated into Solid Lipid Nanoparticles (SLN). *Int. J. Cosm. Sci.* **2001,** *23,* 233–243.
82. Song, C.; Liu, S. A New Healthy Sunscreen System for Human: Solid Lipid Nanopar-ticles as Carrier for 3,4,5-Trimethoxybenzoylchitin and the Improvement by Adding Vitamin E. *Int. J. Biol. Macromol.* **2005,** *36,* 116–119.
83. Choi, C. M.; Berson, D. S. Cosmeceuticals, Seminars in Cutaneous Medicine and Surgery. **2006,** *25* (3) 163–168.
84. Wissing, S. A.; Mäder, K.; Müller, R. H. Solid Lipid Nanoparticles (SLN) as a Novel Carrier System Offering Prolonged Release of the Perfume Allure (Chanel). *Int. Symp. Control Release Bioact. Mater.* **2000,** *27,* 311–312.

85. Müller, R. H.; Petersen, R. D.; Hommoss, A.; Pardeike, J. Nanostructured Lipid Carriers (NLC) in Cosmetic Dermal Products. *Adv. Drug Deliv. Rev.* **2007,** *59*, 522–530.
86. Hassan Hany, M.; El Gazayerly, O. N. Rice Bran Solid Lipid Nanoparticles: Preparation and Characterization. *Int. J. Res. Drug Deliv.* **2011,** *1* (2), 6–9.
87. Pardeike, J.; Hommoss, A.; Muller, R. H. Lipid Nanoparticles (SLN, NLC) in Cosmetic and Pharmaceutical Dermal Products. *Int. J. Pharm.* **2009,** *366*, 170–184.
88. Svenson, S.; Tomalia, D. A. Dendrimers in Biomedical Applications—Reflections on the Field. *Adv. Drug Deliv. Rev.* **2005,** *57*, 2106–2129.
89. L'Oréal. Use of Hyperbranched Polymers and Dendrimers Comprising a Particular Group as Film-Forming Agent, Film-Forming Compositions Comprising Same and Use Particularly in Cosmetics and Pharmaceutics. US Patent 6432423, L'Oréal, 2002.
90. *Self-Tanning Cosmetic Compositions*, L'Oréal, US Patent 6399048, 2002.
91. Adams, G.; Ashton, M. R.; Khoshdel, E. *Hydroxyl-Functionalized Dendritic Macromolecules in Topical Cosmetic and Personal Care Compositions*, Unilever Home & Personal Care, US Patent 6582685, 2002.
92. *Cosmetic or Dermatological Topical Compositions Comprising Dendritic Polyesters*, L'Oréal, US Patent 6287552, 2001.
93. Franzke, M.; Steinbrecht, K.; Clausen, T.; Baecker, S.; Titze, J. *Cosmetic Compositions for Hair Treatment Containing Dendrimers or Dendrimer Conjugates.* US Patent 6068835, Wella Aktiengesellschaft, 2000.
94. Duchene, D.; Wouessidjewe, D.; Poelman, M. C. *Cyclodextrins in Cosmetics: Novel Cosmetic Delivery Systems.* Marcel Dekker Inc.: New York, 1999; pp 275–278.
95. Nacht, S.; Katz, M. *The Microsponge: A Novel Topical Programmable Delivery System*; David, W. O.; Anfon, H. A.; Eds.; Marcel Dekker Inc.: New York, 1992; Vol 42, pp 299–325.
96. Viladot, P. J. L.; Delgado, G. R.; Fernandez, B. A. Lipid Nanoparticle Capsules. European Patent 2549977A2, January 2013.
97. Xu, Z. P.; Zeng, Q. H.; Lu, G. Q.; Yu, A. B. Inorganic Nanoparticles as Carriers for Efficient Cellular Delivery. *Chem. Eng. Sci.* **2006,** *61*, 1027–1040.
98. http://www.nanowerk.com/nanotechnology-in-cosmetics.php (last accessed on January 2016).
99. Gergely, A.; Coroyannakis, L. Nanotechnology in the EU Cosmetics Regulation. *Household and Personal Care Today*, 2009.
100. *Nanocyclic Cleanser Pink.* http://www.nanocyclic.com/Product (last accessed on January 2016).
101. Nohynek, G. J.; et al. Grey Goo on the Skin? Nanotechnology, Cosmetic and Sunscreen Safety. *Crit. Rev. Toxicol.* **2007,** *15*, 215–222.
102. http://www.fda.gov/Cosmetics/GuidanceRegulation/GuidanceDocuments/ucm300886. htm (last accessed on January 2016).
103. Berger, M. *Nanotechnology in Cosmetics—2000 Years Ago...?* http://www.nanowerk. com/spotlight/spotid=791.php (last accessed on January 2016).
104. Oberdorster, G.; Maynard, A.; Donaldson, K.; Castranova, V.; Fitzpatrick, J.; Ausman, K.; Carter, J.; Karn, B.; Kreyling, W.; Lai, D.; Olin, S.; Monteiro-Riviere, N.; Warheit, D.; Yang, H. Principles for Characterizing the Potential Human Health Effects from Exposure to Nanomaterials: Elements of a Screening Strategy. *Part. Fibre Toxicol.* **2005,** *2*, 8.
105. http://www.shb.com/newsevents/2011/NanoCosmeticsBeyondSkinDeep.pdf (last accessed on January 2016).

106. Chaudhry, Q.; Bouwmeester, H.; Hertel, R. F. The Current Risk Assessment Paradigm in Relation to Regulation of Nanotechnologies. In *International Handbook on Regulating Nanotechnologies*; Hodge, G. A., Bowman, D. M., Maynard, A. D., Eds.; Edward Elgar: Cheltenham, 2010; pp 124–143.

107. Nano Science Institute. *Scientific Committee Rules on the Safety of Nanocosmetics.* http://www.nanoscienceinstitute.com/NanoCosmetics.html (last accessed on January 2016).

108. Butz, T. Dermal Penetration of Nanoparticles: What We Know and What We Don't. Cosmetic. Science Conference Proceedings, Munich. *SÖFW J.* **2009,** *135* (4), 8–10.

CHAPTER 14

A DENSITY-FUNCTIONAL STUDY OF Ag-DOPED Cu NANOALLOY CLUSTERS

PRABHAT RANJAN[1], TANMOY CHAKRABORTY[2*], and AJAY KUMAR[1]

[1]Department of Mechatronics, Manipal University, Jaipur 303007, Rajasthan, India

[2]Department of Chemistry, Manipal University, Jaipur 303007, Rajasthan, India

*Corresponding author. E-mail: tanmoy.chakraborty@jaipur.manipal.edu; tanmoychem@gmail.com

CONTENTS

ABSTRACT

The electronic and optical properties of bimetallic Cu_nAg ($n = 1$–8) nanoalloy clusters are investigated using conceptual density functional theory. Geometry optimization of all the clusters are performed with exchange–correlation functional local spin density approximation and basis set LanL2DZ. The computed result reveals that Cu_3Ag cluster has highest HOMO (highest occupied molecular orbital)–LUMO (lowest unoccupied molecular orbital) energy gap in this series. The calculated HOMO–LUMO energy gap shows interesting odd–even oscillation behavior, indicating the clusters with total even number of atoms to be more stable as compared to their neighbor clusters. We have also computed conceptual DFT-based descriptors like electronegativity, hardness, softness, electrophilicity index, and dipole moment of copper–silver clusters. The high value of linear correlation coefficient between HOMO–LUMO energy gap and DFT-based descriptors validates our predicted model. A close agreement between experimental and computed parameters is also reflected from our analysis.

14.1 INTRODUCTION

Since last decade, nanomaterials and nanotechnology have given a new dimension in the research of science and engineering.[1] The combination of two or more atoms from the same group or different group of periodic table at the nanoscale level results in structure called nanoalloy clusters. Due to presence of a large number of quantum mechanical and electronic effects, nanoalloy clusters possess unique electronic and optical properties, which are important for potential technological applications.[2-4] The nanoalloy is classified in terms of size, which have at least one dimension in the range of 1–100 nm. That particular size range exists between the levels of atomic/molecular and bulk material.[5] But, there are still some instances of nonlinear transition of certain physicochemical properties, which may vary not only with respect to size but with composition also.[6,7] There are a number of reports available in which change in electronic, optical, and other physical and chemical properties of clusters with respect to change in size and composition are well described.[1-7] A deep insight into the study of nanoalloy clusters with well-defined size and structure may lead to some other alternatives, which may be used as building blocks for nanodevices.[7,8] The nanoalloy clusters have got immense importance due to its wide range of potential technological and industrial applications especially in the area of biological labeling, solar cells, catalysis, optics, optoelectronics, solid-state physics, nanoscience, etc.[1,3,7–11]

Bimetallic nanoalloy clusters have been investigated from both theoretical and experimental researchers, due to its unique electronic, magnetic, optical, and catalysis properties.[12–18] Jiang et al.[19] have studied geometries, charge distributions, stability, and electronic properties of neutral and cationic copper–silver clusters by DFT method. They reported that silver atoms favor the peripheral position in the clusters. During geometry optimization, they observed that neutral CuAg clusters are planar in nature for larger systems as compare to cationic clusters. Heard et al.[20] have investigated the geometry and electronic properties of eight atoms CuAg and CuAu bimetallic clusters using DFT method. They reported that all the clusters except Au_8 and $CuAu_7$ form 3D geometries. Li et al.[21] have studied geometries and electronic properties of seven atoms CuAg clusters using DFT methods. They have also investigated the optical absorption, Raman spectra, and vibrational spectra of these clusters. The result reveals that intensities of maximum peak of Raman spectra decreasing with increasing Cu atoms, however, intensities of vibrational spectra increases with increasing Cu atoms. Ma et al.[22] have reported electronic and optical properties of Ag_{13}, Cu_{13}, and $Ag_{12}Cu$ clusters using DFT technique. They reported that doping of Cu atom in cluster $Ag_{12}Cu$ makes more electronically active than pure Cu_{13} and Ag_{13} clusters. They have also investigated absorption spectra and density of states of $Ag_{12}Cu$, Cu_{13}, and Ag_{13} clusters.

The detailed study of ground state configurations of nanoalloy clusters and their electronic and optical properties is very much important because its physicochemical properties can be altered through proper control of structure, size, and composition.[23] Group 11 elements of periodic table especially metallic clusters like copper, silver, and gold have filled inner d-orbitals and one unpaired electron in the valence shell.[23–27] The same electronic configuration has been observed experimentally in the case of alkali metal clusters.[28–30] Among the nanoalloy clusters of Group 11 elements, the compound formed between copper and silver is of considerable interest due to its potential technological and industrial applications. The exact location of Ag atoms in the copper–silver core–shell structure is highly important. The position of silver in the cluster has a supervisory effect on the optical properties of such particles as because optical properties are governed by surface plasmon resonance of silver, which is also dependent on its structure.[31] Jankowiak et al.[32] and Piccimin et al.[33] have investigated that bimetallic CuAg nanoalloy clusters, as catalysts may be used to enhance the performance and selectivity. Though a number of computational studies have been performed on copper–silver nanoalloy clusters, a theoretical analysis invoking conceptual density functional theory (CDFT)-based descriptors is still unexplored.

Since last decade density functional theory (DFT) has been dominant method for quantum mechanical computation of periodic systems. Due to its computational friendly behavior, DFT is very much popular to study the many body systems.[8] Super conductivity of metal-based alloys,[34] magnetic properties of nanoalloy clusters,[35,36] quantum fluid dynamics,[37] molecular dynamics,[38] and nuclear physics[39,40] can be extensively studied by DFT methodology. Recently, we have established the importance of CDFT-based descriptors in the domain of nanoalloy clusters.[41–47] The study of DFT has been broadly classified into three subcategories, namely, theoretical, conceptual, and computational.[48–51] The CDFT is highlighted following Parr's dictum "Accurate calculation is not synonymous with useful interpretation. To calculate a molecule is not to understand it."[52]

In this chapter, we have investigated bimetallic Cu_nAg ($n = 1–8$) nanoalloy clusters by using CDFT-based descriptors, namely HOMO–LUMO gap, electronegativity, hardness, softness, electrophilicity index, and dipole moment. An attempt has been made to correlate HOMO–LUMO energy gap of Cu–Ag clusters with their computational counterparts.

14.2 COMPUTATIONAL DETAILS

In this study, we have made an analysis on the bimetallic Cu_nAg ($n = 1–8$) nanoalloy clusters using DFT. Modeling and geometry optimization of all the compounds have been performed using Gaussian 03 software package[53] within theoretical framework of DFT. For geometry optimization, local spin density approximation exchange correlation[54] with basis set LanL2DZ[55] has been adopted. The used computation methodology in this chapter is based on the molecular orbital approach, using linear combination of atomic orbitals. The quadrupole moment of molecule is calculated in terms of analytical integration methodology.

Invoking Koopmans' approximation,[56] we have calculated ionization energy (I) and electron affinity (A) of all the nanoalloys using the following equations:

$$I = -\varepsilon_{HOMO} \tag{14.1}$$

$$A = -\varepsilon_{LUMO} \tag{14.2}$$

Thereafter, using I and A, the conceptual DFT-based descriptors, namely, electronegativity (χ), global hardness (η), molecular softness (S), and electrophilicity index (ω) have been computed. The equations used for such calculations are as given under

$$\chi = -\mu = \frac{I + A}{2} \tag{14.3}$$

where μ represents the chemical potential of the system.

$$\eta = \frac{I - A}{2} \tag{14.4}$$

$$S = \frac{1}{2\eta} \tag{14.5}$$

$$\omega = \frac{\mu^2}{2\eta} \tag{14.6}$$

14.3   RESULTS AND DISCUSSIONS

The bimetallic cluster formed between copper–silver, Cu_nAg ($n = 1$–8) has been studied using DFT. The orbital energies in form of HOMO (highest occupied molecular orbital)–LUMO (lowest unoccupied molecular orbital) gap along with calculated DFT-based descriptors namely electronegativity, hardness, softness, and electrophilicity index and dipole moment have been reported in Table 14.1. We have also computed the quadrupole moment of the same nanoalloy clusters along with different axes, which are represented in Table 14.2. Computed data from Table 14.1 reveal that HOMO–LUMO gaps of the Cu_nAg nanoalloy clusters are maintaining direct relationship with their computed hardness values. This is expected from experimental point of view also, as the frontier orbital energy gap increases, their hardness value increases. As the molecule exhibits highest HOMO–LUMO energy gap, it will be more stable and less reactive. From Table 14.1, it is clear that Cu_5Ag has maximum HOMO–LUMO energy gap, whereas Cu_2Ag has least energy gap. It has been already established by Parr et al.[57] that difference between ionization potential and electron affinity can be used to correlate the optical properties of materials. It is based on the fact that optical properties of materials are interrelated with flow of electrons within the systems which in turn depend on the difference between the distance of HOMO and

LUMO. On that basis, we may conclude that optical properties of instant bimetallic nanoalloy clusters increase with increase of their hardness values. Similarly softness data exhibit an inverse relationship toward the experimental optical properties. The computed dipole moment and electrophilicity index also have an inverse relationship and electronegativity have direct relationship with the HOMO–LUMO energy gap of $Cu_n Ag$ nanoalloy clusters. The linear correlation between HOMO–LUMO energy gap along with their computed softness is lucidly plotted in Figure 14.1.

TABLE 14.1   Computed DFT-Based Descriptors of $Cu_n Ag$; $n = 1$–8 Nanoalloy Clusters.

Species name	HUMO–LUMO gap (eV)	χ (eV)	η (eV)	S (eV)	ω (eV)	Dipole moment (D)
CuAg	1.986	4.558	0.933	0.503	10.458	0.234
Cu_2Ag	0.245	4.122	0.122	4.083	69.392	0.998
Cu_3Ag	1.170	4.503	0.585	0.855	17.332	0.255
Cu_4Ag	0.980	3.864	0.490	1.021	15.241	0.000
Cu_5Ag	2.258	4.530	1.129	0.443	9.088	0.279
Cu_6Ag	1.633	4.136	0.816	0.613	10.478	0.133
Cu_7Ag	1.687	4.598	0.844	0.593	12.535	0.351
Cu_8Ag	1.366	3.181	0.683	0.732	7.407	0.362

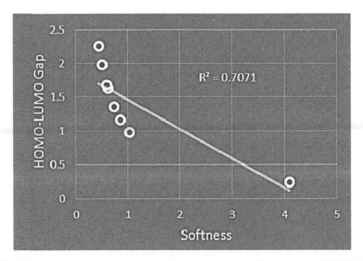

FIGURE 14.1   A linear correlation plot between global softness versus HOMO–LUMO gap.

The quadrupole moment according to Buckingham convention have been also reported in Table 14.2. The quadrupole moment values in different axes are represented in atomic unit (a.u.).

TABLE 14.2 Computed Quadrupole Moment in a.u. of Cu_nAg; $n = 1$–8 Nanoalloy Clusters.

Species name	Quad—xx	Quad—xy	Quad—xz	Quad—yy	Quad—yz	Quad—zz
CuAg	−34.124	0.000	0.000	−34.124	0.000	−26.398
Cu_2Ag	−41.124	2.348	−0.025	−49.389	0.000	−47.638
Cu_3Ag	−60.913	0.000	0.000	−59.944	0.000	−54.630
Cu_4Ag	−74.539	0.000	0.000	−69.551	0.000	−63.661
Cu_5Ag	−87.749	0.000	0.000	−78.037	0.000	−78.581
Cu_6Ag	−95.746	0.002	−0.003	−95.754	−0.007	−103.028
Cu_7Ag	−114.810	0.000	0.000	−99.146	0.000	−100.410
Cu_8Ag	−124.127	−0.036	0.000	−123.947	0.000	−128.697

As per the cluster physics, dissociation energy and second difference of total energy have a marked influence on the relative stability of a particular system. These two energies are highly sensitive quantities and they exhibit pronounced odd–even oscillation behavior for neutral and charged clusters, as a function of cluster size. The similar type of odd–even oscillation behavior is also exhibited by the HOMO–LUMO gap of any particular compound. It has already been reported that cluster with an even number of total atoms possess large HOMO–LUMO energy gap as compared to the clusters with an odd number of total atoms.[58–61] The stability of the even number electronic cluster is actually an outcome of their closed electronic configuration which always produces extra stability. We have reported HOMO–LUMO energy gap as a function of cluster size in Figure 14.2, which shows odd–even oscillation behavior of Cu_nAg nanoalloy clusters.

A comparative analysis has been made between experimental bond length[62–64] and our computed data of the species namely Ag_2, Cu_2, and CuAg. The same is reported in Table 14.3. A close agreement between experimental report and our computed bond length is reflected from Table 14.3. It supports and validates our analysis.

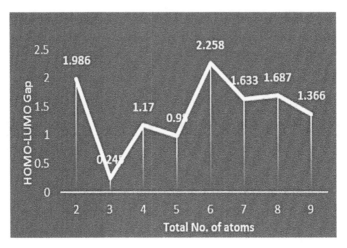

FIGURE 14.2   The size dependence of HOMO–LUMO gap of Cu_nAg ($n = 1$–8) nanoalloy clusters.

TABLE 14.3   The Calculated Bond Length (Å) of Ag_2, Cu_2, and CuAg Species.

Species	Theoretical bond length	Experimental bond length
Ag_2	2.51	2.53[62]
Cu_2	2.21	2.22[63]
CuAg	2.33	2.37[64]

14.4 CONCLUSION

Study of bimetallic nanoalloy clusters is of considerable interest due to its potential technological and industrial applications. In this chapter, we have investigated the cluster formed between copper and silver, the system of Cu_nAg ($n = 1$–8); nanoalloy clusters in terms of CDFT-based descriptors namely HOMO–LUMO energy gap, electronegativity, hardness, softness, electrophilicity index, and dipole moment. Maximum HOMO–LUMO energy gap has been observed in the case of Cu_5Ag cluster, whereas Cu_2Ag has the least energy gap. In this analysis, it is observed that the computed HOMO–LUMO gap runs hand in hand along with the hardness of the clusters. As in absence of any quantitative benchmark, the optical property of Cu_nAg nanoalloy cluster has been assumed to be exactly equivalence of its HOMO–LUMO energy gap. Here, our evaluated data reveal that optical property of these compounds maintains a direct relationship with hardness and inverse relationship with softness. The value of linear regression coefficient between softness and HOMO–LUMO energy gap successfully supports our predicted model. The computed quadrupole moment data of above-mentioned nanoalloy clusters also exhibit the quadrupole charge separation nicely. The calculated bond lengths for the species Cu_2, Ag_2, and CuAg are in agreement with the experimental data. Odd–even oscillation behavior of HOMO–LUMO energy gap as a function of cluster size is also reflected from this study.

KEYWORDS

- **density functional theory**
- **bimetallic nanoalloy**
- **HOMO–LUMO**
- **softness**
- **electrophilicity index**

REFERENCES

1. Zabet-Khosousi, A.; Dhirani, A.-A. Charge Transport in Nanoparticle Assemblies. *Chem. Rev.* **2008,** *108*, 4072–4124.
2. Daniel, M. C.; Astruc, D. Gold Nanoparticles: Assembly, Supramolecular Chemistry, Quantum-Size-Related Properties, and Applications toward Biology, Catalysis, and Nanotechnology. *Chem. Rev.* **2004,** *104*, 293–346.
3. Ghosh, S. K.; Pal, T. Interparticle Coupling Effect on the Surface Plasmon Resonance of Gold Nanoparticles: From Theory to Applications. *Chem. Rev.* **2007,** *107*, 4797–4862.
4. Ghosh Chaudhuri, R.; Paria, S. Core/Shell Nanoparticles: Classes, Properties, Synthesis Mechanisms, Characterization, and Applications. *Chem. Rev.* **2012,** *112*, 2373–2433.
5. Alivisatos, A. P. Semiconductor Clusters, Nanocrystals, and Quantum Dots. *Sci.: New Ser.* **1996,** *271*, 933–937.
6. Kastner, M. A. Artificial Atoms. *Phys. Today* **1993,** *46*, 24–31.
7. Molayem, M.; Grigoryan, V. G.; Springborg, M. Global Minimum Structures and Magic Clusters of Cu_mAg_n Nanoalloys. *J. Phys. Chem. C* **2011,** *115*, 22148–22162.
8. Ismail, R. Theoretical Studies of Free and Supported Nanoalloy Clusters. Ph.D. Thesis, 2012; pp 20–38.
9. Roucoux, A.; Schulz, J.; Patin, H. Reduced Transition Metal Colloids: A Novel Family of Reusable Catalysts? *Chem. Rev.* **2002,** *102*, 3757–3778.
10. Munoz-Flores, B. M.; Kharisov, B. I.; Jimenez-Perez, V. M.; Elizondo Martinez, P.; Lopez, S. T. Recent Advances in the Synthesis and Main Applications of Metallic Nano-alloys. *Ind. Eng. Chem. Res.* **2011,** *50*, 7705–7721.

11. Murray, R. W. Nanoelectrochemistry: Metal Nanoparticles, Nanoelectrodes, and Nano-pores. *Chem. Rev.* **2008,** 108, 2688–2720.
12. Teng, X.; Wang, Q.; Liu, P.; Han, W.; Frenkel, A. I.; Wen; Marinkovic, N.; Hanson, J. C.; Rodriguez, J. A. Formation of Pd/Au Nanostructures from Pd Nanowires via Galvanic Replacement reaction. *J. Am. Chem. Soc.* **2008,** *130*, 1093–1101.
13. Ferrando, R.; Jellinek, J.; Johnston, R. L. Nanoalloys: From Theory to Applications of Alloy Clusters and Nanoparticles. *Chem. Rev.* **2008,** *108*, 845–910.
14. Henglein, A. Physicochemical Properties of Small Metal Particles in Solution: "Micro-electrode" Reactions, Chemisorption, Composite Metal Particles, and the Atom-to-Metal Transition. *J. Phys. Chem.* **1993,** *97*, 5457–5471.
15. Davis, S. C.; Klabunde, K. J. Unsupported Small Metal Particles: Preparation, Reactivity, and Characterization. *Chem. Rev.* **1982,** *82*, 153–208.
16. Lewis, L. N. Chemical Catalysis by Colloids and Clusters. *Chem. Rev.* **1993,** *93*, 2693–2730.
17. Schmid, G. Large Clusters and Colloids. Metals in the Embryonic State. *Chem. Rev.* **1992,** *92*, 1709–1727.
18. Schon, G.; Simon, U. A Fascinating New Field in Colloid Science: Small Ligand-Stabilized Metal Clusters and Possible Application in Microelectronics. *Colloids Polym. Sci.* **1995,** *273*, 101–117.
19. Jiang, Z. Y.; Lee, K. H.; Li, S. T.; Chu, S. Y. Structures and Charge Distributions of Cationic and Neutral Cu_{n-1}Ag Clusters ($n = 2$–8). *Phys. Rev. B* **2006,** *73*, 235423.
20. Heard, C. J.; Johnston, R. L. A Density Functional Global Optimization Study of Neutral 8-Atom Cu–Ag and Cu–Au Clusters. *Eur. Phys. J. D* **2013,** *67*, 34.
21. Li, W.; Chen, F. Structural, Electronic and Optical Properties of 7-Atom Ag–Cu Nano-clusters from Density Functional Theory. *Eur. Phys. J. D* **2014,** *68*, 91.
22. Ma, W.; Chen, F. Optical and Electronic Properties of Cu Doped Ag Clusters. *J. Alloys Compd.* **2012,** *541*, 79–83.
23. Katakuse, I.; Ichihara, T.; Fujita, Y.; Matsuo, T.; Sakurai, T.; Matsuda, H. Mass Distributions of Copper, Silver and Gold Clusters and Electronic Shell Structure: I. *Int. J. Mass Spectrom. Ion Processes* **1985,** *67*, 229–236.
24. Katakuse, I.; Ichihara, T.; Fujita, Y.; Matsuo, T.; Sakurai, T.; Matsuda, H. Mass Distributions of Negative Cluster Ions of Copper, Silver and Gold: I. *Int. J. Mass Spectrom. Ion Processes* **1986,** *74*, 33–41.
25. de Heer, W. A. The Physics of Simple Metal Clusters: Experimental Aspects and Simple Models. *Rev. Mod. Phys.* **1993,** *65*, 611–676.
26. Gantefor, G.; Gausa, M.; Meiwes-Broer, K.-H.; Lutz, H. O. Photoelectron Spectroscopy of Silver and Palladium Cluster Anions. Electron Delocalization *Versus* Localization. *J. Chem. Soc. Faraday Trans.* **1990,** *86*, 2483–2488.
27. Leopold, D. G.; Ho, J.; Lineberger, W. C. Photoelectron Spectroscopy of Mass-Selected Metal Cluster Anions. I. Cu_n^-, $n = 1$–10. *J. Chem. Phys.* **1987,** *86*, 1715–1726.
28. Lattes, A.; Rico, I.; de Savignac, A.; Ahmad-Zadeh Samii, A. Formamide, a Water Substitute in Micelles and Microemulsions xxx Structural Analysis Using a Diels–Alder Reaction as a Chemical Probe. *Tetrahedron* **1987,** *43*, 1725–1735.
29. Chen, F.; Xu, G. Q.; Hor, T. S. A. Preparation and Assembly of Colloidal Gold Nanoparticles in CTAB-Stabilized Reverse Microemulsion. *Mater. Lett.* **2003,** *57*, 3282–3286.
30. Taleb, A.; Petit, C.; Pileni, M. P. Optical Properties of Self-Assembled 2D and 3D Superlattices of Silver Nanoparticles. *J. Phys. Chem. B* **1998,** *102*, 2214–2220.
31. Langlois, C.; Wang, Z. W.; Pearmain, D.; Ricolleau, C.; Li, Z. Y. HAADF-STEM Imaging of CuAg Core–Shell Nanoparticles. *J. Phys.: Conf. Ser.* **2010,** *241*, 012043–012047.

32. Jankowiak, J. T. A.; Barteau, M. A. Ethylene Epoxidation over Silver and Copper–Silver Bimetallic Catalysts: I. Kinetic and Selectivity. *J. Catal.* **2005**, *236*, 366–378.

33. Piccinin, S.; Zafeiratos, S.; Stampfl, C.; Hansen, T. W.; Hävecker, M.; Teschner, D.; et al. Alloy Catalyst in a Reactive Environment: The example of Ag–Cu Particles for Ethylene Epoxidation. *Phys. Rev. Lett.* **2010**, *104*, 035503-1–035503-4.

34. Wacker, O. J.; Kummel, R.; Gross, E. K. U. Time-Dependent Density-Functional Theory for Superconductors. *Phys. Rev. Lett.* **1994**, *73*, 2915–2918.

35. Illas, F.; Martin, R. L. Magnetic Coupling in Ionic Solids Studied by Density Functional Theory. *J. Chem. Phys.* **1998**, *108*, 2519–2527.

36. Gyorffy, B.; Staunton, J.; Stocks, G. In Fluctuations in Density Functional Theory: Random Metallic Alloys and Itinerant Paramagnets. *Series B* **1995**, *337*, 461–464 [Proceedings of a NATO advanced study institute on Density Functional Theory, August 16–27, 1993, Italy].

37. Kümmel, S.; Brack, M. Quantum Fluid-Dynamics from Density Functional Theory. *Phys. Rev. A* **2001**, *64*, 022506.

38. Car, R.; Parrinello, M. Unified Approach for Molecular Dynamics and Density Functional Theory. *Phys. Rev. Lett.* **1985**, *55*, 2471–2474.

39. Koskinen, M.; Lipas, P.; Manninen, M. Shapes of Light Nuclei and Metallic Clusters. *Nucl. Phys. A* **1995**, *591*, 421–434.

40. Schmid, R. N.; Engel, E.; Dreizler, R. M. Density Functional Approach to Quantum Hydrodynamics: Local Exchange Potential for Nuclear Structure Calculations. *Phys. Rev. C* **1995**, *52*, 164–169.

41. Ranjan, P.; Dhail, S.; Venigalla, S.; Kumar, A.; Ledwani, L.; Chakraborty, T. A Theoretical Analysis of Bi-metallic (Cu–Ag)$_{n=1-7}$ Nano-alloy Clusters Invoking DFT Based Descriptors. *Mater. Sci.—Pol.* **2015**, *33*, 719.

42. Ranjan, P.; Venigalla, S.; Kumar, A.; Chakraborty, T. Theoretical Study of Bi-metallic Ag$_m$Au$_n$ ($m + n = 2$–8) Nano Alloy Clusters in Terms of DFT Methodology. *New Front. Chem.* **2014**, *23*, 111–122.

43. Ranjan, P.; Kumar, A.; Chakraborty, T. Computational Investigation of Ge-Doped Au Nanoalloy Clusters. *IOP Conf. Ser.: Mater. Sci. Eng.* **2016**, *149*, 012172.

44. Ranjan, P.; Kumar, A.; Chakraborty, T. A Theoretical Analysis of Bi-metallic AgAu$_n$ ($n = 1$–7) Nano-alloy Clusters Invoking DFT Based Descriptors. In *Research Methodology in Chemical Sciences: Experimental and Theoretical Approaches*; CRC & Apple Academic Press: Boca Raton, FL, 2016; pp 337–346. ISBN: 9781771881272.

45. Ranjan, P.; Kumar, A.; Chakraborty, T. Computational Study of AuSin ($n = 1$–9) Nano-alloy Clusters Invoking DFT Based Descriptors. *AIP Conf. Proc.* **2016**, *1724*, 020072.

46. Ranjan, P.; Kumar, A.; Chakraborty, T. Theoretical Analysis: Electronic and Optical Properties of Gold–Silicon Nanoalloy Clusters. *Mat. Today Proc.* **2016**, *3*, 1563–1658.

47. Venigalla, S.; Dhail, S.; Ranjan, P.; Jain, S.; Chakraborty, T. Computational Study about Cytotoxicity of Metal Oxide Nanoparticles Invoking Nano-QSAR Technique. *New Front. Chem.* **2014**, *23*, 123–130.

48. Parr, R. G.; Yang, W. Density-Functional Theory of the Electronic Structure of Molecules. *Annu. Rev. Phys. Chem.* **1995**, *46*, 701–728.

49. Kohn, W.; Becke, A. D.; Parr, R. G. Density Functional Theory of Electronic Structure. *J. Phys. Chem.* **1996**, *100*, 12974–12980.

50. Liu, S.; Parr, R. G. Second-Order Density-Functional Description of Molecules and Chemical Changes. *J. Chem. Phys.* **1997**, *106*, 5578–5586.

51. Ziegler, T. Approximate Density Functional Theory as a Practical Tool in Molecular Energetics and Dynamics. *Chem. Rev.* **1991,** *91,* 651–667.
52. Geerlings, P.; De Proft, F. Chemical Reactivity as Described by Quantum Chemical Methods. *Int. J. Mol. Sci.* **2002,** *3,* 276–309.
53. Gaussian 03, Revision C.02; Frisch, M. J.; Trucks, G. W.; Schlegel, H. B.; Scuseria, G. E.; Robb, M. A.; Cheeseman, J. R.; Montgomery, Jr.; J. A.; Vreven, T.; Kudin, K. N.; Burant, J. C.; Millam, J. M.; Iyengar, S. S.; Tomasi, J.; Barone, V.; Mennucci, B.; Cossi, M.; Scalmani, G.; Rega, N.; Petersson, G. A.; Nakatsuji, H.; Hada, M.; Ehara, M.; Toyota, K.; Fukuda, R.; Hasegawa, J.; Ishida, M.; Nakajima, T.; Honda, Y.; Kitao, O.; Nakai, H.; Klene, M.; Li, X.; Knox, J. E.; Hratchian, H. P.; Cross, J. B.; Bakken, V.; Adamo, C.; Jaramillo, J.; Gomperts, R.; Stratmann, R. E.; Yazyev, O.; Austin, A. J.; Cammi, R.; Pomelli, C.; Ochterski, J. W.; Ayala, P. Y.; Morokuma, K.; Voth, G. A.; Salvador, P.; Dannenberg, J. J.; Zakrzewski, V. G.; Dapprich, S.; Daniels, A. D.; Strain, M. C.; Farkas, O.; Malick, D. K.; Rabuck, A. D.; Raghavachari, K.; Foresman, J. B.; Ortiz, J. V.; Cui, Q.; Baboul, A. G.; Clifford, S.; Cioslowski, J.; Stefanov, B. B.; Liu, G.; Liashenko, A.; Piskorz, P.; Komaromi, I.; Martin, R. L.; Fox, D. J.; Keith, T.; Al-Laham, M. A.; Peng, C. Y.; Nanayakkara, A.; Challacombe, M.; Gill, P. M. W.; Johnson, B.; Chen, W.; Wong, M. W.; Gonzalez, C.; Pople, J. A. *Gaussian 03, Revision C.02*; Gaussian, Inc.: Wallingford, CT, 2004.
54. Vosko, S. H.; Wilk, L.; Nusair, M. Accurate Spin-Dependent Electron Liquid Correlation Energies for Local Spin Density Calculations: A Critical Analysis. *Can. J. Phys.* **1980,** *58,* 1200–1211.
55. Wang, H. Q.; Kuang, X. Y.; Li, H. F. Density Functional Study of Structural and Electronic Properties of Bimetallic Copper–Gold Clusters: Comparison with Pure and Doped Gold Clusters. *Phys. Chem. Chem. Phys.* **2010,** *12,* 5156–5165.
56. Parr, R. G.; Yang, W. *Density Functional Theory of Atoms and Molecules*; Oxford University Press: Oxford, 1989.
57. Parr, R. G. Absolute Electronegativity and Hardness Correlated with Molecular Orbital Theory. *Proc. Nat. Acad. Sci. USA* **1986,** *83,* 8440–8441.
58. Ping, D. L.; Yu, K. X.; Peng, S.; Ru, Z. Y.; Fang, L. Y. A Comparative Study of Geometries, Stabilities and Electronic Properties between Bimetallic Ag_nX (X = Au, Cu; n = 1–8) and Pure Silver Clusters. *Chin. Phys. B* **2012,** *21,* 043601–043613.
59. Hakkinen, H.; Landman, U. Gold Clusters (AuN, $2 \leq N \leq 10$) and their Anions. *Phys. Rev.* **2000,** *62,* 2287–2290.
60. Li, X. B.; Wang, H. Y.; Yang, X. D.; Zhu, Z. H.; Tang, Y. J. Size Dependence of the Structures and Energetic and Electronic Properties of Gold Clusters. *J. Chem. Phys.* **2007,** *126,* 084505.
61. Jain, P. K. A DFT-Based Study of the Low-Energy Electronic Structures and Properties of Small Gold Clusters. *Struct. Chem.* **2005,** *16,* 421–426.
62. Beutel, V.; Kramer, H. G.; Bhale, G. L.; Kuhn, M.; Weyers, K.; Demtroder, W. High-Resolution Isotope Selective Laser Spectroscopy of Ag_2 Molecules. *J. Chem. Phys.* **1993,** *98,* 2699–2708.
63. Balbuena, P. B.; Derosa, P. A.; Seminario, J. M. Density Functional Theory of Copper Clusters. *J. Phys. Chem. B* **1999,** *103,* 2830–2840.
64. Bishea, G. A.; Marak, N.; Morse, M. D. Spectroscopic Studies of Jet-Cooled CuAg. *J. Chem. Phys.* **1991,** *95,* 5618–5629.

CHAPTER 15

DESIGN, DEVELOPMENT, AND OPTICAL PROPERTIES OF FUNCTIONAL ACTIVE METHACRYLATE POLYMER/ZnO NANOCOMPOSITES

S. MOHAMMED SAFIULLAH, I. PUGAZHENTHI, and K. ANVER BASHA*

P.G. & Research Department of Chemistry, C. Abdul Hakeem College (Autonomous), Melvisharam, Tamil Nadu 632509, India

Corresponding author. E-mail: kanverbasha@gmail.com

CONTENTS

ABSTRACT

There is a crucial need to design a functionally active polymer hybrid for the protection of material structure that are exposing to harmful ultraviolet (UV) radiation. In this chapter, a poly(pyridine-4-yl-methyl) methacrylate ZnO nanocomposite (PPyMMA/ZnO) was designed and developed by in situ solution polymerization. The ZnO nanoparticles (ZnONPs) were chemically modified by oleic acid (OA-ZnO) and incorporated during the solution polymerization for homogeneous dispersion. The Fourier transform infrared spectroscopy and X-ray diffraction studies confirmed that the obtained nanocomposite is homogeneous with good compatibility between the two counterparts. The morphological change arises owing to the addition of OA-ZnO were observed by using field emission scanning electron microscope and transmission electron microscope. An optical study was carried out to test the optical properties of PPyMMA/ZnO (2%, 5%, and 5%), which reveals that 2% ZnONPs loading exhibits an excellent UV shielding properties.

15.1 INTRODUCTION

Due to the anthropogenic activities, the thinning of the ozone layer has been increased enormously, which leads to permit the dosage of ultraviolet (UV) radiation on the earth's surface.[1,2] Several studies have proven that the UV radiation could cause severely adverse impact on the human beings.[2,3] The exposure of UV radiation on materials may accelerate/cause significant degradation. The UV radiation damage is responsible for the discoloration of dyes and pigments, weathering, yellowing of plastics, loss of gloss, mechanical properties, etc.[4] Therefore, designing and development of UV-shielding materials attracts significant research interests in materials science.[5,6] An appropriate polymer–inorganic nanocomposite can signify a synergetic association between UV-absorbing inorganic materials and polymeric matrix, which is considered to be a superior system for designing bulk, flexible, and UV-shielding coatings that are suitable for practical applications.[6]

Researchers have focused on the preparation of polymeric hybrid coatings by incorporating some wide band gap semiconductor fillers like TiO_2,[7] ZnS,[8] and ZnO[6] into the polymer matrix. Recently, the influence of the copolymer films in the ZnO nanoparticle's (ZnONP) optical properties was studied by Pizarro et al.[9] Zhang et al. fabricated visible light traversing and UV-shielding ZnO quantum dots–PMMA nanocomposite films.[10] For optical applications, transparency of the material is an important point of concern.

And hence, the literature is concerning only about PMMA-based nanocomposites, but the design of material for anticorrosive application needs modification. The PMMA is not an effective choice for the anticorrosive application due to its poor adhesiveness.

The study of functional polymer nanocomposites is a rising interest in material engineers, to design the smart materials for industrial applications. The engineering of polymeric hybrid for the protection of exposed materials from UV radiation is essential; this can be achieved using functionally active polymer with optically induced nanofiller. Thus, there is a demand to design a functionally active polymer hybridized with ZnONPs for UV shielding applications.

The main point of concern is to design a methacrylate polymer with electron rich functional group (pyridine). The pyridine-4-yl-methyl methacrylate is an electron rich monomer that possesses strong electron acceptor character. The benefit of functionalizing a 6-membered heterocyclic moiety in the polymeric sides are (1) the incorporation of pyridine in the polymeric structures may enhance the charge-transfer processes upon exposure to UV radiation and (2) the increased van der waals forces of a 6-membered pyridine moiety imparts better electrostatic forces to the finished polymer that improves the adhesion. And hence, the harmful UV radiation may not influence the polymeric materials to undergo deterioration. In the present work, an attempt has been taken to report the optical properties of an electrically active methacrylate polymer nanocomposite named poly{(pyridine-4-yl-methyl) methacrylate}/ZnO (PPyMMA/ZnO).

15.2 MATERIALS AND METHOD

Zinc acetate and sodium hydroxide were purchased from Sigma Aldrich, India. Pyridine-4-methanol and methacrylic acid were obtained from Merck, India, and distilled under reduced pressure before use. Azobisisobutyronitrile (AIBN, Sigma Aldrich, India) was crystallized from ethanol at 50°C. Tetrahydrofuran (THF) acquired from Merck was dried by sodium metal before use. All the other chemicals were procured from Merck and purified by standard methods.

15.2.1 PREPARATION OF ZNO NANOPARTICLES

2.2 g (25 mmol) of ZnOAc was dissolved in 200 mL of double-distilled water and stirred well for about 20 min. To this, 8 g NaOH in 300-mL double-distilled water was added drop wise under vigorous stirring until the solution became homogeneous. The solution was digested at 70°C for 2 h, filtered, and dried for 1 h at room temperature. The obtained precipitate was calcined for 4 h at 400°C.[11]

15.2.2 *SURFACE MODIFICATION OF ZnO NANOPARTICLES BY OLEIC ACID*

The surface modification of ZnONPs was carried out with the oleic acid.[12] Initially, 50 mL of 2% solution of oleic acid in ethanol was prepared. To this solution, 0.5 g of ZnONPs was added and stirred at 50°C. After 4 h, the contents were centrifuged (1×10^6 rpm) and the grafted ZnONPs (OA-ZnO) were collected, washed with ethanol followed by acetone (5×30 mL) and dried under reduced pressure.

15.2.3 *SYNTHESIS OF PYMMA*

To synthesize PyMMA, an esterification of pyridine-4-methanol with meth-acrylic acid was carried out by utilizing P_2O_5/SiO_2 as a dried solid support without using a solvent.[13] A quantity of 500 mg P_2O_5 was stirred with 1 g of SiO_2 under dry atmosphere at room temperature. To this, a 1:1 ratio of 4-pyrdine methanol and methacrylic acid was added. The reaction mixture was stirred for about 2 h with moisture protection at room temperature. The organic mixture was washed with 2% sodium bicarbonate solution followed by distilled water and dried over sodium sulfate. On evaporating the solvent under reduced pressure, a pure yellow-colored viscous PyMMA liquid was obtained.

15.2.4 *PREPARATION OF PPYMMA/ZNO*

PPyMMA/ZnO was prepared by adding 10 mmol of (pyridine-4-yl)methyl methacrylate and 2, 5, and 10 wt% of OA-ZnO in THF solvent. After soni-cation for 30 min, polymerization was carried out at 60°C using AIBN as radical initiator under N_2 atmosphere. After cooling, white-colored amor-phous powder was obtained by the addition of hexane, it was further repre-cipitated from chloroform to get the pure product (Scheme 15.1a). Same experimental procedure was followed (Scheme 15.1b) to synthesize pure PPyMMA polymer without ZnONPs.

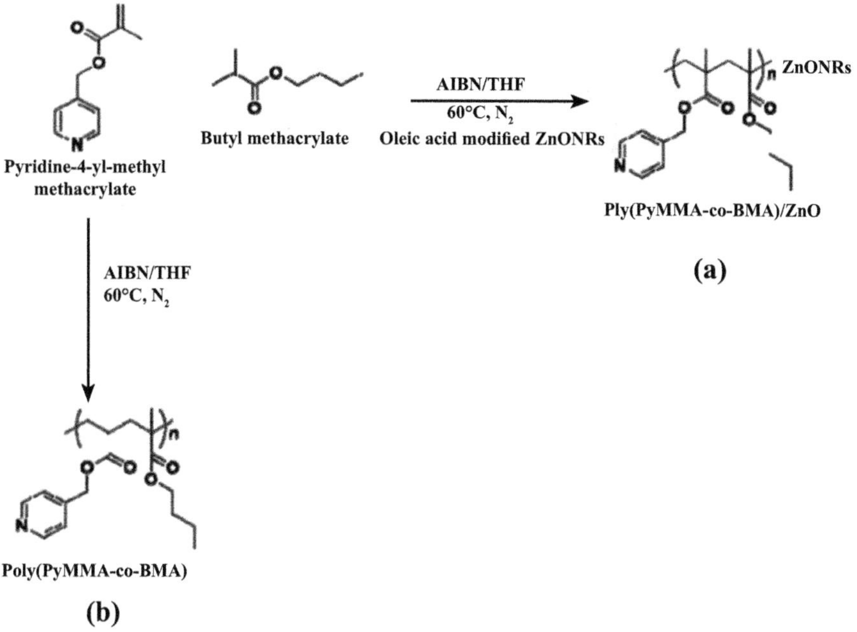

SCHEME 15.1 Synthesis of (a) PPyMMA/ZnO and (b) PPyMMA.

15.2.5 INSTRUMENTATION AND CHARACTERIZATION

The PPyMMA and its ZnO nanocomposites were characterized by Fourier transform infrared spectroscopy (FTIR) spectroscopy (Shimadzu IR Affinity-1S Spectrometer) and X-ray diffraction (XRD) [Bruker-D8 Advanced X-ray diffractometer with Cu $K\alpha$ (1.5418 Å)]. The morphology of the PPyMMA and its ZnO nanocomposites were studied using field emission scanning electron microscope (FESEM) (HITACHI SU6600). Transmission electron microscope (TEM) (FEI-TECNAI G2-20 TWIN 200 kV) images of PPyMMA/ZnO were taken to know the OA-ZnO distribution. The Bruker EDX with LN2 detector was used to identify the chemical compositions of the PPyMMA and its ZnO nanocomposites. The optical studies were carried out for all the samples with the film thickness of 50–60 µm. The UV–visible transmission spectra were recorded on JASCO-V 670 spectrophotometer in the wavelength range 300–800 nm to investigate the optical properties of nanocomposite materials.

15.3   RESULTS AND DISCUSSION

15.3.1   FTIR CHARACTERIZATION

The FTIR spectrum (Fig. 15.1a) of ZnONPs shows a broadband with low intensity at 3232.55 cm^{-1}, which corresponds to the vibration mode of –OH indicates the presence of moisture on the ZnONPs. An appearance of band at 420 cm^{-1} is due to Zn–O bond. Figure 15.1b illustrates the FTIR spectra of OA-grafted ZnONPs. The peaks at 1568 and 1454 cm^{-1} were assigned to symmetric and asymmetric C=O stretching of Zn-oleate, which is not observed in Figure 15.1a. The FTIR spectrum of PyMMA (Fig. 15.1c) displays a band at 1724.3cm^{-1} assigned to –C=O stretching vibration of ester group. Aromatic C=C stretching was found at 1444 cm^{-1}. The C–N stretching was identified at 1388 cm^{-1}. The peaks at 1249 and 1143 cm^{-1} were attributed to the C–O stretching. The C–H out-of-plane bending vibration of the

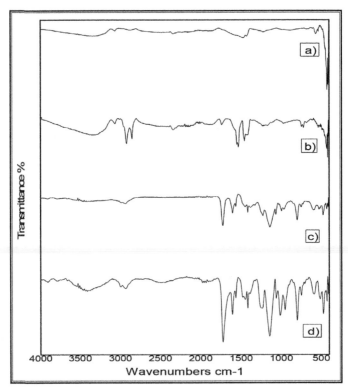

FIGURE 15.1   FTIR spectra of (a) ZnONPs, (b) OA-ZnO, (c) PPyMMA, and (d) PPyMMA/ZnO.

aromatic ring was observed at 796 cm^{-1}. The peak at 590 cm^{-1} was owing to the C–C out-of-plane bending vibration of the aromatic ring. The FTIR spectrum of PPyMMA/ZnO (Fig. 15.1d) shows the same PPyMMA characteristic peaks with an additional band at 498 cm^{-1}, correspond to Zn–O. But, there is an evidence for the decrease in peak intensity due to the addition of OA-ZnO in the PPyMMA. This observation clearly explains about the good compatibility arises between the counterparts (OA-ZnO and PPyMMA), which facilitates the charge transfer between the OA-ZnO and PPyMMA.

15.3.2 XRD STUDIES

The crystallinity of pristine ZnONPs was analyzed by X-ray diffractogram (Fig. 15.2a). The sharp diffraction peaks at $2\theta = 32$ (1 0 0), 34 (0 0 2), 36 (1 0 1), 47 (1 0 2), 56 (1 1 0), and 62 (1 0 3) demonstrate the crystallinity and hexagonal wurtzite structure of ZnONPs, which is in accordance with the standard JCPDS file no. 36-1451. To know the effect of oleic acid treatment on the crystalline size of nanoparticles, the XRD pattern of OA-ZnO was taken (Fig. 15.2b). The diffraction peaks of these two profiles are consistent with typical wurtzite structure of ZnO. The XRD patterns of the pure PPyMMA and PPyMMA/ZnO were recorded to study the effect of OA-ZnO incorporation in the PPyMMA.

The XRD pattern (Fig. 15.2d), showing a broad noncrystalline peak at 10°–30°, confirms the amorphous nature of PPyMMA. The presence of characteristic diffraction peaks of ZnONPs in Fig. 15.2c confirms the incorporation of it. The XRD patterns of PPyMMA/ZnO show the co-existence of broad amorphous peak (10°–30°) of the PPyMMA and crystalline peaks (30°–80°) of ZnONPs.

15.3.3 MORPHOLOGICAL STUDIES

FESEM micrographs of freshly synthesized ZnONPs and OA-ZnO provide the information about the rod-shaped morphology. It is interesting to note that the morphology of OA-ZnO (Fig. 15.3b) is similar to ZnO nanorods (Fig. 15.3a). No appreciable change was observed in the morphology of ZnO nanorods before and after grafting with oleic acid. The FESEM study reveals that the surface of PPyMMA is homogeneous and smooth (Fig. 15.3c). The addition of OA-ZnO in PPyMMA creates a rough and heterogeneous surface (Fig. 15.3d), which indicate that the incorporation of surface modified

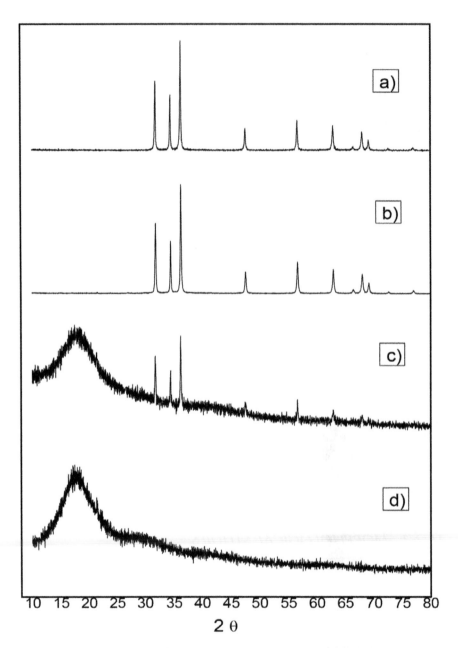

FIGURE 15.2   XRD pattern of (a) ZnONPs, (b) OA-ZnO, (c) PPyMMA/ZnO (2%), and (d) PPyMMA.

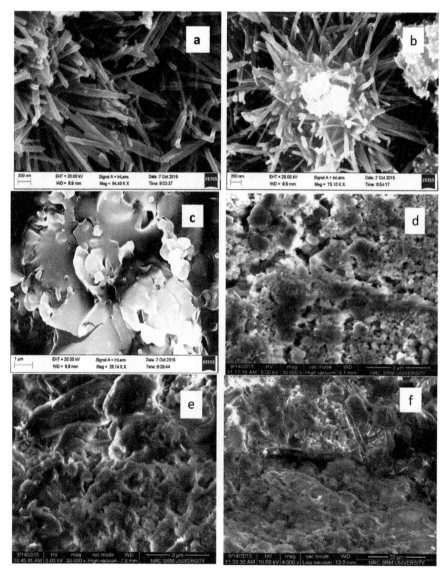

FIGURE 15.3 FESEM images of (a) ZnONPs, (b) OA-ZnO, (c) PPyMMA, and PPyMMA/ZnO, (d) 2%, (e) 5%, and (f) 10%.

nanoparticle had a significant influence on the morphology of polymer. The compatibility between the polymer and OA-ZnO is good, which may be the reason for the significant morphological changes and it was supported by TEM micrographs (Fig. 15.4). The uniform dispersion of well-separated

OA-ZnO nanorods with an average size of 25 nm was observed in Figure 15.4a for the PPyMMA/ZnO (2 wt%), whereas other fractions were also dispersed evenly, but little agglomeration was seldom seen in nanodimensions (Fig. 15.4b and c).

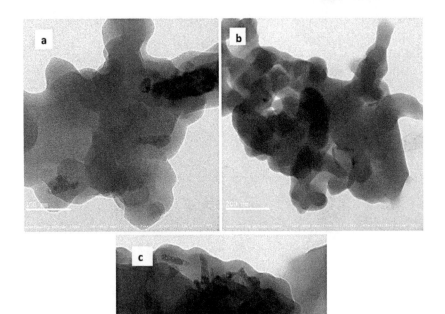

FIGURE 15.4   TEM images of PPyMMA/ZnO (a) 2%, (b) 5%, and (c) 10%.

15.3.4   EDX ANALYSIS

The EDX spectra of OA-ZnO have been shown (Fig. 15.5b). The peaks corresponding to C, Zn, and O indicate the anchoring of OA on the surface of ZnO nanorods. The comparison of EDX spectra (Fig. 15.5c and d) of the polymer and its nanocomposite reveal the peaks corresponding to Zn and O appeared in PPyMMA/ZnO. These results clearly express the successive incorporation of ZnONPs into the PPyMMA matrix.

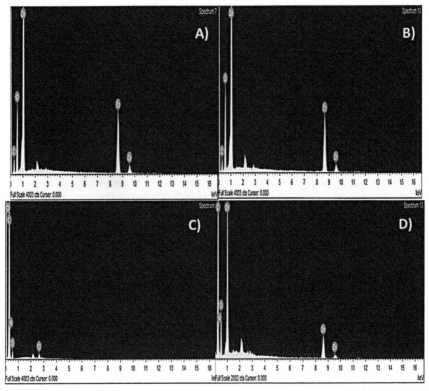

FIGURE 15.5   EDX spectra of (a) ZnONPs, (b) OA-ZnO, (c) PPyMMA, and (d) PPyMMA/ZnO (2%).

15.3.5   OPTICAL PROPERTIES

15.3.5.1   UV ABSORBANCE

To investigate the optical properties, the absorption spectra of the PPyMMA and PPyMMA/ZnO with different percentages (10, 5, 2 wt%) of ZnO loading were taken and shown in Figure 15.6. The absorption spectrum of pure PPyMMA has shown one broad absorbance band at 260 nm, which is assigned to the $\pi \rightarrow \pi^*$ electronic transitions of pyridine moiety. The absorbance spectra of the PPyMMA/ZnO revealed an additional absorbance band at 370 nm corresponding to ZnO along with a fundamental peak of the PPyMMA. A blue shift was observed in ZnO nanocomposite compare to the ZnO bulk material (380 nm) at room temperature by 10 nm,[14] which may be due to the quantum confinement effect, that is, due to the reduction in the crystallite size.[15–17] From Figure 15.6, the absorbance of the nanocomposite increases with the increase of ZnONPs content. This is due to the absorption of incident radiation by free electrons of ZnONPs and get excited to the PPyMMA.[18] It is well known that the pyridine moiety possesses strong electron acceptor character. It is concluded that the PPyMMA/ZnO can improve the UV weatherability of the polymer film and the undercoats. As a result, it can be potentially applied to UV-shielding materials such as fibers and coatings.[19]

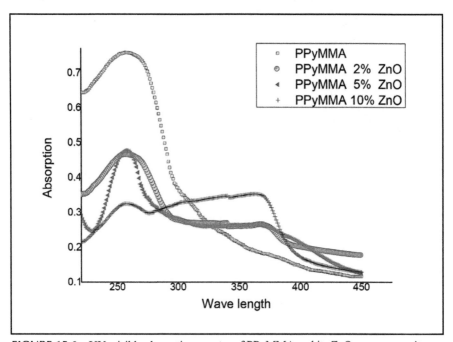

FIGURE 15.6   UV–visible absorption spectra of PPyMMA and its ZnO nanocomposite.

15.3.5.2   UV TRANSMITTANCE

The transmittance spectra of PPyMMA and its ZnO nanocomposite films with different ZnO filler contents at two different wavelengths, that is, at 372 and 550 nm is shown in (Fig. 15.7). Increase in ZnO weight% (0, 2, 5, 10 wt%) decreases the transmittance of the resultant nanocomposite at UV region.[19] Hence, the increase in ZnONPs percentage in the nanocomposite, increases the UV absorption light at 350–400 nm. Among the various fractions of filler, the 2-wt% ZnONPs loading in PPyMMA showed 45% of transmittance in UV region that infers that the composite has strong UV-blocking efficiency. Furthermore, it was well known that absorption bands of ZnONPs were not exist at the visible region 400–800 nm; hence, it is suggested that the loss of transparency in this visible region (550 nm) was caused only by the light scattering.[20,21]

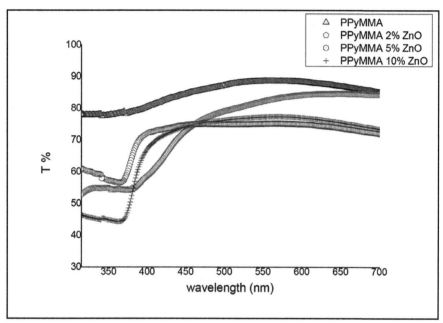

FIGURE 15.7 UV–visible transmission spectra of PPyMMA and its ZnO nanocomposite.

The PPyMMA with 2% ZnONPs has 50% transmittance (>50% more absorption) in UV region and better transparency >82%. The transparency of PPyMMA/ZnO (2%) was found to be close as noticed for bare PPyMMA (87%) at 550 nm in visible region, which suggested that PPyMMA/ZnO (2%) has better optical homogeneity (Fig. 15.4a). The transmittance in the UV wave length range was very low indicating that the prepared film could be used as UV-blocking materials. The observed higher transmittance in the vis–near IR region indicates that the prepared nanocomposites could be used as window layer for solar cells as well.[22]

15.3.5.3 OPTICAL ENERGY GAP

Figure 15.8 demonstrates the relationship between $(ahv)^2$ and photon energy (hv) of PPyMMA and PPyMMA/ZnO nanocomposites.

$$(\alpha h v)^2 = A(h v - E_g)$$

where α is the absorption coefficient, h is the incident photon energy, A is a constant, and E_g is the optical energy gap. The optical energy gap was estimated from the extrapolation of linear portion of the graph to the photon energy x axis. The E_g values of PPyMMA and its ZnO nanocomposites were tabulated in Table 15.1. It is evident from Table 15.1 that the optical energy gap of PPyMMA and PPyMMA/ZnO (2 wt% ZnO) was found to be 4.10 and 4.02 eV, respectively, whereas for the higher weight percentage of ZnO (5 and 10 wt%) were 3.73 and 3.49 eV, respectively. From Table 15.1, it is apparent that the optical energy gap value reduces from 4.10 eV for pure PPyMMA to 3.49 eV for PPyMMA/ZnO nanocomposite. This decrease may be attributed to the formation of chemical bonding between the chains of PPyMMA and OA-ZnO[23] that is the main cause for localized states generation. The low value of PPyMMA/ZnO (10%) indicates that there is a charge transfer between the highest occupied molecular orbital of ZnONPs and the lowest unoccupied molecular orbital energy bands of PPyMMA that makes the lower energy transitions feasible.[23]

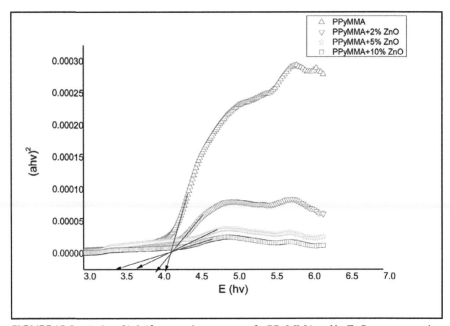

FIGURE 15.8   A plot of $(\alpha h v)^2$ versus photon energy for PPyMMA and its ZnO nanocomposite.

TABLE 15.1 Optical Data of PPyMMA and its ZnO Nanocomposites.

Samples	Optical energy gap (eV)
PPyMMA	4.10
PPyMMA/ZnO (2 wt%)	4.02
PPyMMA/ZnO (5 wt%)	3.73
PPyMMA/ZnO (10wt%)	3.49

15.4 CONCLUSION

In summary, we have synthesized a functional polymer composite made up of poly(pyridine-4-yl-methyl) methacrylate and ZnO nanorods. The incorporation and homogeneous dispersion of ZnO nanorods into PPyMMA matrix was confirmed by using FTIR, XRD, SEM, and TEM. The studies confirm that the surface modification of ZnONPs leads to enhance the compatibility of nanoparticle with polymer. The nanocomposites showed high absorption and low transmittance in UV spectrum due to the behavior of ZnONPs. The optical energy gap of pure polymer has more value than polymer/ZnO nanocomposites. The relation between the optical energy gap with the ZnONP's weight percentage in nanocomposites is inversely proportional. The results showed that polymer/ZnO nanocomposites exhibited a good UV screening effect since their absorption at wavelengths below 350 nm. Hence, the prepared nanocomposite is a promising candidate for UV-shielding coating materials in marine applications is our future concern.

ACKNOWLEDGMENT

The authors would like to thank the Defence Research and Development Organization (DRDO) INDIA, Ref. No. ERIP/ER/1204672/M/01/1525 for funding.

KEYWORDS

- **poly(pyridine-4-yl-methyl) methacrylate**
- **nanocomposites**
- **ZnO nanoparticles**
- **ultraviolet shielding**
- **band gap energy**

REFERENCES

1. Ries, G.; Heller, W.; Puchta, H.; Sandermann, H.; Seidlitz, H. K.; Hohn, B. *Nature* **2000**, *406*, 98–101.
2. Li, S.; Toprak, M. S.; Jo, Y. S.; Dobson, J.; Kim, D. K.; Muhammed, M. *Adv. Mater.* **2007**, *19*, 4347–4352.
3. Eita, M.; Wågberg, L.; Muhammed, M. *ACS Appl. Mater. Interfaces* **2012**, *4*, 2920–2925.
4. Gu, H. *Mater. Des.* **2008**, *29*, 1476–1479.
5. Yan, L.; Chouw, N.; Jayaraman, K. *Mater. Des.* **2015**, *71*, 17–25.
6. Wong, T.-T.; Lau, K.-T.; Tam, W.-Y.; Leng, J.; Etches, J. A. *Mater. Des.* **2014**, *56*, 254–257.
7. Dutta, K.; Manna, S.; De, S. *Synth. Met.* **2009**, *159*, 315–319.
8. Zhang, Y.; Wang, X.; Liu, Y.; Song, S.; Liu, D. *J. Mater. Chem.* **2012**, *22*, 11971–11977.
9. Pizarro, G. C.; Marambio, O. G.; Jeria-Orell, M.; Oyarzún, D. P.; Geckeler, K. E. *Mater. Des.* **2016**, *111*, 513–521.
10. Zhang, Y.; Zhuang, S.; Xu, X.; Hu, J. *Opt. Mater.* **2013**, *36*, 169–172.
11. Akinci, A.; Sen, S.; Sen, U. *Composites B: Eng.* **2014**, *56*, 42–47.
12. Devikala, S.; Kamaraj, P.; Arthanareeswari, M. *Chem. Sci. Trans.* **2013**, *2* (S_1) S129–S134.
13. Sannakki, B.; Nivrtirao, E. *Asian J. Chem.* **2011**, *23*, 5566.
14. Hemalatha, K.; Rukmani, K.; Suriyamurthy, N.; Nagabhushana, B. *Mater. Res. Bull.* **2014**, *51*, 438–446.
15. Tang, E.; Cheng, G.; Pang, X.; Ma, X.; Xing, F. *Colloid Polym. Sci.* **2006**, *284*, 422–428.
16. Fujihara, S.; Naito, H.; Kimura, T. *Thin Solid Films* **2001**, *389*, 227–232.
17. Nyffenegger, R. M.; Craft, B.; Shaaban, M.; Gorer, S.; Erley, G.; Penner, R. M. *Chem. Mater.* **1998**, *10*, 1120–1129.
18. Khan, M.; Chen, M.; Wei, C.; Tao, J.; Huang, N.; Qi, Z.; Li, L. *Appl. Phys. A* **2014**, *117*, 1085–1093.
19. Shanshool, H. M.; Yahaya, M.; Yunus, W. M. M.; Abdullah, I. Y. *J. Mater. Sci.: Mater. Electron.* **2016**, *27*, 9804–9811.
20. Yabe, S.; Yamashita, M.; Momose, S.; Tahira, K.; Yoshida, S.; Li, R.; Yin, S.; Sato, T. *Int. J. Inorg. Mater.* **2001**, *3*, 1003–1008.
21. Demir, M. M.; Koynov, K.; Akbey, Ü.; Bubeck, C.; Park, I.; Lieberwirth, I.; Wegner, G. *Macromolecules* **2007**, *40*, 1089–1100.
22. Sugumaran, S.; Bellan, C. *Optik—Int. J. Light Electron Optics* **2014**, *125*, 5128–5133.
23. Abdelghany, A. M.; Abdelrazek, E. M.; Badr, S. I.; Morsi, M. A. *Mater. Des.* **2016**, *97*, 532–543.

PART V
Selected Topics

CHAPTER 16

SUPERCONDUCTORS, SUPERCONDUCTIVITY, BCS THEORY, AND ENTANGLED PHOTONS FOR QUANTUM COMPUTING

FRANCISCO TORRENS[1*] and GLORIA CASTELLANO[2]

[1]Institut Universitari de Ciència Molecular, Universitat de València, Edifici d'Instituts de Paterna, PO Box 22085, E-46071 València, Spain

[2]Departamento de Ciencias Experimentales y Matemáticas, Facultad de Veterinaria y Ciencias Experimentales, Universidad Católica de Valencia San Vicente Mártir, Guillem de Castro-94, E-46001 València, Spain

*Corresponding author. E-mail: torrens@uv.es

CONTENTS

ABSTRACT

Quantum simulators are controllable quantum systems that can be use to simulate other quantum systems. What basic interaction between the electrons is responsible for the superconducting behavior? What minimum Hamiltonian does describe the phenomenon of high-temperature superconductivity? A quantum simulator checks various Hamiltonian candidates for relevant phases. Such systems can be *superconductors*.

16.1 INTRODUCTION

Cirac and Zoller proposed questions (Qs) and hypothesis (H) on high-temperature (HT) superconductivity.[1]

Q2. What basic interaction between the electrons (e^-) is responsible for SC behavior?

Q3. What minimum Hamiltonian does describe the phenomenon of HT superconductivity?

H12. A quantum simulator checks the various Hamiltonian candidates for relevant phases.

In earlier publications, fractal hybrid-orbital analysis,[2,3] resonance,[4] molecular diversity,[5] periodic table of the elements (PTE),[6,7] law, property, information entropy, molecular classification, simulators,[8–11] and labor risk prevention and preventive healthcare at work with nanomaterials[12,13] were reviewed. In the present report, the aim is to understand *superconductors* (SCs) and entangled photons for quantum computing.

16.2 SUPERCONDUCTORS

The SCs form a special group of materials with a high electric conductivity.[14] The electric resistance of all metals decays monotonously with decaying temperature (cf. Fig. 16.1). In certain metals and alloys, however, the electric resistance drops abruptly almost to zero at a definite critical temperature T_c, that is, the material becomes an SC. Superconductivity is found in 30 elements and 1000 alloys. The SC properties are shown by many alloys with the structure composed of an ordered solid solution and intermediate phases (σ, Laves, etc.). At common temperatures, however, the electric conductivity of these substances is not high.

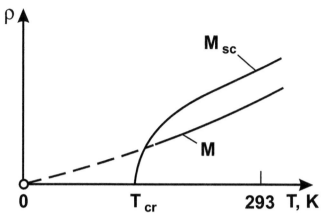

FIGURE 16.1   Changes of metals (*M*)/superconductors (*M*$_{sc}$) electric resistance at cryogenic temperatures.

   The change of a metal to SC state is associated with a phase transformation. A new phase state can be characterized by that free e⁻ cease to interact with ions of the crystal lattice but start to interact with one another. As a result, e⁻ with antiparallel spins are paired. The resultant spin moment of paired e⁻ is equal to zero and an SC changes to a diamagnetic. All e⁻ pairs occupy lower energy levels where they cease to experience thermal scattering, since the energy that an e⁻ pair can obtain on interaction with ions of the crystal lattice is too low to sustain the scattering. The SC state can be destroyed not only by heating but also in a strong magnetic field (MF) or on passage of a high electric current (at certain critical values of MF intensity and current density). Among all elements capable of changing to SC state, Nb presents the highest T_c = 9.17 K. The SC alloys that found application (65/35BT) are characterized by a high content of Nb. Alloy Grade 65BT contains 22–26% Ti, 63–68% Nb, and 8.5–11.5% Zr and presents T_c = 9.7 K. At a temperature T = 4.2 K, the critical current density is 2.8 × 10⁶ A·m⁻² and the critical MF intensity, (6–7.2) × 10⁶ A·m⁻¹. Alloy Grade 35BT (60–64% Ti, 33.5–36.5% Nb, and 1.7–4.3% Zr) features an elevated brittleness, because of which wires made of the alloy are cast in a Cu matrix. Both 65/35BT alloys are employed in windings of powerful generators, high-capacity magnets (e.g., magnetic-cushion trains), and tunnel diodes for electronic computers. The ability of diamagnetic supercomputers to repulse MFs is used in magnetic pumps, which can generate MFs of enormous intensity and cryogenic gyroscopes. If a gyroscope armature is made of an SC material, it floats, as it were, in MF and is mounted without supports and bearings, which eliminates friction and increases largely instrument service life.

16.3　SUPERCONDUCTIVITY

Onnes (1911) examined supercooled metals properties.[15] He cooled He till the temperature in which it becomes liquid at 4.2 K. On bathing metals in $He_{(l)}$, he investigated how their electrical behavior altered. When he placed an Hg test tube in $He_{(l)}$, the metal's electric resistance suddenly dropped. Hg, a liquid at room temperature (RT, 300 K), becomes solid at 4 K, conducting perfectly: its resistance is zero.[16] $Hg_{(s)}$ is an *SC*. Other metals (e.g., Pb, Nb, Rh) were also SCs, although common materials used for cables at RT (e.g., Cu, Ag, Au) are not. Pb becomes SC at 7.2 K, and Nb/Rh presents a characteristic T_c, below which resistance disappears. Electric currents that flow via SCs never stop. Currents can go over a supercooled Pb ring during years, without losing energy. However, at RT, they soon demean themselves. In SCs, resistance is so low that currents could circulate during thousands of millions years without demeaning themselves. Quantum rules hinder them from losing energy: no viable states exist by which they could do it.

16.3.1　EXPLAINING SCs

Bardeen, Cooper, and Shrieffer[17,18] published SC *Bardeen–Cooper–Schrieffer (BCS) theory*, which described how e⁻ movements in an SC material become coordinated in such a way that they act as a system whose behavior is described via wave equations. Metals consist of a cations lattice surrounded by a sea of e⁻, which are free to move via the lattice and produce electric currents. However, they must overcome strengths that refuse to their movement. At RT, atoms are not motionless. They shake around. The e⁻ in movement must avoid the cations that struggle and spread when striking versus them, which collisions cause electric resistance, which stops the current and loses energy. At supercold temperatures, the cations do not shake so much. The e⁻ move at larger distances without bounding on them. The T_c varies versus SC-material atomic mass. If it were because of e⁻ properties, this was not the case, as all e⁻ are the same. Hg-heavy isotopes, for example, present a slightly lower T_c. All metal lattices must be involved: heavy ions move like e⁻. Theory of BCS assumes that e⁻ go hand-in-hand and begins a *dance*. The proper lattice vibrations lend their rhythm to the waltz of e⁻, which form lax couples (*Cooper's pairs*), which movements are linked. The e⁻ are *fermions*, which normally Pauli's exclusion principle hinders to meet in the same quantum state. However, when they are matched, e⁻ behave like *bosons* and inhabit similar states. The set energy decays. An energy *bandgap* above these acts as a damper. At low temperature

(LT), e^- present not enough energy to free themselves and push via the lattice. They avoid the collisions that cause resistance. Theory of BCS predicts that superconductivity breaks if e^- acquire enough energy to jump the forbidden band, which size is a direct function of T_c. In addition to present a zero resistance, SCs show another strange property: they cannot contain an MF, which was discovered by Meissner and Ochsenfeld (Meissmer's effect, 1933). The SC expels MF when creating currents on its surface, which eliminate what there would be inside if it was a normal conductor.

16.3.2   HEATING UP

It began as a race (1960s) to try finding new SC types. Physicists wanted to find SCs with high T_c, which could be more widely used. $He_{(l)}$ is difficult to produce and maintain. $N_{(l)}$, at 77 K, is much easier to handle and produce. Physicists searched for materials that could work at temperatures that could be reached with $N_{(l)}$. The SC materials that function at RT are the last objective, although people are already far from reaching it. It was found that SC alloys (e.g., Nb/Ti, Nb/Sn) superconducted at temperatures somewhat higher (10, 18 K) than the original metals, which were used to produce SC cables, which were used to build potent magnets, which could be used for particles accelerators. Josephson's later prediction led to a series of new devices. He discovered that one could pass current via a sandwich of two SCs separated by a thin insulating layer. Electric energy lays across the sandwich filler via a quantum tunnel, forming a Josephson junction (JJ), which is sensitive enough to measure tiny MFs, a thousand of millions times lesser than Earth's one. Bednorz and Müller (1986) discovered types of ceramics, which could superconduct at 30 K, which supposed a great advance.[19] They were formed by a Ba/La/Cu/O mixture (cuprates), which was unexpected, because ceramics are generally used for insulators at RT (e.g., as insulators in electric towers, substations). In 1987, a ceramic that contains Y instead of La became SC at 90 K, which overcame $N_{(l)}$ limit, which made it economically feasible of being used and opened a route to find other high-T_c SCs. Nowadays, 130 K were overcome, but none is useful at RT.

16.4   BCS THEORY

BCS theory (1957) is the first microscopic theory of superconductivity since its discovery in 1911.[17,18]

16.5 KOSTERLITZ–THOULESS TRANSITION

Berezinskii–Kosterlitz–Thouless (BKT) transition is a phase transition in the two-dimensional (2D) *XY* model. It is a transition from bound vortex–antivortex pairs at LTs to unpaired vortices and antivortices at some T_c. The transition is named after condensed matter physicists Vadim Berezinskii, John M. Kosterlitz, and David J. Thouless. The BKT transitions can be found in several 2D systems in condensed matter physics that are approximated by the *XY* model, for example, Josephson junction arrays and thin disordered SC granular films. More recently, the term has been applied by the 2D SC insulator transition community to the pinning of Cooper pairs in the insulating regime, because of similarities with the original vortex BKT transition.

16.6 BKT TRANSITION IN PROXIMITY-COUPLED SUPERCONDUCTING ARRAYS

Newrock group showed BKT vortex unbinding transition in triangular planar arrays of proximity-coupled Pb/Sn junctions.[20] Resistive-transition temperature dependence and current–voltage (*I–V*) characteristics nonlinear features are reliable with theories of topological ordering in 2D SCs.

16.7 MICRO-LEDS OUTPUT ENTANGLED PHOTONS FOR QUANTUM COMPUTING

Scalability and foundry compatibility (e.g., conventional Si-based integrated computer processors) in developing quantum technologies are major challenges versus research (cf. Fig. 16.2). Pelucchi group introduced a quantum photonic technology with the potential to enable the large-scale fabrication of semiconductor-based, site-controlled, scalable arrays of electrically driven sources of polarization-entangled photons (whose actions and states are linked), which encode quantum information.[21] Sources design is based on quantum dots (QDs) grown in micrometer-sized pyramidal recesses along the crystallographic direction (1 1 1)B, which theoretically ensures high QDs symmetry (requirement for bright entangled-photon emission). A selective electric injection scheme in the nonplanar structures allowed a high density of light-emitting diodes, with some producing entangled photon pairs that violate Bell's inequality. Compatibility with semiconductor fabrication technology, good reproducibility, and lithographic position control make the devices attractive candidates for integrated photonic circuits for quantum information processing.

FIGURE 16.2   A quantum simulator: optical network.

16.8   QUANTUM COMMUNICATIONS AND COMPUTING

Broadband quantum light sources with many frequency modes in a single waveguide showed potential for scalable quantum state generation.[22]

16.9   CONCLUDING REMARK

From the present information, the following concluding remark can be drawn.

Quantum simulation constitutes an exciting field that contains great promises for the future. However, short-term goals should be clearly defined; for example, to find systems that satisfy the above-exposed criteria and, in particular, to demonstrate in the laboratory the simulation of a quantum many-body system, in which a great-scale entanglement (already shown) take part, which could not be represented with classic means. Such systems can be superconductors.

ACKNOWLEDGMENTS

Francisco Torrens belongs to the Institut Universitari de Ciència Molecular, Universitat de València. Gloria Castellano belongs to the Departamento de Ciencias Experimentales y Matemáticas, Facultad de Veterinaria y Ciencias Experimentales, Universidad Católica de Valencia *San Vicente Mártir*. The authors thank support from Generalitat Valenciana (Project No. PROMETEO/2016/094) and Universidad Católica de *Valencia San Vicente Mártir* (Project No. PRUCV/2015/617).

KEYWORDS

- **quantum simulators**
- **quantum systems**
- **Hamiltonian**
- **superconductors**
- **superconductivity**

REFERENCES

1. Cirac, J. I.; Zoller, P. Goals and Opportunities in Quantum Simulation. *Nat. Phys.* **2012**, *8*, 264–266.
2. Torrens, F. Fractals for Hybrid Orbitals in Protein Models. *Complexity Int.* **2001**, *8*, 01-1–13.
3. Torrens, F. Fractal Hybrid-Orbital Analysis of the Protein Tertiary Structure. *Complexity Int.* (in press).
4. Torrens, F.; Castellano, G. Resonance in Interacting Induced-Dipole Polarizing Force Fields: Application to Force-Field Derivatives. *Algorithms* **2009**, *2*, 437–447.
5. Torrens, F.; Castellano, G. Molecular Diversity Classification via Information Theory: A Review. *ICST Trans. Complex Syst.* **2012**, *12* (10–12), e4-1–8.
6. Torrens, F.; Castellano, G. Reflections on the Nature of the Periodic Table of the Elements: Implications in Chemical Education. In *Synthetic Organic Chemistry*; Seijas, J. A., Vázquez Tato, M. P., Lin, S. K., Eds.; MDPI: Basel, Switzerland, 2015; Vol. 18, pp 8-1–15.
7. Putz, M. V., Ed. *The Explicative Dictionary of Nanochemistry*; Apple Academic–CRC: Waretown, NJ (in press).
8. Torrens, F.; Castellano, G. Reflections on the Cultural History of Nanominiaturization and Quantum Simulators (Computers). In *Sensors and Molecular Recognition*;

Laguarda Miró, N., Masot Peris, R., Brun Sánchez, E., Eds.; Universidad Politécnica de Valencia: València, Spain, 2015; Vol. 9, pp 1–7.

9. Torrens, F.; Castellano, G. Ideas in the History of Nano/Miniaturization and (Quantum) Simulators: Feynman, Education and Research Reorientation in Translational Science. In *Synthetic Organic Chemistry*; Seijas, J. A., Vázquez Tato, M. P., Lin, S. K., Eds.; MDPI: Basel, Switzerland, 2016; Vol. 19, pp 1–16.

10. Torrens, F.; Castellano, G. Nanominiaturization and Quantum Computing. In *Sensors and Molecular Recognition*; Costero Nieto, A. M., Parra Álvarez, M., Gaviña Costero, P., Gil Grau, S., Eds.; Universitat de València: València, Spain, 2016; Vol. 10, pp 31-1-5.

11. Torrens, F.; Castellano, G. Nanominiaturization, Classical/Quantum Computers/Simulators, Superconductivity and Universe. In *Methodologies and Applications for Analytical and Physical Chemistry*; Haghi, A. K., Thomas, S., Palit, S., Main, P., Eds.; Apple Academic–CRC: Waretown, NJ (in press).

12. Torrens, F.; Castellano, G. *Book of Abstracts, Certamen Integral de la Prevención y el Bienestar Laboral*, València, Spain, September 28–29, 2016; Generalitat Valenciana–INVASSAT: València, Spain, 2016; p 3.

13. Torrens, F.; Castellano, G. Nanoscience: From a Two-dimensional to a Three-dimensional Periodic Table of the Elements. In *Methodologies and Applications for Analytical and Physical Chemistry*; Haghi, A. K., Thomas, S., Palit, S., Main, P., Eds.; Apple Academic–CRC: Waretown, NJ (in press).

14. Arzamasov, B., Ed. *Materials Science*; MIR: Moscow, 1989.

15. Onnes, H. K. Further experiments with liquid helium. C. On the change of electric resistance of pure metals at very low temperatures, *etc*. IV. The resistance of pure mercury at helium temperatures. *Comm. Phys. Lab. Univ. Leiden* **1911,** *1911*(120b), 1–1.

16. Baker, J. *50 Quantum Physics Ideas You Really Need to Know*; Quercus: London, UK, 2013.

17. Bardeen, J.; Cooper, L. N.; Schrieffer, J. R. Microscopic Theory of Superconductivity. *Phys. Rev.* **1957,** *106*, 162–164.

18. Bardeen, J.; Cooper, L. N.; Schrieffer, J. R. Theory of Superconductivity. *Phys. Rev.* **1957,** *108*, 1175–1204.

19. Bednorz, J. G.; Müller, K. A. Possible High-T_c Superconductivity in the Ba−La−Cu−O System. *Z. Phys. B: Condens. Matter* **1986,** *64*, 189–193.

20. Resnick, D. J.; Garland, J. C.; Boyd, J. T.; Shoemaker, S.; Newrock, R. S. Kosterlitz-Thouless Transition in Proximity-Coupled Superconducting Arrays. *Phys. Rev. Lett.* **1981,** *47*, 1542 -1545.

21. Chung, T. H.; Juska, G.; Moroni, S. T.; Pescaglini, A.; Gocalinska, A.; Pelucchi, E. Selective Carrier Injection Into Patterned Arrays of Pyramidal Quantum Dots for Entangled Photon Light-Emitting Diodes. *Nat. Photon.* **2016,** *10*, 782–787.

22. Kues, M.; Reimer, C.; Roztocki, P.; Moss, D.; Morandotti, R. Quantum Communications and Computing. *Photon. Spectra* **2017,** *51* (1), 76–82.

CHAPTER 17

CUPRATE-BASED HIGH-TEMPERATURE SUPERCONDUCTOR

SHILNA K. V. and SWAPNA S. NAIR*

Department of Physics, School of Mathematical and Physical Science, Central University of Kerala, Kasaragod, Kerala 671314, India

Corresponding author. E-mail: swapnasharp@gmail.com

CONTENTS

ABSTRACT

High-temperature superconductivity in cuprates is one of the emanant phenomena in strongly correlated electron systems. Focus on cuprates sharply increased with the remarkable breakthrough of the 1986 discovery of lanthanum barium copper oxide. After that many cuprates were identified, among these, bismuth strontium calcium copper oxide are the "kings" of superconductors. It is an important candidate due to absence of toxic elements like Hg, Tl, and Pb. These materials become superconducting at the temperature of liquid nitrogen which can be produced relatively cheaply.

Even after 20 years of their discovery, their peculiar properties continued to surprise the physicists and forced them to develop new theoretical tools and innovative experiments to understand their fundamental properties better. This chapter will present recent reports on cuprates including their salient features and applications and also more about a cuprate-based high-temperature superconductor.

17.1 INTRODUCTION

Superconductivity is a quantum mechanical phenomenon whereby certain materials loses its electrical resistance and expulse magnetic fields when cooled below a particular temperature, known as critical temperature T_c. In 1908, H. Kamerlingh Onnes initiated the field of low-temperature physics by liquefying helium, after 3 years, he discovered the phenomenon of superconductivity for helium. In 1911, he found that below 4.15 K, the DC resistance of mercury dropped to zero and also observed superfluid transition of helium at 2.2 K. In subsequent decades, other superconducting materials were discovered. In 1913, the element lead was found to superconduct at 7 K. Later, Niobium was discovered to have a transition temperature of 9.2 K. Henceforth, great efforts have been devoted by physicists for the study of superconductivity. In 1933, Meissner and Ochenfeld discovered that when a sphere is cooled below its transition temperature in a magnetic field it excludes the magnetic flux. This phenomenon later came to be known as the Meissner effect.

The basic properties of superconductors:

A superconductor has several main macroscopic characteristics such as

- zero resistance,
- critical field and critical current density,

- the Meissner effect,
- the Josephson effect,
- the isotope effect,
- anomalous specific heat capacity, and
- abnormal infrared electromagnetic absorption.

Bardeen, Cooper, and Schrieffer developed an effective microscopic theory of superconductivity called as the BCS theory in 1957. In condensed-matter physics, BCS theory is applicable for conventional superconductors where electron–phonon interaction occurs and inapplicable in unconventional superconductors, where electron–electron interaction plays an important role in electron pairing.

During the mid-1980s, a new research area "High-temperature superconductivity" developed.[1,2] high temperature superconductors (HTSCs) represent a new class of materials having extraordinary superconducting and magnetic properties and great potential for technological, scientific as well as industrial applications such as loss-less transmission lines, power generation, computers with reduced size and power consumption, superconducting magnets, levitating trains to medical diagnostics.[3–5] HTSCs have a critical temperature above 30 K with a high critical temperature, higher critical magnetic field, and current density. The first HTSC was discovered in 1986, by Karl Muller and Johannes Bednorz, and was subsequently awarded the Nobel Prize in Physics. Until the discovery of Fe-based superconductors in 2008, the term HTSCs was used interchangeably with cuprate superconductor. The best known HTSCs are the compounds of copper and oxygen such as

- bismuth strontium calcium copper oxide (BSCCO);
- yttrium barium copper oxide (YBCO);
- thallium barium calcium copper oxide (TBCCO); and
- mercury barium calcium copper oxide (HBCCO).

After the discovery of yttrium-based cuprate with T_c = 93 K, BSCCO system at a transition temperature of above 105 K was also reported.[6] In the same year, TBCCO were found to show superconductivity at 120 K.[7] The critical temperatures of mercury-based cuprates are found to be 133 K.[3,8] Iron-based superconductors are compounds of iron and a pnictogen such as arsenic or phosphorus, which has the second highest T_c behind cuprates. Another family includes carbon-based oxycarbonates having T_c value 117 K.[9]

HTSCs are considered as unconventional superconductors which cannot be explained by BCS theory. The first theoretical description of HTSC was proposed by P. W. Anderson using the resonating valence bond theory. Some other studies are done based on the d-wave pair symmetry and spin-fluctuation theory for these superconductors. The study of high-temperature superconductivity is still in progress. At present, the mechanism of formation of Cooper pairs is unknown. As mentioned earlier, the phenomenon of superconductivity occurs usually at low temperatures, which makes it difficult to apply to real-life situations. Superconductors that have higher critical temperature which is close to room temperature will have immense applications. A superconductor with higher values of critical temperature can obtain zero resistance at a more easily attainable temperature has huge advantage. This can be achieved by improving the quality and composition of the existing materials and or synthesizing new materials.

Due to its unique properties, BSCCO is the widest used cuprate-based HTSC. The discovery of superconductivity in BSCCO system by Maeda et al.[6] and Raveau[10] resulted in the identification of three superconducting phases Bi-2201, Bi-2212, and Bi-2223.[11–14] Here, we describe a Bi-based superconductor having an acceptable T_c which may have potential in the development of an inexpensive and economical HTSC compound.

17.2   CUPRATES

Cuprate superconductors are generally considered to be quasi-two-dimensional materials. The structure of cuprate superconductors adopts oxygen deficient distorted multilayered perovskite structure. This structure is termed as defect perovskite structure.[15] The CuO_2 interplane coupling is very weak, where the superconducting phenomenon varies. The bond type between Cu and O is covalent. Copper is present in a partial oxidized mixed state of Cu^{2+}–Cu^{3+}. More layers of CuO_2 imply higher T_c, since superconductivity takes place between these layers. The current carriers of the HTSC are the holes or electrons doped into the two-dimensional CuO_2 plane. Carrier doping into this Mott insulating state induces high T_c superconductivity.

17.2.1   MATERIALS

Important classes of cuprates are given below:

BSCCO has three superconducting phases, having general formula $Bi_2Sr_2Ca_{n-1}Cu_nO_{2n+4+x}$, they are Bi-2201, Bi-2212, and Bi = 2223 with transition temperatures T_c = 20 K (n = 1, 2201 phase), 85 K (n = 2, 2212 phase), and 110 K (n = 3, 2223 phase). The only difference between two consecutive phases is the addition of a double $Ca-Cu-O_2$ in the unit cell. BSSCO shows more anisotropy than other cuprates; it is fascinating to study the properties. Numerous works are in progress for the synthesis of new superconducting phases by different methods using BSCCO as the parent material.[16-18]

17.3.1 CRYSTAL STRUCTURE

LBCO: Lanthanum barium copper oxide is the first oxide-based HTSC which is an insulating material having T_c equal to 35 K. They shows tetragonal crystal structure with lattice parameters $a = b = 3.78, c = 13.2$.

YBCO: This is the materials having T_c around 90 K and was the first material to break the liquid nitrogen temperature. YBCO systems are most ordered crystals. The unit cell of $YBa_2Cu_3O_7$ consists of three pseudocubic elementary perovskite unit cells.

TBCCO: Generally, TBCCO system shows transition temperature ranging from 85 to 125 K. TBCCO phases such as 2201, 2212, and 2223 exhibits T_c of 85, 110, and 125, respectively.

HBCCO: Till today, HBCCO system is found to demonstrate highest T_c among the members of HTSC family. They shows transition temperature ranging from 94 K (1201 phase), 128 K (1212 phase), and 134 K (1223 phase).

17.3 BSCCO

A new category of superconductors, Bi–Sr–Ca–Cu–O has been considered to be one of the most promising candidates because of the absence of rare earth elements and other toxic elements like Hg, Tl, and Pb. The use of rare earth elements has raised concern, as they are toxic to the environment. Because of the absence of these elements, BSCCO are important category. Transition temperature of the BSCCO system range from 20 to 110 K. Elements used in the manufacture of BSCCO are reportedly of low cost and also can be made effortlessly. Moreover, it has high T_c and stability in superconducting properties with respect to the oxygen loss and fast degradation in air emphasize their relevance, compared to YBCO.

The existence of Bi-based phase depends on the layered structure. Lanthanum, barium, strontium, or other atoms present in the neighboring layers stabilize the structure. The structure of cuprates may be described by the number of CuO_2 planes, the apical oxygen, and the metal–oxygen charge reservoir.

The CuO_2 planes vary from single-layer to an infinite-layer structure among the different families of cuprates. Cuprates crystal structure causes large anisotropy in superconducting properties; as a result, the electrical conduction is highly anisotropic. Higher conductivity is observed parallel to the CuO_2 plane than in the perpendicular direction. The crystallographic unit cell of BSCCO structure contains three layers: reservoir layer, superconducting layer, and insulating layer. Reservoir layer (i.e., SrO and BiO layer) that reserves the electron and just above it contains superconducting layer (i.e., CuO). Superconductivity is seen between the alternating multilayers of CuO_2 planes. In these CuO layers, the doping can be possible to induce the superconducting property. Above this, it contains insulating layer (CaO and CuO), where CaO and CuO plane form the Josephson junction between them. The CuO_2 planes are present in between Bi–O double layers; deformation of the weak Bi–O planes affects the charge reservoir distribution in the layered HTSC and hence lowers the transition temperature.

In our synthesis, we prefer BSCCO 2212 phase as its T_c is considerably high and it is easy to synthesize as they are thermodynamically stable over a wide range of temperature and within the stoichiometric range, as compared to 2223 phase.

17.3.2 SYNTHESIS OF BSCCO

Cuprate materials are brittle ceramics which are expensive to manufacture. There are several methods of preparation for BSCCO, solid-state reaction route, coprecipitation, pyrolysis, melt-process method, and sol–gel technique.

Some synthesis methods and characterization methods of the BSCCO system adopted by different researchers are discussed:

Bock et al., Solid State Communication, Vol. 72, 453–458 (1989)

Melt process method was adopted by Bock et al.[19] with Bi_2O_3, SrO, CaO, and CuO as starting materials and T_c of the system was found to be 85 K. The precursors were mixed together and heated in an aluminum crucible up to 1000–1100°C and cooled in a controlled mode.

Tampieri et al., Material Chemistry and Physics, Vol. 34, 157–161 (1993)

The samples prepared by Tampieri et al.[20] using high-purity commercial powders of Bi, Cu, Pb, Sr, and Ca by solid-state synthesis route. Here, powders were wet mixed in alcohol with polythelene balls, then calcinated and sintered at 780 and 850°C.

Hanjin Lim et al.,[21] Journal of Materials Science, Vol. 31, 2349–2352 (1996)

Solid-state method with precursors Bi_2O_3, $SrCO_3$, $CaCO_3$, and CuO powders were mixed in the strochiometric ratio and then ball milled with isopropanol and ZrO_2 grinding media.

M. O. Petropoulou et al., Journal of Thermal Analysis, Vol. 52, 903–914 (1998)

T_c of the samples prepared by M. O. Petropoulou *et al.*[22] were found to be 77 and 85 K, respectively, using solid-state method. In single-step method, the starting materials were Bi_2O_3, $SrCo_3$, $CaCo_3$, and CuO and were mixed in the molar ratio and calcinated at 830–845°C. In two-step process, $SrCo_3$, $CaCo_3$, and CuO were mixed in the molar ratio and the mixture was heated at 955°C. Then, Bi_2O_3 was added and heated at 845°C. The prepared sample contains impurity phase of 2201.

Tampieri et al., Physica C 306, 21–33 (1998)

Sample preparation by three different methods such as solid state, pyrolysis, and sol–gel reaction were used by Tampieri et al.[23] Here, metallic oxides were used as the starting materials. Sol–gel method resulted in single phase. Phase confirmation of samples prepared by solid state and pyrolysis technique was found to contain CaO, CuO, and BSCCO 2201 impurity phases.

Fruth et al., Journal of the European Ceramic Society, Vol. 24, 1827–1830 (2004)

Fruth et al.[24] prepared glass ceramic superconductors in the Bi–Sr–Ca–Cu–O by melt-quenching method in the presence of some additive (Pb, B). The studied glasses were characterized by scanning electron microscopy (SEM), X-ray diffraction (XRD), thermogravimetric and differential thermal analysis (DTA/TGA), and differential scanning calorimetry. The glass-transition temperatures were found to range in the 380–430 domains. The samples doped with low amount of B present obvious transit.

Patel et al., Superconductor Science and Technology, Vol. 18, 317–324 (2005)

Patel et al.[25] reported the characterization of attrition milled powder of Bi-2212 phase with $T_c = 87$ K, after two high-temperature treatment for 60 h. Critical current density J_c was reported to be 10^6 A cm^{-2} at 4 K measured by magnetic field response properties. This single phase BSCCO was prepared using oxides and carbonates of Bi, Sr, Ca, and Cu as precursor and was identified by powder XRD. More recently, preparation of ceramic BSCCO 2223 and 2212 phases by sol–gel method using nitrate solutions and its characterization by DTA and IR is also reported.

Arshad et al., Journal of Thermal Analysis and Calorimetry, Vol. 89, 595–600 (2006)

Sol–gel synthesis route was utilized by Arshad et al.[26] with starting material as all metallic nitrate. The resultant powder was subjected to thermal treatment at 800, 845, and 860°C.

Kanungo et al.[27], Doctoral Dissertation (2013)

Kanungo reported the characterization results of Gd-doped BSCCO with different concentrations such as 1%, 3%, and 5%. Their characterization are also done by various methods like XRD, SEM, temperature dependence of resistivity, and current (I)–voltage (V) measurement. Effect of doping on the value of critical current density T_c is analyzed. The preparation of lead free BSCCO-2212 phase is done by solid-state reaction route.

17.3.3   SOL–GEL TECHNIQUE

Synthesis technique is a solution chemistry process, modified sol–gel technique even for the mass scale production because of its cost efficiency, the ease of preparation, shorter heat treatment which gives good corrosion resistant homogeneous less particle sized resultant product having better phase purity. Also, this offers probable tenability of T_c with respect to particle size of the materials.

Nitrates of bismuth, strontium, calcium, and copper were used as the starting materials for the synthesis of BSCCO in the molar ratio 1:1:1:2.5%. Excess bismuth was added to compensate for the evaporation during the thermal treatment. The precursors in desired molar ratio were dissolved in ethylene glycol and stirred continuously for an hour. The obtained light-blue color homogenous sol was heated at 60°C for an hour. The resultant product is a gel. The temperature was raised to 150°C so that the solvent gets evaporated and final get is self-ignited to obtain a dark residue. The collected residue was ground using an agate mortar and pestle for an hour and the prepared sample was sintered.

A higher annealing temperature produces a higher T_c, but high-temperature treatment leads to the evaporation of some constituent of the system. Such a loss in Bi allows other stacking alternatives. Physical properties of the material are affected by these structural defects. From earlier works, it is found that some stacking faults produce lower or higher T_c phase. A deficiency in the Bi content would lead to an increase in the average valance of copper, thereby changing the T_c. Also T_c value depressed due to substitution at the Cu site in the CuO_2 planes than any substitution. Increasing sintering time should lead to the change in electron correlation suppression and will produce superconductivity. It has been reported that T_c of the material is related to the sintering temperature and time, variations in oxygen stoichiometry and the presence of Cu(III) (Figure 17.1).

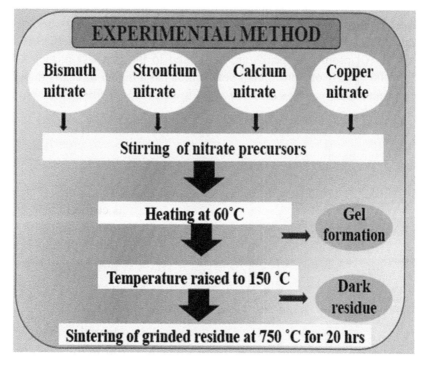

FIGURE 17.1   Sol–gel synthesis technique.

17.3.4   PROPERTIES

17.3.4.1   STRUCTURAL PROPERTIES

XRD is used for the structural characterization of materials and each crystalline solid has a unique characteristic XRD pattern. Phase confirmation of the sintered powder sample is done by X-ray diffractometer (Model: Rigaku, Miniflex, Japan) with filtered 0.154 Cu $K\alpha$ radiation.

Transition temperature of different cuprate compounds had different values, depending on the main components of the system. With the change in constituent element of cuprates, it showed different T_c values. In addition, sintering like influencing factors also produced major variations in T_c. Dou et al. studied about the variation of T_c with the sintering time and concluded that T_c of $Tl_2Ba_2Ca_2Cu_3O_{10+\delta}$ decreases with the increase in sintering time.[28] They demonstrated that decrease in T_c takes place due to the loss of thallium during the long-period sintering treatment. Moreover, they presented after effects of decrease in T_c, it allowed some crack formation which resulted from excessive mechanical deformation. To explain the effect of sintering conditions, samples were sintered at different temperatures and subjected to characterization. Now the structural measurements are discussed. Prominent peaks of the XRD spectra are indexed by matching it with the JCPDS/ICDP database. All prominent peaks are compared to the BSCCO phase. XRD pattern shows only one prominent peak which has a higher intensity than the other peaks. XRD patterns are shown in Figures 17.2–17.5.

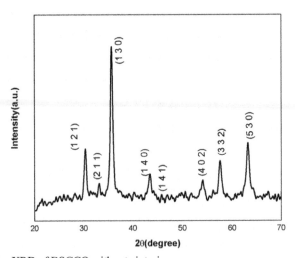

FIGURE 17.2   XRD of BSCCO without sintering.

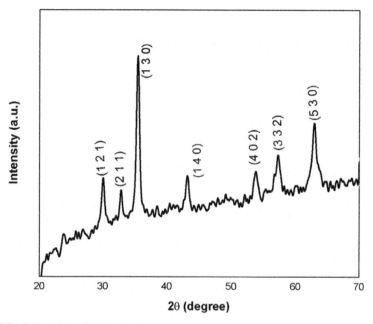

FIGURE 17.3 XRD of BSCCO sintered at 500°C for 6 h.

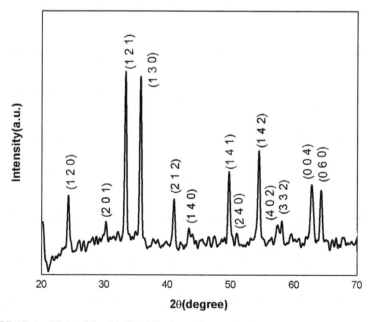

FIGURE 17.4 XRD of the 6-h (900°C) sintered sample of BSCCO.

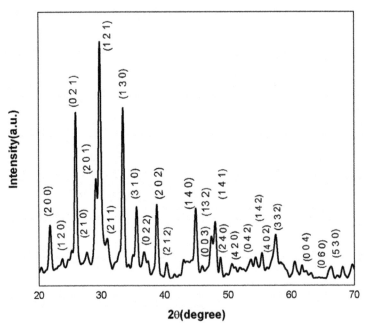

FIGURE 17.5　XRD plot of the 20-h (750°C) sintered sample.

17.3.4.2　MAGNETIC PROPERTIES

The HTSC comes under the category of Type-II superconductors and shows a gradual change in transition temperature as a function of the magnetic field. A vibrating sample magnetometer or vibrational sample magnetometer (VSM) is a scientific instrument that measures magnetic properties. Magnetic measurements of the powder sample were performed using Quantum Design Versalab VSM. The magnetic field and temperature range is about −30–30 T and 50–300 K, respectively. Temperature dependence of magnetization was measured in constant magnetic field of 100 Oe. Magnetic data of unsintered and 20 h (750°C) sintered samples are plotted. The critical temperature of the sintered sample is obtained as 72 K (Figures 17.6–17.8).

The superconducting properties are indirectly probed the magnetic transition studies. The results are performed M–T measurements. It shows the temperature dependence of magnetization for BSCCO and the transition temperature. Normally, both magnetic and resistivity transitions in a superconductor will be at the same temperature.

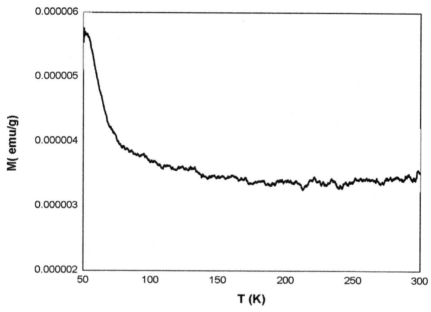

FIGURE 17.6   *M–T* measurement showing no transition.

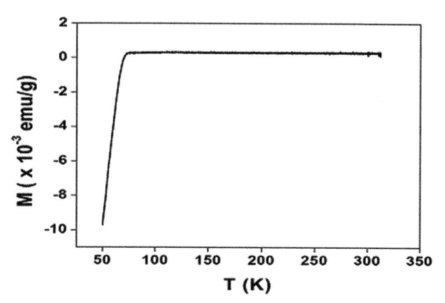

FIGURE 17.7   *M–T* curve of the sintered BSCCO powder.

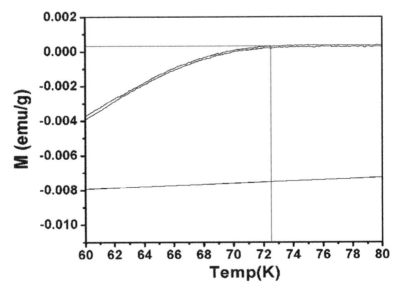

FIGURE 17.8   A typical graph showing magnetic transition.

17.3.4.3   DIELECTRIC PROPERTIES

The dielectric constant is the ratio of the electrical permittivity of the material to the permittivity of free space. It is a measure of how concentrated the electric flux is in the material. Dielectric constant can be measured using an LCR meter. Generally, this electronic device is used for the measurement of inductance, capacitance, and resistance of a material. The precision impedance analyzer (WAYNE KERR, 6500B) was used for the measurement of capacitance of the silver-pasted sample pellet. Dielectric constants of the materials were calculated from the obtained data. The variation of dielectric constant with applied frequency was studied. The frequency is varied from 0 to 100.

Dielectric data plotted against applied frequency and is given by Figure 17.9. At sufficiently low frequencies, the total contribution to the dielectric constant can be from electronic, ionic, and dipolar oscillations. However, at intermediate frequencies, dipolar frequencies and interfacial polarizations cannot follow the applied frequencies and they will lag behind. Dielectric constant approaches the saturation for higher applied frequencies. Here, only the electronic contribution to the polarization can stay with the applied frequency.

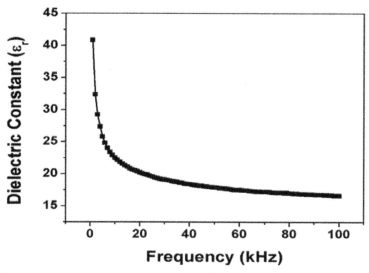

FIGURE 17.9   Dielectric data plotted against applied frequency.

17.3.5   MECHANISM

The origin of superconductivity in HTSCs is one of the major unsolved problems for theoretical physicist. Cuprates behave in a complex manner because of a strong electron correlation. Central mechanism of superconductivity is considered to be electron–phonon interaction or spin fluctuations. Cuprates shows strong magnetic interactions in the form of antiferromagnetic (AF) spin correlations. Unusual magnetic properties of copper oxides at its normal state are due to low-lying spin fluctuations. The AF spin fluctuation is absolutely effective mechanism in high T_c cuprates. AF interaction and its fluctuation may be a source of electron pairing in cuprates. Cuprates are unconventional superconductors having the $d_{x^2-y^2}$ wave function.[29] Also Monthoux and Pines[30] suggested that transition to a superconducting state can occur by exchange of AF paramagnon induced by the retarded interaction between planar quasi-particles for weak interactions. Harlingen reported the possibility of the d-wave superconductivity in the Hubbard model had using the random-phase approximation or the fluctuation exchange approximation.[30–36] Swarovski discussed that the study of the typical Fermi surface of current carriers in CuO_2 plane gives details of phase transition to AF state.

   The pseudogap problem is the heart of cuprates because it is the fundamental property of underdoped copper oxides. For HTSC, pseudogap state corresponds to the maximum critical temperature. The pseudogap derives into the superconducting gap and appears due to short-range order fluctuations. For underdoped samples, the pseudogap formation was clearly for temperatures higher than T_c.[34] It is important that some signs of pseudogap existence were observed for slightly overdoped samples also.

17.3.6 APPLICATIONS

Superconductors have remarkable applications because of its ability to carry an electric charge without any loss across an infinite distance and also block out magnetic fields. The practical applications of conventional superconductors are limited due to their very low operating temperature. At the same time, the discovery of high T_c materials extends the feasible applications of superconductors. The lack of a permanent, renewable, environment friendly source of energy is the main challenges faced by science and humanity. The invention of materials which provide zero-energy loss during power transmission would be a great breakthrough and thereby reducing the ever-increasing demand for energy production. The other prospective application of superconductivity is lossless power transmission, that is, a superconducting wire can transmit a DC current without any loss and an AC current with an extremely small loss.

Application of superconductors includes electrical, magnetic, medical application, and application of Josephson's effect. Mainly, these materials are used in powerful superconducting electromagnet-based magnetic levitating trains, magnetic resonance imaging (MRI), superconducting quantum interference devices (SQUID), magnetoencephalography, microwave devices and resonators to high-energy physics experiments, sensors, etc.[3]

Applications of these materials are grouped as

1. Electrical applications
    a) Power transmission
    b) Electrical generators and transformers
    c) Sensitive electrical equipment
    d) Superconducting microchip (storage element)
    e) Switching device (e.g., cryotron)
    f) Fault current limiters

2. Magnetic applications
    a) Magnetic levitation (magnetic levitation transport)
    b) Storing electrical power and magnetic flux

3. Medical applications
    a) MRI
    b) Nuclear magnetic resonance scanning
    c) Diagnosis of brain tumor and defective cells (detection of magnetic signals from brain, heart, etc.)
    d) Magneto-hydrodynamic power generation

4. Applications of Josephson devices
    a) Magnetic sensors
    b) Bolometers
    c) Photon detectors
    d) Gradiometers
    e) Oscilloscopes
    f) Decoders
    g) Analogue to digital converters
    h) Oscillators
    i) Microwave amplifiers
    j) Sensors for biomedical, scientific, and defense purposes
    k) Digital circuit development for integrated circuits
    l) Microprocessors
    m) Random access memories (RAMs)

5. Applications of SQUIDS
    a) Storage device for magnetic flux
    b) Study of earthquakes
    c) Removing paramagnetic impurities

6. Telecommunications
    a) Efficient filters in cellular telephone towers
    b) Separate signals of individual phone calls

17.4 CONCLUSIONS

Superconductivity is a field which has immense applications in the area of industry as well as science. When it comes to the case of high-temperature superconductivity, the relevance multiplies. High-temperature superconductivity can in fact quench the thirst for the never-ending increase of power consumption and pressure on the available energy sources. The current research indicates that cuprate-based BSCCO will serve as a good candidate in the field of eco-friendly HTSC.

KEYWORDS

- **HTSC**
- **strongly correlated electron systems**
- **cuprates**
- **BSCCO**
- **sol–gel**

REFERENCES

1. Bednorz, J. G.; Muiller, K. A. Possible High T_c Superconductivity in the Ba–La–Cu–O System Z. *Phys. B: Condens. Matter* **1986**, *64*, 189–193.

2. Wu, M. K.; Ashburn, J. R.; Torng, C. J.; Hor, P. H.; Meng, R. L.; Gao, L.; Huang, Z. J.; Wang, Y. Q.; Chu, C. W. Superconductivity at 93 K in a New Mixed-Phase Y–Ba–Cu–O Compound System at Ambient Pressure. *Phys. Rev. Lett.* **1987**, *58*, 908–910.

3. Malik, M. A.; Malik, B. A. High Temperature Superconductivity: Materials, Mechanism and Applications. *Bulg. J. Phys.* **2014**, *41*, 305–314.

4. Homa, D.; Liang, Y.; Pickrell, G. Superconducting Fiber. *Appl. Phys. Lett.* **2013**, *103*, 082601–082604.

5. Camarasa-Gomez, M.; Di Marco, A.; Hekking, F. W.; Winkelmann, C. B.; Courtois, H.; Giazotto, F. Superconducting Cascade Electron Refrigerator. *Appl. Phys. Lett.* **2014**, *104*, 192601.

6. Maeda, H.; Tanaka, Y.; Fukutumi, M.; Asano, T. A New High-T_c Oxide Superconductor without a Rare Earth Element. *Jpn. J. Appl. Phys.* **1988**, *16*, L209–L210.

7. Sheng, Z. Z.; Hermann, A. M. Bulk Superconductivity at 120 K in the Ti–Ca/Ba–Cu–O System. *Nature* **1988**, *332*, 138–139.

8. Schilling, A.; Cantoni, M.; Guo, J. D.; Ott, H. R. Superconductivity above 130 K in the Hg–Ba–Ca–Cu–O System. *Nature* **1993**, *363*, 56–58.

9. Kawashima, T.; Matsui, Y.; Takayama-Muromachi, E. New Oxycarbonate Superconductors $(Cu_0, 5C_0, 5) Ba_2Ca_{n-1}Cu_nO_{2n+3}$ (n = 3, 4) Prepared at High Pressure. *Phys. C: Supercond.* **1994**, *224*, 69–74.

10. Michel, C.; Hervieu, M.; Borel, M. M.; Grandin, A.; Deslandes, F.; Provost, J.; Raveau, B. Superconductivity in the Bi–Sr–Cu–O System. In *Ten Years of Superconductivity: 1980–1990*; Springer: Netherlands, 1987; pp 300–302.

11. Tallon, J. L.; Buckley, R. G.; Gilberd, P. W.; Presland, M. R.; Brown, I. W. M.; Bowden, M. E.; Christian, L. A.; Goguel, R. High-T_c Superconducting Phases in the Series $Bi_{2.1}(Ca, Sr)_{n+1}Cu_nO_{2n+4+\delta}$. *Nature* **1988**, *333*, 153–156.

12. Subramanian, M. A.; Torardi, C. C.; Calabrese, J. C.; Gopalakrishnan, J.; Morrissey, K. J.; Askew, T. R.; Flippen, R. B.; Chowdhry, U.; Sleight, A. W. A New High-Temperature Superconductor: $Bi_2Sr_3^-\cdot Ca\cdot Cu_2O_{8+\delta}$. *Science* **1988**, *239*, 1015–1017.

13. Torardi, C. C.; Parise, J. B.; Subramanian, M. A.; Gopalakrishnan, J.; Sleight, A. W. Oxygen Nonstoichiometry in Copper-Oxide Based Superconductors and Related

Systems: Structure of Nonsuperconducting $Bi_2Sr_{3-x}Y_xCu_2O_{8+y}$ ($x \approx 0.6–1.0$). *Phys. C: Supercond.* **1989**, *157*, 115–123.

14. Majewski, P. BiSrCaCuO High-T_c Superconductors. *Adv. Mater.* **1994**, *6*, 460–469.

15. Tarascon, J. M.; McKinnon, W. R; LePage, Y.; Stoffel, N.; Giroud, M. Preparation, Structure and Properties of the Superconducting $Bi_2Sr_2Ca_{n-1}Cu_nO_y$ with n = 1, 2 and 3. *Phys. Rev. B* **1988**, *38*, 8885–8892.

16. Swain, S. *Studies On Superconductor/Nanocomposite of BSCCO/BiFeO*, Msc. Dissertation, National Institute of Technology, India, 2011.

17. Hazen,R. M.; Prewitt, C. T.; Angel, R. J.; Ross, N. L.; Finger, L. W.; Hadidiacos, C. G.; Veblen, D. R.; Heaney, P. J.; Hor, P. H.; Meng, R. L.; Sun, Y. Y.; Wang, Y. Q.; Xue, Y. Y.; Huang, Z. J.; Gao, L.; Bechtold, J.; Chu, C. W. Superconductivity in the High-T, Bi–Ca–Sr–Cu–O System: Phase Identification. *Phys. Rev. Lett.* **1988**, *60*, 1174–1177.

18. Raykova, R.; Raycheva, T.; Bogdanov, B.; Pashev, P.; Hristov, Y. Sol–Gel Synthesis and Thermal Characterization of the Batches of BSCCO System. *Научни Трудове на Русенския Университет* **2014**, *5*, 59–63.

19. Bock, J.; Preisler, E. Preparation of Single Phase 2212 Bismuth Strontium Calcium Cuprate by Melt Processing. *Solid State Commun.* **1989**, *72*, 453–458.

20. Tampieri,A.;Landi,E.;Bellosi,A.KineticStudyoftheFormationofBi₁·8PbO·2Sr₂Ca₂Cu₃Oₓ Ceramic Superconductor. *Mater. Chem. Phys.* **1993**, *34*, 157–161.

21. Lim, H.; Byrne, J. G. Effect of Precursor History on Synthesis of High-T_c BPSCCO Superconductor. *J. Mater. Sci.* **1996**, *31*, 2349–2352.

22. Petropoulou, M. O; Argyropoulou, R; Tarantilis, P; Kokkinos, E; Ochsenkühn, E. M; Parissakis, G. Comparison of the Oxalate Co-precipitation and the Solid State Reaction Methods for the Production of High Temperature Superconducting Powders and Coatings. *J. Mater. Process. Technol.* **2002**, *127*, 122–128.

23. Tampieri, A.; Calestani, G.; Celotti, G.; Masini, R.; Lesca, S. Multi-Step Process to Prepare Bulk BSCCO (2223) Superconductor with Improved Transport Properties. *Phys. C: Supercond.* **1998**, *306*, 21–33.

24. Fruth, V.; Popa, M.; Ianculescu, A.; Stir, M.; Preda, S.; Aldica, G. High-T_c Phase Obtained in the Pb/Sb Doped Bi–Sr–Ca–Cu–O System. *J. Eur. Ceram. Soc.* **2004**, *24*, 1827–1830.

25. Patel, R. H.; Nabialek, A.; Niewczas, M. Characterization of Superconducting Properties of BSCCO Powder Prepared by Attrition Milling. *Supercond. Sci. Technol.* **2005**, *18*, 317–324.

26. Arshad, M.; Qureshi, A; Masud, K.; Qazi, N. Production of BSCCO Bulk High T_c Superconductors by Sol–Gel Method and their Characterization by FTIR and XRD Techniques. *J. Therm. Anal. Calorim.* **2006**, *89*, 595–600.

27. Kanungo, S. *Synthesis and Characterization of Gd Doped BSCCO-2212*, Doctoral Dissertation, National Institute of Technology, India, 2013.

28. Dou, S. X.; Liu, H. K.; Bourdillon, A. J.; Tan, N. X.; Savvides, N.; Andrikidis, C.; Sorrell, C. C. Processing, Characterisation and Properties of the Superconducting Tl–Ba–Ca–Cu–O System. *Supercond. Sci. Technol.* **1988**, *1*, 83.

29. Garnier, V.; Caillard, R.; Sotelo, A.; Desgardin, G. Relationship among Synthesis, Microstructure and Properties in Sinter-Forged Bi-2212 Ceramics. *Phys. C: Supercond.* **1999**, *319*, 197–208.

30. Monthoux, P.; Pines, D. Spin-Fluctuation-Induced Superconductivity in the Copper Oxides: A Strong Coupling Calculation. *Phys. Rev. Lett.* **1992**, *69*, 961–964.

31. Moriya, T.; Ueda, K. Spin Fluctuations and High Temperature Superconductivity. *Adv. Phys.* **2000,** *49,* 555–606.
32. Tohyama, T. Recent Progress in Physics of High-Temperature Superconductors. *Jpn. J. Appl. Phys.* **2012,** *51,* 0100041–01000412.
33. Van Harlingen, D. J. Phase-Sensitive Tests of the Symmetry of the Pairing State in the High-Temperature Superconductors—Evidence for dx_2-y_2 Symmetry. *Rev. Mod. Phys.* **1995,** *67,* 515–535.
34. Sadovski, M. V. Pseudogap in High-Temperature Superconductors. *Phys. Usp.* **2001,** *5,* 515–539.
35. Schriffer, J. R. Approaches to the Theory of High Temperature Superconductivity. *Phys. B* **1999,** *259,* 433–439.
36. Timusk, T.; Statt, B. The Pseudogap in High-Temperature Superconductors: An Experimental Survey. *Rep. Prog. Phys.* **1999,** *62,* 61–122.

CHAPTER 18

BASICS OF CARBOHYDRATES

RAJEEV SINGH[1] and ANAMIKA SINGH[2*]

[1]*Department of Environment Studies, Satayawati College, University of Delhi, Delhi, India*

[2]*Department of Botany, Matreyi College, University of Delhi, Delhi, India*

Corresponding author. E-mail: arjumika@gmail.com

CONTENTS

ABSTRACT

In this chapter a concise review on the basics of carbohydrates is presented. There are three major classes of carbohydrates: monosaccharides, oligosaccharides, and polysaccharides.

18.1   INTRODUCTION

Carbohydrates are saccharides and the word "saccharide" is derived from the Greek *sakcharon*, meaning "sugar," and commonly known as "staff of life." Carbohydrates are the most abundant biomolecule and it has series of compounds of carbon, hydrogen, and oxygen in which the atoms of the latter two elements are in the ratio of 2:1 (as in water). Its empirical formula $C_m(H_2O)_n$ (where m could be different from n). Carbohydrates are actually hydrates of carbon. Carbohydrates act as energy source, fuels, storage, metabolic products, and intermediates. Carbohydrates are also acting as genetic material in the form of ribose and deoxyribose sugars (DNA and RNA). Carbohydrates have aldehyde or ketone groups with multiple hydroxyl groups.

There are three major classes of carbohydrates: monosaccharides, oligosaccharides, and polysaccharides.

18.2   MONOSACCHARIDES

Monosaccharides are sugars, with single polyhydroxy aldehyde or ketone units. Six-carbon D-glucose sugar (dextrose) is the most abundant monosaccharide. Monosaccharides which are having more than four carbons tend to have cyclic structures. Monosaccharides generally contain three to nine carbon atoms. These may be linked together to form a large variety of oligosaccharide structures. Glucose and fructose are the six-carbon monosaccharides having five hydroxyl groups. Most of the hydroxyl groups are attached to the carbons having chiral centers (four different groups attached to single carbon) and it gives a specific characteristic to all monosaccharide, known as sterioisomers. The monosaccharides molecules are generally unbranched carbon chains and all the carbon atoms are linked together by single bonds. One of the carbon atoms is double-bonded to an oxygen atom to form a carbonyl group while other carbon atoms have a hydroxyl group. If the aldehyde (carbonyl) group is at an end of the carbon chain, the monosaccharide is known as aldose and if the ketone (carbonyl) group is at any other position the monosaccharide is a ketose.

18.3 DISACCHARIDES

Disaccharides are sugars having two monosaccharide units. Most abundant disaccharide is sucrose (cane sugar), consists of the six-carbon sugars D-glucose and D-fructose. All monosaccharides and disaccharides have names ending with the suffix "-ose."

18.3.1 IMPORTANT DISACCHARIDES

Two monosaccharides reacts to form a disaccharide molecule and loss of water molecule occurs; this reaction is dehydration reaction (loss of water) (Fig. 18.1). Disaccharides joined together through an "oxygen bridge." The carbon–oxygen bonds are called glycosidic bonds. The most common disaccharides are sucrose, lactose, and maltose.

$$2C_6H_{12}O_6 \rightarrow C_{12}H_{22}O_{11} + H_2O$$

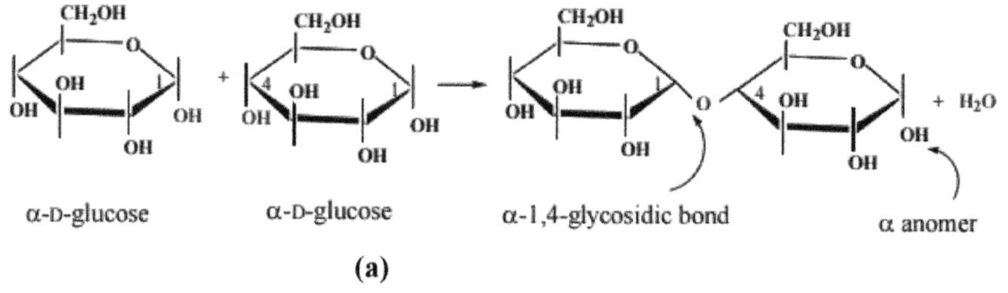

α-D-glucose α-D-glucose α-1,4-glycosidic bond α anomer

(a)

FIGURE 18.1 The artwork showing disaccharide formation and dehydration reaction.

18.3.1.1 SUCROSE

Sucrose, commonly known as table sugar, is the most abundant disaccharide in the biological system. Sucrose contains an α-D-glucose unit and a β-D-fructose unit joined by α,β(1→2) glycosidic linkage. Sucrose is a nonreducing sugar because its glycosidic bond involves both anomeric carbons; therefore, there is no free aldehyde group (Fig. 18.2a).

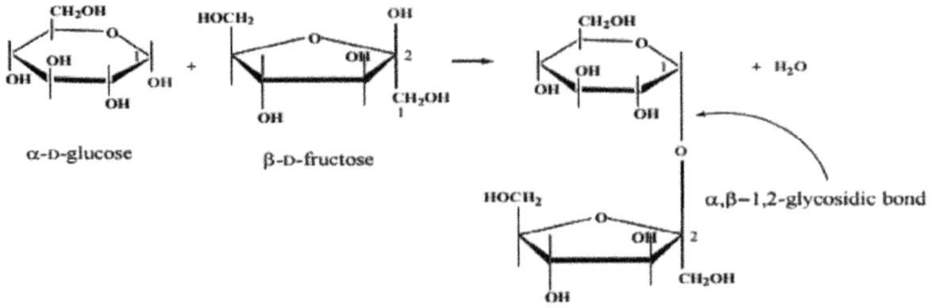

(a)

18.3.1.2 LACTOSE

Lactose, commonly known milk sugar, is made up of a β-D-galactose unit and a D-glucose unit joined by a β(1→4) glycosidic linkage (Fig. 18.2b). Lactose is reducing sugar (glucose ring on the right has a free anomeric carbon that can open to give an aldehyde).

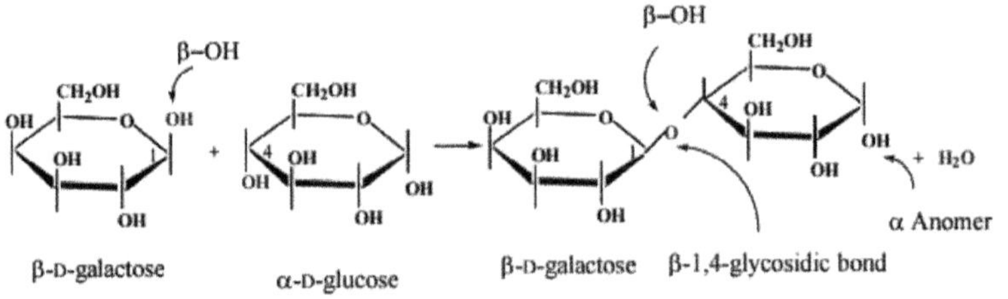

(b)

18.3.1.3 MALTOSE

Maltose (malt sugar) is made up of two D-glucose units. The glycosidic linkage between two glucose units is an (1→4) linkage (Fig. 18.2c). Maltose is a reducing sugar (glucose ring on the right can open to give an aldehyde).

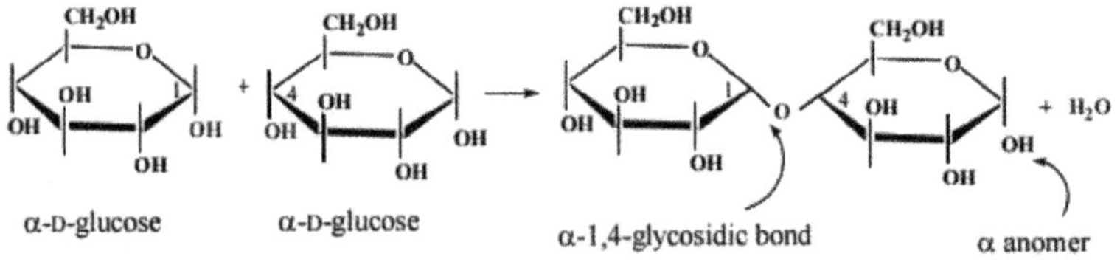

(c)

FIGURE 18.2 The artworks showing disaccharide formation and types of bonds in disaccharides: (a) sucrose, (b) lactose, and (c) maltose.

18.4   OLIGOSACCHARIDES

Oligosaccharides are short chains of monosaccharide units, joined together by glycosidic bonds. Term oligosaccharide is derived from Greek *olígos*, "a few," and *sácchar*, "sugar" is a saccharide polymer containing a small number (typically 2–10 of simple sugars (monosaccharides). Oligosaccharides can have many functions including cell recognition, cell binding, and in immune response. Generally, they are attached to either *N*- or *O*-linked to compatible amino acid side chains in proteins or to lipid moieties. *N*-linked oligosaccharides are found attached to asparagine through beta linkage to the nitrogen of amine. Alternately, *O*-linked oligosaccharides are generally attached to threonine or serine on the alcohol group of the side chain.

The process of *N*-linked glycosylation occurs cotranslationally or concurrently while the proteins are being translated. It is added cotranslationally and *N*-linked glycosylation helps to determine the folding of polypeptides due to the hydrophilic nature of sugars. They are small carbohydrates which are formed by condensation of two to nine monosaccharides. *N*-linked oligosaccharides usually has the oligosaccharide linked to the amide nitrogen of the Asn residue, in the sequence Asn–X–Ser/Thr. X can be any amino acid except for proline (though it is rare to see Asp, Glu, Leu, or Trp). Oligosaccharides that participate in *O*-linked glycosylation are attached to threonine or serine on the alcohol group of the side chain. *O*-linked glycosylation occurs in the Golgi apparatus, in which monosaccharide units are added to a complete polypeptide chain. Cell surface proteins and extracellular proteins are *O*-glycosylated. Glycosylation sites in *O*-linked oligosaccharides are specified only in the secondary and tertiary structures of the polypeptide, which will dictate where glycosyltransferases will add sugars. Both glycoproteins and glycolipids have a covalently attached carbohydrate attached to their respective molecule. They are very abundant on the surface of the cell, and their interactions contribute to the overall stability of the cell. Carbohydrate is very important as it is used for the classification of blood group. The blood groups are classified on the basis of antigen found at surface of red blood cells (RBC); these antigens are simple chains of sugars (oligosaccharides). Type of oligosaccharide attached at the surface RBC determines the person's blood group (Table 18.1).

18.4.1 OLIGOSACCHARIDE AND ITS ROLE IN BLOOD GROUP DETERMINATION

Carbohydrate is very important as it is used for the classification of blood group. The blood groups are classified on the basis of antigen found at surface of RBC; these antigens are simple chains of sugars (oligosaccharides). Type of oligosaccharide attached at the surface RBC determines the person's blood group (Table 18.1).

TABLE 18.1 Showing Types of Blood Groups Due to Different Oligosaccharide Attached at RBC.

S. No	Antigen	Blood group	Can receive blood
1	O type	O	O
2	A type	A	A, O
3	B type	Or B	B, O
4	AB type	AB	A, B, O

18.5 POLYSACCHARIDES

Polysaccharides are sugar polymers containing more than 20 or so monosaccharide units, and some have hundreds or thousands of units. A polymer is a large molecule composed of many small, repeating structural units that are identical. The repeating structural units are called monomers. Most of the carbohydrates found in nature found as polysaccharides. Polysaccharides are carbohydrates formed by more than nine monosaccharide units linked by glycosidic bonds. They are polymers of high molecular weight. Polysaccharides, also called glycans, differ from each other by mean of monosaccharide units, length of their chains, types of bonds linking the units, and in the branching pattern. For the synthesis of polysaccharide, there is no template required (different from protein synthesis), but it depends upon enzymatic action that catalyze the polymerization of monomeric units. Glycoconjugates are carbohydrates linked to many proteins and lipids and it helps in cell–cell interaction. It is a key role of carbohydrate and because of tremendous structural diversity. For each type of monosaccharide to be added to the growing polymer, there is a separate enzyme, and each enzyme acts only when the enzyme that inserts the preceding subunit has acted. The alternating action of several enzymes produces a polymer with a precisely repeating sequence, but the exact length varies from molecule to molecule, within a general size class. Depending upon the monomeric units, polysaccharide is homo or heteropolysaccharide. Homopolysaccharides are having single type of monomeric unit, while heteropolysaccharides contain two or more different kinds of monomeric units. Bacterial cell envelope (the peptidoglycan) is a heteropolysaccharide. The rigid layer is built from two alternating monosaccharide units. While in animal tissues, extracellular space is occupied by different types of heteropolysaccharide.

18.5.1 STARCH

It is the primary storage polysaccharide in the plants and the most common forms of starch is amylose and amylopectin. Starch molecules are heavily hydrated because they have many exposed hydroxyl groups easily available to form hydrogen bonds with water. Starch is a tightly coiled helical structure stabilized by hydrogen bonds. Starch contains two types of glucose polymer, amylose and amylopectin. The former consists of long, unbranched chains of D-glucose units connected by (α1→4) linkages (Fig. 18.3a). Most of the starch contains 20–25% amylose and 75–80% amylopectins. It completely hydrolyzes into amylose and amylopectins and produces only D-glucose. The glycosidic linkages joining successive glucose residues in amylopectin chains are (α1→4), but the branch points, occurring every 24–30 residues are (α1→6) linkages. An amylose molecules is a continuous chains of D-glucose units joined by α1,4-glycosidic linkage. There are no branch points in the molecule which may contain of as many as 4000 D-glucose units, while amylopectin is a branched-chain polysaccharide larger than amylose. Amylopectin molecules, on the average, consist of several thousand -D-glucose units joined by -1,4-glycosidic linkage (Fig. 18.3b). The molecular masses routinely approach 1 million or more. Wheat, rice, corn, and potatoes are the important sources of starch.

Amylose

α-1,4-glycosidic bonds

(a)

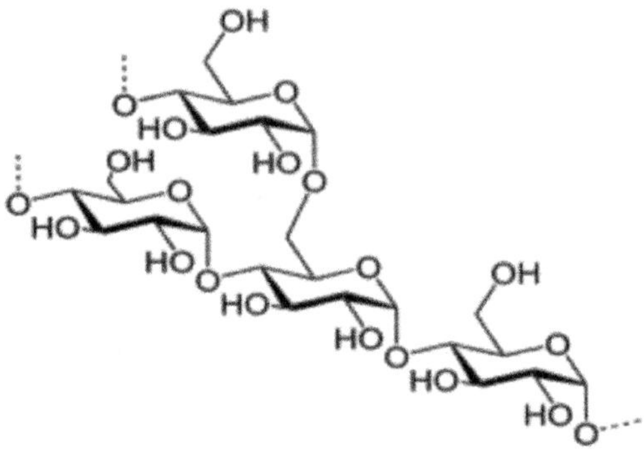

(b)

FIGURE 18.3   (a) amylase and (b) amylopectin.

18.5.2   GLYCOGEN

Glycogen is the main glucose storage molecule in animal cells. Structurally, glycogen is very similar to amylopectin (-D-glucose units joined by -1,4-glycosidic linkages), but glycogen is a highly branched structure. Similar to starch, it is highly hydrated as large number of exposed hydroxyl group is present to form hydrogen bonding. Glycogen occurs intracellularly as large clusters or granules. Glycogen is a polymer of $(\alpha 1 \rightarrow 4)$-linked subunits of glucose, with $(\alpha 1 \rightarrow 6)$-linked branches and it is a compact structure. Like starch, glycogen is a tightly bound structure stabilized by hydrogen bonding. Glycogen is abundant in the liver and also found in skeletal muscle. In liver cells, glycogen is found in the form of large granules, which are themselves clusters of smaller granules composed of single, highly branched glycogen molecules with an average molecular weight of several million.

18.5.3   CELLULOSE

Cellulose is the most abundant organic molecule found in nature. Cellulose is a fibrous, tough, water-insoluble substance, is found in the cell walls of plants. Structurally, it is long, unbranched, D-glucose polymer in which the glucose units are linked by $\beta(1 \rightarrow 4)$ glycosidic bonds (Fig. 18.4), because cellulose is a linear, unbranched homopolysaccharide of 10,000–15,000 D-glucose units, linked by $(\beta 1 \rightarrow 4)$ glycosidic bonds. In cellulose, the glucose residues have the β configuration (Fig. 18.4). Due to this, cellulose has very different three-dimensional structures and physical properties.

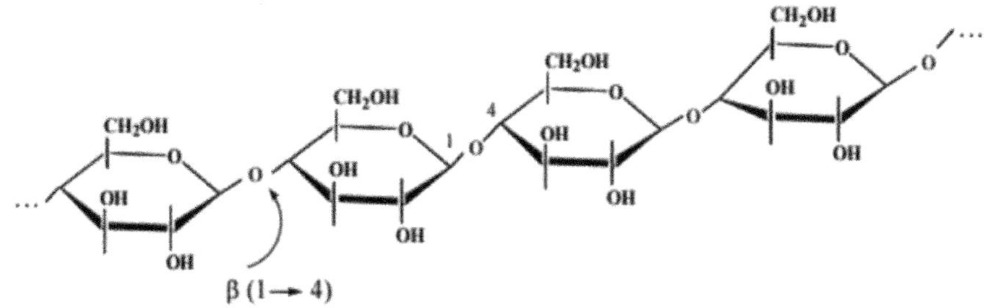

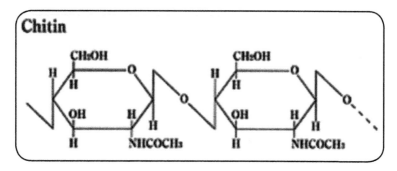

FIGURE 18.4   The artwork showing structure of cellulose.

18.5.4   CHITIN

It is the second most abundant polysaccharide. It is a linear homopolysaccharide composed of *N*-acetyl-D-glucosamine residues in β linkage (Fig. 18.5). It is almost similar to cellulose, but the only difference from cellulose is the replacement of a hydroxyl group at C-2 with an acetylated amino group. Chitin forms extended fibers and found commonly in vertebrate animals. Chitin is the principal component of the hard exoskeletons of nearly a million species of arthropods, for example, insects, lobsters, and crabs, and is probably the second most abundant polysaccharide, next to cellulose, in nature.

FIGURE 18.5   The artwork showing structure of chitin.

Bacterial Cell Wall Contains a Heteropolysaccharide: The bacterial cell wall has an important component which gives rigidity to the cell and it is a heteropolymer of alternating (β1→4)-linked *N*-acetylglucosamine and *N*-acetylmuramic acid units linked by glycosidic linkage. These linear polymers lie side by side in the cell wall in many numbers and are cross-linked by short peptides. Its number depends on the bacterial species. Lysozyme is the enzyme which degrades these cross-linked polymers and it hydrolyzes the glycosidic bond between *N*-acetylglucosamine and *N*-acetylmuramic acid and kills bacterial cells. Lysozyme is present in tears, presumably a defense against bacterial infections of the eye. It is also produced by certain bacterial viruses to ensure their release from the host bacteria, an essential step of the viral infection cycle.

1. Glycosaminoglycans: It is a heteropolysaccharides and it is a linear polymer composed of repeating disaccharide units). One of the two monosaccharides is always either *N*-acetylglucosamine or *N*-acetylgalactosamine, the other is in most cases, a uronic acid, usually glucuronic acid. When glycosaminoglycan chains are attached to a protein molecule, the compound is known as "proteoglycans." Some of the examples are hyaluronic acid, chondroitin sulfate, and heparin.

Hyaluronic acid is a disaccharide containing repeating units of *N*-acetyl glucosamine and D-glucuronic acid linked together by β-1,3 linkage. These heteropolysacchrides are clear, highly viscous solutions, and serve as lubricants in the synovial fluids of joints and provide a cushioning effect. Chondroitin sulfate is polysaccharide found in cartilage and contains alternating units of D-glucuronic acid and *N*-acetyl-D-galactosamine linked by β-1,3 linkage. Heparin is made up of D-glucuronate sulfate/L-idurunote sulfate and *N*-sulphoglucosamine-6-sulfate linked by α(1→4) glycosidic bonds. Almost 90% of uronic acids are iduronic acids. It is present in liver, lungs, spleen, monocytes, etc.; it is an anticoagulant agent.

18.6 ANALYSIS OF CARBOHYDRATES

18.6.1 MOLISCH'S TEST

Molisch's reagent is 10% alcoholic solution of α-naphthol. This is a common chemical test to detect the presence of carbohydrates. Carbohydrates undergo dehydration by sulfuric acid to form furfural (furfuraldehyde) that reacts with α-naphthol to form a violet-colored product.

18.6.2 FEHLING'S TEST

This is an important test to detect the presence of reducing sugars. Fehling's solution A is copper sulfate solution and Fehling's solution B is potassium sodium tartrate. On heating, carbohydrate reduces deep blue solution of copper(II) ions to red precipitate of insoluble copper oxide.

$$\underset{\text{Glucose}}{\overset{\displaystyle CHO}{\underset{\displaystyle CH_2OH}{\overset{|}{\underset{|}{(CHOH)_4}}}}} + 2Cu(OH)_2 + NaOH \xrightarrow[\text{ions}]{\text{Tartrate}} \underset{\text{Sod. Salt of gluconic acid}}{\overset{\displaystyle COO^-Na^+}{\underset{\displaystyle CH_2OH}{\overset{|}{\underset{|}{(CHOH)_4}}}}} + 3H_2O + \underset{\text{Red ppt.}}{Cu_2O}$$

18.6.3 BENEDICT'S TEST

Benedict's test distinguishes reducing sugar from nonreducing sugar. Benedict's reagent contains blue copper(II) ions (Cu^{2+}, cupric ions) that are reduced to copper(I) ions (Cu^+, cuprous ions) by carbohydrates. These ions form precipitate as red-colored cuprous copper(I) oxide.

$$CuSO_4 \rightarrow Cu^{2+} + SO_4^{2-}$$

$$Cu^{2+} + \text{Reducing sugar} \rightarrow Cu^+$$

$$Cu^+ \rightarrow Cu_2O$$

18.6.4 TOLLEN'S TEST

Tollen's reagent is ammoniacal silver nitrate solution. On reacting with carbohydrate, elemental silver precipitates out of the solution, occasionally onto the inner surface of the reaction vessel. This produces silver mirror on the inner wall of the reaction vessel.

$$AgNO_3 + NH_4OH \rightarrow NH_4NO_3 + AgOH$$

$$2AgOH \rightarrow Ag_2O + H_2O$$

$$Ag_2O + 2NH_4OH \rightarrow \underset{\text{(Soluble)}}{2[Ag(NH_3)_2]OH} + 3H_2O$$

$$\underset{\text{(Glucose)}}{\overset{\displaystyle CH_2OH}{\underset{\displaystyle CHO}{\overset{|}{\underset{|}{(CHOH)_4}}}}} + Ag_2O \xrightarrow{NH_4OH} \underset{\text{(Gluconic acid)}}{\overset{\displaystyle COOH}{\underset{\displaystyle COOH}{\overset{|}{\underset{|}{(CHOH)_4}}}}} + \underset{\text{(Silver mirror)}}{2Ag}$$

18.6.5   IODINE TEST

Iodine test is used to detect the presence of starch. Iodine is not much soluble in water so iodine solution is prepared by dissolving iodine in water in presence of potassium iodide. Iodine dissolved in an aqueous solution of potassium iodide reacts with starch to form a starch/iodine complex which gives characteristics blue-black color to the reaction mixture (Fig. 18.6).

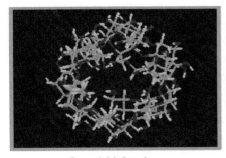
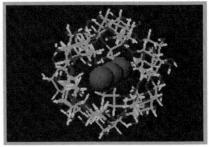

Starch Molecule   Iodine slides into starch molecule to
give blue - black starch - iodine complex

FIGURE 18.6   The artwork showing starch molecules and its reaction with iodine.

POINTS TO REMEMBER

- Sugars, also known as saccharides, contain aldehyde or ketone groups with two or more hydroxyl groups.
- Monosaccharides show steriochemical properties due to presence of one or more chiral carbon.
- Monosaccharides are commonly found as unbranched or cyclic structures. Cyclic structures are represented as a Haworth perspective formula.
- Sterioisomerism, mutarotation, and epimer formation is the important feature of monosaccharides.
- Different monomers joints together through glycosidic bonds and are forming disaccharides.
- Oligosaccharides are short polymers of several monosaccharides joined by glycosidic bonds.
- Monomers joints together to form polymers and are homopolysaccharides or heteropolysacchrides.

- Starch and glycogen are important reserve food and are homopolysacchride.
- The homopolysaccharides cellulose, chitin, are structural units.
- Cellulose is the most abundant polysaccharide in biosphere.

KEYWORDS

- **carbohydrates**
- **saccharide**
- **carbons**
- **sterioisomers**
- **oligosaccharides**

REFERENCES

1. Nelson, D. L.; Cox, M. M. *Lehninger Principles of Biochemistry*, 5th ed.; W H Freeman & Co, 2008. ISBN: 978-0-716-77108-1.
2. Gilbert, H. F. *Basic Concepts in Biochemistry: A Student's Survival Guide*; McGraw-Hill Education/Medical, 1999. *ISBN*-10: 0071356576.
3. Rodwell, V. *Harpers Illustrated Biochemistry*, 30th ed.; McGraw-Hill Education/Medical, 2015. *ISBN*-10: 0071825347.

INDEX

Printed and bound by CPI Group (UK) Ltd, Croydon, CR0 4YY
23/10/2024
01777704-0012